Oxford Applied Mathematics
and Computing Science Series

General Editors
J. Crank, H. G. Martin, D. M. Melluish

G. D. SMITH

Brunel University

Numerical solution of partial differential equations

FINITE DIFFERENCE METHODS

SECOND EDITION

CLARENDON PRESS · OXFORD

Oxford University Press, Walton Street, Oxford OX2 6DP

OXFORD LONDON GLASGOW NEW YORK
TORONTO MELBOURNE WELLINGTON CAPE TOWN
IBADAN NAIROBI DAR ES SALAAM LUSAKA ADDIS ABABA
KUALA LUMPUR SINGAPORE JAKARTA HONG KONG TOKYO
DELHI BOMBAY CALCUTTA MADRAS KARACHI

First edition published in Oxford Mathematical Handbooks 1965
Second edition 1978 *Reprinted with corrections* 1979

British Library Cataloguing in Publication Data

Smith, Gordon Dennis
  Numerical solution of partial differential
  equations. – 2nd ed. – (Oxford applied mathematics
  and computing series).
  1. Differential equations, Partial – Numerical
  solutions
  I. Title
  515'.353   QA374   78-40239

  ISBN 0-19-859625-1
  ISBN 0-19-859626-X Pbk.

Reproduced from copy supplied
printed and bound in Great Britain
by Billing and Sons Limited
Guildford, London, Oxford, Worcester

Preface to the first edition

THIS book has been written primarily as a textbook for students with no previous knowledge of numerical methods whatsoever, and will, I hope, provide a bridge to the growing number of advanced treatises on the numerical solution of partial differential equations. Because of its purpose, the various methods of solution and analysis have been illustrated through a series of worked examples and, in addition, every chapter except the first includes a fairly large number of exercises, with worked solutions, that have often been used to extend the theory as well as to amplify points in the text.

The book is intended for students taking degree and similar courses in mathematics, physics and engineering. It aims to introduce students to the finite-difference methods in use today for solving partial differential equations, and is self-contained in that no finite-difference calculus is assumed, all basic finite-difference formulae being derived through Taylor's series. I have not hesitated, however, to make occasional asides for the benefit of students familiar with this calculus. This self-containment will, I hope. give the book a wider appeal than that already mentioned, and make it of value to some postgraduate workers as well as undergraduate workers in many fields of science and engineering.

In the first chapter I have attempted to give an overall picture of the finite-difference method of solution in relation to the various types of partial differential equations, and have developed some basic formulae needed for the remainder of the book. This descriptive summary will, I trust, give students a sense of perspective of the subsequent field of study before they get involved in the concomitant detail. Chapter 2 illustrates the simplicity and flexibility of finite-difference methods through the numerical solution of several heat-flow problems, but leaves the investigation of convergence and stability to the third chapter where these topics are dealt with both descriptively and analytically. The theoretical analyses of convergence and stability have been deliberately isolated in this way so as to enable students with no knowledge of matrix algebra to read Chapters 2, 4 and 5 without interruption by sections they might the Courant-Friedrichs-Lewy condition for convergence. The propagation of discontinuities in initial data is also discussed at greater length. The chapter on elliptic equations has undergone the most change and now includes the standard work on the SOR iterative method, a section on ordering vectors that is new to undergraduate textbooks, and recent methods for large systems of linear equations.

Finally, I should like to take this opportunity to express my appreciation to the Vice-Chancellor and Council of Brunel University for approving my application for sabbatical leave. Without it I could not have undertaken the preparation of this second edition. I should also like to thank those of my colleagues who so willingly shared the burden of my absence. In particular it gives me tremendous pleasure to express my gratitude to Dr. N. Papamichael for his thoughtful and constructive criticisms of my initial drafts and for his permission to use parts of his M.Tech dissertation on consistent orderings (Brunel University). I am also indebted to Dr. J. Gregory for many helpful discussions and to Professor L. Fox for comments upon weaknesses in the first edition and several useful suggestions with regard to this revised edition.

Brunel University   GORDON D. SMITH
May, 1977

Preface to the second edition

IN THIS second edition substantial additions have been made to most chapters, outmoded material has been eliminated, and all the work on iterative methods, which previously spread over the whole book, has been put together in Chapter 5. Each chapter also incorporates a large number of new exercises with solutions. As before, only finite-difference methods of approximation have been dealt with. In this second edition, however, unlike the first, I have not hesitated to use matrix algebra and well-known notations and results from finite-difference calculus when their use has made the presentation of new work simpler and more precise. This should present no difficulties to most readers as the mathematics needed is that normally covered by the end of the second year of undergraduate courses in mathematics, physics and engineering. The initial basic work in each chapter can still be studied by students with no previous knowledge of numerical methods.

When this book was first written it was quite normal to use the same symbol for the solution of the difference equation and the solution of the differential equation, except when it was essential to distinguish between them. Nowadays it is customary to distinguish between them at all times and this has been done in the new sections of the second edition. The resulting mixture of old and new practices is unfortunate but is most unlikely to give difficulties to any reader.

With regard to the changes the chapter on parabolic equations has been altered the least, the original work merely being supplemented by a few miscellaneous methods for improving accuracy and a section on non-linear parabolic equations. Chapter 3 has been considerably improved by reconsidering the concept of stability and relating it to convergence and consistency via Lax's equivalence theorem. Useful theorems on eigenvalues have also been added, together with the analytical solutions of homogeneous difference equations. The chapter on hyperbolic equations has been augmented by the theory and numerical work associated with first-order equations, by the Lax-Wendroff and Wendroff formulae and otherwise feel they must understand before proceeding further. Matrix notation has been used occasionally in Chapter 5, but the essential ideas of this chapter have been explained without reference to matrix algebra.

Finally, I make no apology for a small amount of repetition of some topics, mainly iterative methods, in two or three chapters, as this has kept Chapters 2, 4 and 5 relatively independent of one another and should allow students to read them in whatever order they please without excessive cross-referencing.

I am deeply indebted to Professor J. Crank for the useful criticisms and valuable suggestions he made on reading the original and revised manuscripts and am extremely grateful to Ian Parker for programming and computing almost all the numerical examples and exercises. I should also like to thank Mrs. M. Richmond for the speed and accuracy with which she typed the manuscript, Mr. C. C. Ritchie for reading the final typescript and Mr. E. Lovelock for the diagrams. I am also pleased to acknowledge my debt to Dr. J. Topping and Dr. E. T. Goodwin for permission to study at the Mathematics Division of the National Physical Laboratory as a Guest-Worker during 1958, where my interest in this subject was first generated, and to Mr. E. L. Albasiny and Dr. D. W. Martin of the National Physical Laboratory for an extremely stimulating set of lectures they gave on partial differential equations at Brunel.

In conclusion I warmly thank the officials of Oxford University Press for the careful attention they have given to the preparation of this book.

Brunel,   GORDON D. SMITH
March, 1964

Contents

1 Introduction and finite-difference formulae

THE mathematical formulation of most problems in science involving rates of change with respect to two or more independent variables, usually representing time, length or angle, leads either to a partial differential equation or to a set of such equations. Special cases of the two dimensional second-order equation

$$a\frac{\partial^2\phi}{\partial x^2}+b\frac{\partial^2\phi}{\partial x\,\partial y}+c\frac{\partial^2\phi}{\partial y^2}+d\frac{\partial\phi}{\partial x}+e\frac{\partial\phi}{\partial y}+f\phi+g=0,$$

where a, b, c, d, e, f and g may be functions of the independent variables x and y and of the dependent variable ϕ, occur more frequently than any other because they are often the mathematical form of one of the conservation principles of physics.

For reasons that are given in Chapter 4 this equation is said to be *elliptic* when $b^2-4ac<0$, *parabolic* when $b^2-4ac=0$, and *hyperbolic* when $b^2-4ac>0$.

Two-dimensional elliptic equations

These equations, of which the best known are Poisson's equation

$$\frac{\partial^2\phi}{\partial x^2}+\frac{\partial^2\phi}{\partial y^2}+g=0$$

and Laplace's equation

$$\frac{\partial^2\phi}{\partial x^2}+\frac{\partial^2\phi}{\partial y^2}=0,$$

are generally associated with equilibrium or steady-state problems. For example, the velocity potential for the steady flow of incompressible non-viscous fluid satisfies Laplace's equation and is the mathematical way of expressing the idea that the rate at which such fluid enters any given region is equal to the rate at which it leaves it. Similarly, the electric potential V associated with a two-dimensional electron distribution of charge density ρ satisfies Poisson's equation

$\partial^2 V/\partial x^2 + \partial^2 V/\partial y^2 + \rho/\varepsilon = 0$, where ε is the dielectric constant. This is the partial differential equation form of the well-known theorem by Gauss which states that the total electric flux through any closed surface is equal to the total charge enclosed.

The *analytical* solution of a two-dimensional elliptic equation is a function of the space co-ordinates x and y which satisfies the partial differential equation *at every point* of the area S inside a plane *closed* curve C and satisfies certain conditions *at every point* on this boundary curve C (Fig. 1.1). The function ϕ, for instance, from which we can calculate the displacements and shear stresses within a long solid elastic cylinder in a state of torsion satisfies

$$\frac{\partial^2 \phi}{\partial x^2} + \frac{\partial^2 \phi}{\partial y^2} + 2 = 0$$

at every point of a right cross-section, and has a constant value round the perimeter of the cross-section. Similarly, the steady motion of incompressible viscous fluid through a straight uniform tube can be found from a function that satisfies Laplace's equation at every point of the cross-section and equals $\frac{1}{2}(x^2 + y^2)$ at each point on the boundary.

The condition that the dependent variable must satisfy round the boundary curve C is termed the boundary condition.

To the present, only a limited number of special types of elliptic equations have been solved analytically and the usefulness of these solutions is further restricted to problems involving shapes for which

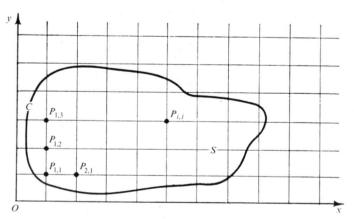

Fig. 1.1

the boundary conditions can be satisfied. This not only eliminates all problems with boundary curves that are undefined in terms of equations, but also many for which the boundary conditions are too difficult to satisfy even though the equations for the boundary curves are known. In such cases approximation methods, whether analytical or numerical in character, are the only means of solution, apart from the use of analogue devices. Analytical approximation methods often provide extremely useful information concerning the character of the solution for critical values of the dependent variables but tend to be more difficult to apply than the numerical methods, and will not be discussed in this book. Of the numerical approximation methods available for solving differential equations those employing finite-differences are more frequently used and more universally applicable than any other. Before outlining these methods however, the reader should be aware of the manner in which the term 'approximation method' is used. Finite-difference methods are approximate in the sense that derivatives *at a point* are approximated by difference quotients *over a small interval*, i.e., $\partial \phi / \partial x$ is replaced by $\delta \phi / \delta x$ where δx is small, but the solutions are *not* approximate in the sense of being crude estimates. The data of the problems of technology are invariably subject to errors of measurement, besides which, all arithmetical work is limited to a finite number of significant figures and contains rounding errors, so even analytical solutions provide only approximate numerical answers. Finite-difference methods generally give solutions that are either as accurate as the data warrant or as accurate as is necessary for the technical purposes for which the solutions are required. In both cases a finite-difference solution is as satisfactory as one calculated from an analytical formula. In future, all non-analytical approximation methods will be called numerical methods.

They are not of course restricted to problems for which no analytical solutions can be found. The numerical evaluation of an analytical solution is often a laborious task, as can be seen by inspecting the solution of the torsion problem for a rectangular cross-section defined by $x = \pm a$, $y = \pm b$, namely

$$\phi = b^2 - y^2 - 32b^2\pi^{-3} \sum_{n=0}^{\infty} \frac{(-1)^n}{(2n+1)^3} \operatorname{sech} \frac{(2n+1)\pi a}{2b}$$
$$\times \cosh \frac{(2n+1)\pi x}{2b} \cos \frac{(2n+1)\pi y}{2b},$$

and numerical methods generally provide adequate numerical solutions more simply and efficiently. This is certainly so with finite-difference methods for solving partial differential equations.

In these methods, (Fig. 1.1), the area of integration of the elliptic equation, i.e. the area S bounded by the closed curve C, is overlayed by a system of rectangular meshes formed by two sets of equally spaced lines, one set parallel to Ox and the other parallel to Oy, and an approximate solution to the differential equation is found at the points of intersection $P_{1,1}$, $P_{1,2}, \ldots, P_{i,j}, \ldots$ of the parallel lines, which points are called mesh points. (Other terms in common use are pivotal, nodal, grid, or lattice points.) This solution is obtained by approximating the partial differential equation over the area S by n *algebraic* equations involving the values of ϕ at the n mesh points internal to C. The approximation consists of replacing each derivative of the partial differential equation at the point $P_{i,j}$ (say) by a finite-difference approximation in terms of the values of ϕ at $P_{i,j}$ and at neighbouring mesh points and boundary points, and in writing down for each of the n internal mesh points the algebraic equation approximating the differential equation. This process clearly gives n algebraic equations for the n unknowns $\phi_{1,1}$, $\phi_{1,2}, \ldots$ $\phi_{i,j}, \ldots$. Accuracy can usually be improved either by increasing the number of mesh points or by including 'correction terms' in the approximations for the derivatives.

Parabolic and hyperbolic equations

Problems involving time t as one independent variable lead usually to parabolic or hyperbolic equations.

The simplest parabolic equation, $\partial u/\partial t = \kappa \partial^2 u/\partial x^2$, derives from the theory of heat conduction and its solution gives, for example, the temperature u at a distance x units of length from one end of a thermally insulated bar after t seconds of heat conduction. In such a problem the temperatures at the *ends* of a bar of length l (say) are often known for all time. In other words, the *boundary conditions* are known. It is also usual for the temperature distribution along the bar to be known at some particular instant. This instant is usually taken as zero time and the temperature distribution is called the *initial condition*. The solution gives u for values of x between 0 and l and values of t from zero to infinity. Hence the area of integration S in the x-t plane (Fig. 1.2), is the infinite area bounded by the x axis

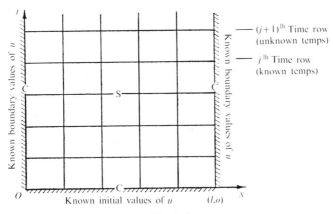

Fig. 1.2

and the parallel lines $x = 0$, $x = l$. This is described as an *open* area because the boundary curves marked C do not constitute a closed boundary in any finite region of the x-t plane.

Application of finite-difference methods of solution to parabolic equations are no different from their application to elliptic equations in so far as the integration of the differential equation over S is approximated by the solution of algebraic equations. The structure of the algebraic equations is different however in that they propagate the solution forward from one time row to the next in a step-by-step fashion.

Hyperbolic equations generally originate from vibration problems, or from problems where discontinuities can persist in time, such as with shock waves, across which there are discontinuities in speed, pressure and density. The simplest hyperbolic equation is the one-dimensional wave equation $\partial^2 u/\partial t^2 = c^2 \partial^2 u/\partial x^2$, giving, for example, the transverse displacement u at a distance x from one end of a vibrating string of length l after a time t. As the values of u at the ends of the string are usually known for all time (the boundary conditions) and the shape and velocity of the string are prescribed as zero time (the initial conditions), it is seen (Fig. 1.2), that the solution is similar to that of a parabolic equation in that the calculation of u for a given x and t, $(0 \leqslant x \leqslant l)$, entails integration of the equation over the open area S bounded by the open curve C. Although hyperbolic equations can be solved numerically by finite-difference methods, those involving only two independent variables,

x and t say, are more usually dealt with by the method of characteristics. This method finds special curves in the x-t plane, called characteristic curves, along which the solution of the *partial* differential equation is reduced to the integration of an *ordinary* differential equation. This ordinary equation is generally integrated by numerical methods.

In conclusion it is worth noting that whereas changes to the shape of the area of integration or to the boundary and initial conditions of partial differential equations often make their analytical solutions impossible, such changes do not fundamentally affect finite-difference methods although they sometimes necessitate rather complicated modifications to the methods.

Finite-difference approximations to derivatives

When a function u and its derivatives are single-valued, finite and continuous functions of x, then by Taylor's theorem,

$$u(x+h) = u(x) + hu'(x) + \tfrac{1}{2}h^2 u''(x) + \tfrac{1}{6}h^3 u'''(x) + \ldots \qquad (1.1)$$

and

$$u(x-h) = u(x) - hu'(x) + \tfrac{1}{2}h^2 u''(x) - \tfrac{1}{6}h^3 u'''(x). \ldots \qquad (1.2)$$

Addition of these expansions gives

$$u(x+h) + u(x-h) = 2u(x) + h^2 u''(x) + O(h^4), \qquad (1.3)$$

where $O(h^4)$ denotes terms containing fourth and higher powers of h. Assuming these are negligible in comparison with lower powers of h it follows that,

$$u''(x) = \left(\frac{d^2 u}{dx^2}\right)_{x=x} \simeq \frac{1}{h^2}\{u(x+h) - 2u(x) + u(x-h)\}, \qquad (1.4)$$

with a leading error on the right-hand side of order h^2.

Subtraction of equation (1.2) from equation (1.1) and neglect of terms of order h^3 leads to

$$u'(x) = \left(\frac{du}{dx}\right)_{x=x} \simeq \frac{1}{2h}\{u(x+h) - u(x-h)\}, \qquad (1.5)$$

with an error of order h^2.

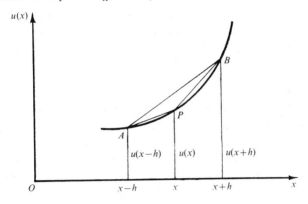

Fig. 1.3

Equation (1.5) clearly approximates the slope of the tangent at P by the slope of the chord AB, and is called a *central-difference* approximation. We can also approximate the slope of the tangent at P by either the slope of the chord PB, giving the *forward-difference* formula,

$$u'(x) \simeq \frac{1}{h}\{u(x+h) - u(x)\}, \qquad (1.6)$$

or the slope of the chord AP giving the *backward-difference* formula

$$u'(x) \simeq \frac{1}{h}\{u(x) - u(x-h)\}. \qquad (1.7)$$

Both (1.6) and (1.7) can be written down immediately from equations (1.1) and (1.2) respectively, assuming second and higher powers of h are negligible. This shows that the leading errors in these forward and backward-difference formulae are both $O(h)$.

Notation for functions of several variables

Assume u is a function of the independent variables x and t. Subdivide the x-t plane into sets of equal rectangles of sides $\delta x = h$, $\delta t = k$, as shown in Fig. 1.4, and let the co-ordinates (x, t) of the

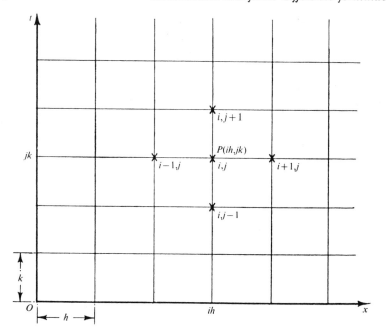

Fig. 1.4

representative mesh point P be

$$x = ih; \quad t = jk,$$

where i and j are integers.

Denote the value of u at P by

$$u_P = u(ih, jk) = u_{i,j}.$$

Then by equation (1.4),

$$\left(\frac{\partial^2 u}{\partial x^2}\right)_P = \left(\frac{\partial^2 u}{\partial x^2}\right)_{i,j} \simeq \frac{u\{(i+1)h, jk\} - 2u\{ih, jk\} + u\{(i-1)h, jk\}}{h^2}.$$

i.e.

$$\left(\frac{\partial^2 u}{\partial x^2}\right)_{i,j} \simeq \frac{u_{i+1,j} - 2u_{i,j} + u_{i-1,j}}{h^2}, \tag{1.8}$$

with a leading error of order h^2. Similarly,

$$\left(\frac{\partial^2 u}{\partial t^2}\right)_{i,j} \simeq \frac{u_{i,j+1} - 2u_{i,j} + u_{i,j-1}}{k^2}, \tag{1.9}$$

with a leading error or order k^2.

With this notation the forward-difference approximation for $\partial u/\partial t$ at P is

$$\frac{\partial u}{\partial t} \simeq \frac{u_{i,j+1} - u_{i,j}}{k}, \tag{1.10}$$

with a leading error of $O(k)$.

2 Parabolic equations

Transformation to non-dimensional form

THE computational stage of all numerical methods for solving problems of any complexity generally involves a great deal of arithmetic. It is usual therefore to arrange, whenever possible, for one solution to suffice for a variety of different problems. This can be done by expressing all equations in terms of non-dimensional variables. Then all problems with the same non-dimensional mathematical formulation can be dealt with by means of one solution. For example, the oscillation of a pendulum in a viscous medium and the discharge of electricity from a capacitance through a resistance and inductance are different problems physically, but identical mathematically when expressed in terms of non-dimensional variables. The problems need not, of course, be dimensionally different, but merely variations of the same type of problem, as we would have with the calculation of the periods of oscillation of springs of different lengths l supporting different masses m and having different stiffnesses s. A single solution of the corresponding non-dimensional equation would allow us to solve a wide variety of spring problems because a single parameter ξ, say, would replace some combination of l, m and s.

This non-dimensionalizing process is illustrated below with the parabolic equation

$$\frac{\partial U}{\partial T} = \kappa \frac{\partial^2 U}{\partial X^2}, \quad \kappa \text{ constant}, \tag{2.1}$$

the solution of which gives the temperature U at a distance X from one end of a thin uniform rod after a time T. (This assumes the rod is heat-insulated along its length so that temperature changes occur through heat conduction along its length and heat transfer at its ends.) Let L represent the length of the rod and U_0 some particular temperature such as the maximum or minimum temperature at zero time. Put

$$x = \frac{X}{L} \text{ and } u = \frac{U}{U_0}.$$

Then

$$\frac{\partial U}{\partial X} = \frac{\partial U}{\partial x}\frac{dx}{dX} = \frac{\partial U}{\partial x}\frac{1}{L}$$

and

$$\frac{\partial^2 U}{\partial X^2} = \frac{\partial}{\partial X}\left(\frac{\partial U}{\partial X}\right) = \frac{\partial}{\partial x}\left(\frac{1}{L}\frac{\partial U}{\partial X}\right)\frac{dx}{dX} = \frac{1}{L^2}\frac{\partial^2 U}{\partial x^2},$$

so equation (2.1) transforms to

$$\frac{\partial(uU_0)}{\partial T} = \frac{\kappa}{L^2}\frac{\partial^2(uU_0)}{\partial x^2},$$

i.e.

$$\frac{1}{\kappa L^{-2}}\frac{\partial u}{\partial T} = \frac{\partial^2 u}{\partial x^2}.$$

Writing $t = \kappa T/L^2$ and applying the function of a function rule to the left side yields

$$\frac{\partial u}{\partial t} = \frac{\partial^2 u}{\partial x^2} \qquad\qquad (2.2)$$

as the non-dimensional form of (2.1).

It should be noted that the number representing the length of the rod is 1.

An explicit method of solution

By equations (1.10) and (1.8) one finite-difference approximation to

$$\frac{\partial u}{\partial t} = \frac{\partial^2 u}{\partial x^2} \qquad\qquad (2.3)$$

is

$$\frac{u_{i,j+1} - u_{i,j}}{k} = \frac{u_{i+1,j} - 2u_{i,j} + u_{i-1,j}}{h^2},$$

where

$$x = ih, (i = 0, 1, 2, \ldots),$$

and

$$t = jk, (j = 0, 1, 2, \ldots).$$

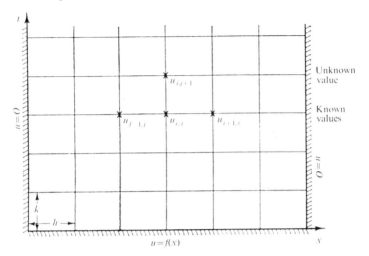

Fig. 2.1

This can be written as

$$u_{i,j+1} = u_{i,j} + r(u_{i-1,j} - 2u_{i,j} + u_{i+1,j}) \tag{2.4}$$

where $r = \delta t/(\delta x)^2 = k/h^2$, and gives a formula for the unknown 'temperature' $u_{i,j+1}$ at the $(i, j+1)$th mesh point in terms of known 'temperatures' along the jth time row (Fig. 2.1). Hence we can calculate the unknown pivotal values of u along the first time row, $t = k$, in terms of known boundary and initial values along $t = 0$, then the unknown pivotal values along the second time row in terms of the calculated pivotal values along the first, and so on. A formula such as this which expresses *one* unknown pivotal value directly in terms of known pivotal values is called an explicit formula.

Example 2.1

As a numerical example let us solve (2.4) given that the ends of the rod are kept in contact with blocks of melting ice and that the initial temperature distribution in non-dimensional form is

$$\begin{aligned}(a)\ &u = 2x, &0 \leqslant x \leqslant \tfrac{1}{2}, \\ (b)\ &u = 2(1-x), &\tfrac{1}{2} \leqslant x \leqslant 1.\end{aligned} \tag{2.5}$$

In other words we are seeking a numerical solution of $\partial u/\partial t = \partial^2 u/\partial x^2$ which satisfies

(i) $u = 0$ when $x = 0$ and 1 for all t. (The boundary conditions)

(ii) $u = 2x$ for $0 \leqslant x \leqslant \frac{1}{2}$,
and $u = 2(1-x)$ for $\frac{1}{2} \leqslant x \leqslant 1$. $\Big\} t = 0$. (The initial condition)

(This initial temperature distribution could be obtained by heating the centre of the rod for a long time and keeping the ends in contact with the ice.)

For $\delta x = h = \frac{1}{10}$, the initial values and boundary values are as shown in Table 2.1. The problem is symmetric with respect to $x = \frac{1}{2}$ so we need the solution only for $0 \leqslant x \leqslant \frac{1}{2}$.

TABLE 2.1

$x = 0$	0·1	0·1	0·3	0·4	0·5	0·6	
$j = 0$	0	0·2	0·4	0·6	0·8	1·0	0·8	x
$j = 1$	0							
$j = 2$	0							
$j = 3$	0							
$j = 4$	0							
t								

Case 1

Take $\delta x = h = \frac{1}{10}$, $\delta t = k = \frac{1}{1000}$, so $r = k/h^2 = \frac{1}{10}$. Equation (2.4) then reads as

$$u_{i,j+1} = \tfrac{1}{10}(u_{i-1,j} + 8u_{i,j} + u_{i+1,j}). \tag{2.6}$$

For pencil and paper calculations the relationship between these four function values is represented very conveniently by the 'molecule' in Fig. 2.2. The numbers in the 'atoms' are the multipliers of the function values at the corresponding mesh points.

Application of equation (2.6) to the data of Table 2.1 is shown in Table 2.2, and readers are recommended to check some of the calculations, remembering that the values of u at $x = \frac{4}{10}$ and $\frac{6}{10}$ are equal because of symmetry. (Increasing values of t, i.e. of j, are

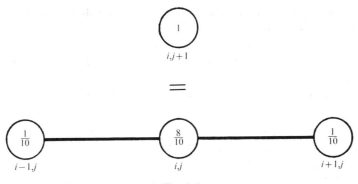

Fig. 2.2

shown moving downwards for convenience of calculation.) As examples,

$$u_{5,1} = \tfrac{1}{10}\{0\cdot8 + (8 \times 1) + 0\cdot8\} = 0\cdot9600.$$
$$u_{4,2} = \tfrac{1}{10}\{0\cdot6 + (8 \times 0\cdot8) + 0\cdot96\} = 0\cdot7960.$$

TABLE 2.2

	$i = 0$ $x = 0$	$i = 1$ $0\cdot1$	$i = 2$ $0\cdot2$	$i = 3$ $0\cdot3$	$i = 4$ $0\cdot4$	$i = 5$ $0\cdot5$	$i = 6$ $0\cdot6$
$(j = 0)t = 0\cdot000$	0	0·2000	0·4000	0·6000	0·8000	1·0000	0·8000
$(j = 1)$ 0·001	0	0·2000	0·4000	0·6000	0·8000	0·9600	0·8000
$(j = 2)$ 0·002	0	0·2000	0·4000	0·6000	0·7960	0·9280	0·7960
$(j = 3)$ 0·003	0	0·2000	0·4000	0·5996	0·7896	0·9016	0·7896
$(j = 4)$ 0·004	0	0·2000	0·4000	0·5986	0·7818	0·8792	0·7818
$(j = 5)$ 0·005	0	0·2000	0·3999	0·5971	0·7732	0·8597	0·7732
$(j = 10)$ 0·01	0	0·1996	0·3968	0·5822	0·7281	0·7867	0·7281
$(j = 20)$ 0·02	0	0·1938	0·3781	0·5373	0·6486	0·6891	0·6486

The analytical solution of the partial differential equation satisfying these conditions is

$$u = \frac{8}{\pi^2} \sum_{n=1}^{\infty} \frac{1}{n^2} (\sin \tfrac{1}{2}n\pi)(\sin n\pi x)\exp(-n^2\pi^2 t).$$

Comparison of this solution with the finite-difference one at $x = 0.3$, as given below, shows that the finite-difference solution is reasonably accurate. The percentage error is the difference of the solutions expressed as a percentage of the analytical solution of the partial differential equation.

TABLE 2.3

	Finite-difference solution $(x = 0.3)$	Analytical solution $(x = 0.3)$	Difference	Percentage error
$t = 0.005$	0·5971	0·5966	0·0005	0·08
$t = 0.01$	0·5822	0·5799	0·0023	0·4
$t = 0.02$	0·5373	0·5334	0·0039	0.7
$t = 0.10$	0·2472	0·2444	0·0028	1·1

The comparison at $x = 0.5$ is not quite so good because of the discontinuity in the initial value of $\partial u/\partial x$, from $+2$ to -2, at this point (Equation 2.5). Inspection of Table 2.4 shows, however, that the effect of this discontinuity dies away as t increases.

TABLE 2.4

	Finite-difference solution $(x = 0.5)$	Analytical solution $(x = 0.5)$	Difference	Percentage error
$t = 0.005$	0·8597	0·8404	0·0193	2·3
$t = 0.01$	0·7867	0·7743	0·0124	1·6
$t = 0.02$	0·6891	0·6809	0·0082	1·2
$t = 0.10$	0·3056	0·3021	0·0035	1·2

It can be proved analytically that when the boundary values are constant the effect of discontinuities in initial values and initial derivatives upon the solution of a parabolic equation decreases as t increases. (See Chapter 3, exercise 20.)

An examination of Tables 2.19 and 2.21 given in exercise 1 at the end of this chapter shows that the same finite-difference solution for a problem in which the initial function and all its derivatives are continuous is very close indeed to the solution of the partial differential equation.

Richtmyer, reference 31, has shown for this particular finite-difference scheme that when the initial function and its first $(p-1)$ derivatives are continuous and the pth derivative ordinarily discontinuous (i.e., changes by finite jumps), then the difference between the solution of the partial differential equation and a convergent solution of the difference equation is of order $(\delta t)^{(p+2)/(p+4)}$, for small δt.

In this example, $p = 1$, so the difference is of order $(\delta t)^{\frac{3}{5}}$. As $(0 \cdot 001)^{\frac{3}{5}} = 0 \cdot 016$, it is seen that the finite-difference solution is actually better than the estimate indicates, a feature common to most error estimates. When all the derivatives are continuous, $p \to \infty$, and the error is of order δt.

Case 2

Take $\delta x = h = \frac{1}{10}$, $\delta t = k = \frac{5}{1000}$, so $r = k/h^2 = 0 \cdot 5$. Then equation (2.4) gives

$$u_{i,j+1} = \tfrac{1}{2}(u_{i-1,j} + u_{i+1,j}), \tag{2.7}$$

and the solution obtained by applying this finite-difference equation to the boundary and initial values is recorded in Table 2.5.

TABLE 2.5

$i = 0$	1	2	3	4	5	6
$x = 0$	0·1	0·2	0·3	0·4	0·5	0·6
$T = 0 \cdot 000$ 0	0·2000	0·4000	0·6000	0·8000	1·0000	0·8000
$0 \cdot 005$ 0	0·2000	0·4000	0·6000	0·8000	0·8000	0·8000
$0 \cdot 010$ 0	0·2000	0·4000	0·6000	0·7000	0·8000	0·7000
$0 \cdot 015$ 0	0·2000	0·4000	0·5500	0·7000	0·7000	0·7000
$0 \cdot 020$ 0	0·2000	0·3750	0·5500	0·6250	0·7000	0·6250
$\vdots$						
$0 \cdot 100$ 0	0·0949	0·1717	0·2484	0·2778	0·3071	0·2778

TABLE 2.6

	Finite-difference solution $(x = 0 \cdot 3)$	Analytical solution $(x = 0 \cdot 3)$	Difference	Percentage error
$t = 0 \cdot 005$	0·6000	0·5966	0·0034	0·57
$t = 0 \cdot 01$	0·6000	0·5799	0·0201	3·5
$t = 0 \cdot 02$	0·5500	0·5334	0·0166	3·1
$t = 0 \cdot 1$	0·2484	0·2444	0·0040	1·6

It is seen that this finite-difference solution is not quite as good an approximation to the solution of the partial differential equation as the previous one; nevertheless it would be adequate for most technical purposes.

Case 3

Take $\delta x = \frac{1}{10}$, $\delta t = \frac{1}{100}$, so $r = \delta t/(\delta x)^2 = 1$. Then equation (2.4) gives

$$u_{i,j+1} = u_{i-1,j} - u_{i,j} + u_{i+1,j}, \qquad (2.9)$$

and the solution of this finite-difference scheme is as below.

TABLE 2.7

	$i = 0$	1	2	3	4	5	6
	$x = 0$	0·1	0·2	0·3	0·4	0·5	0·6
$t = 0\cdot00$	0	0·2	0·4	0·6	0·8	1·0	0·8
0·01	0	0·2	0·4	0·6	0·8	0·6	0·8
0·02	0	0·2	0·4	0·6	0·4	1·0	0·4
0·03	0	0·2	0·4	0·2	1·2	−0·2	1·2
0·04	0	0·2	0·0	1·4	−1·2	2·6	−1·2

Considered as a solution of the partial differential equation this is obviously meaningless, although it is, of course, the correct solution of equation (2.9) with respect to the initial values and boundary values given.

These three cases clearly indicate that the value of r is important and it will be proved in Chapter 3 that this explicit method is valid only when $0 < r \leqslant \frac{1}{2}$. (The conditions that must be satisfied for a valid expansion are dealt with both descriptively and analytically in Chapter 3 under the headings of convergence, stability and consistency. Any reader who would prefer to have an introduction to these concepts at this stage could do so by reading the descriptive treatments of these topics as they are independent of the remainder of this chapter.)

The graphs opposite compare the analytical solution of the partial differential equation (shown as continuous curves) with the finite-difference solution (shown by dots) for values of r just below and above $\frac{1}{2}$, and the same number of time-steps.

Crank–Nicolson implicit method

Although the explicit method is computationally simple it has one serious drawback. The time step $\delta t = k$ is necessarily very small

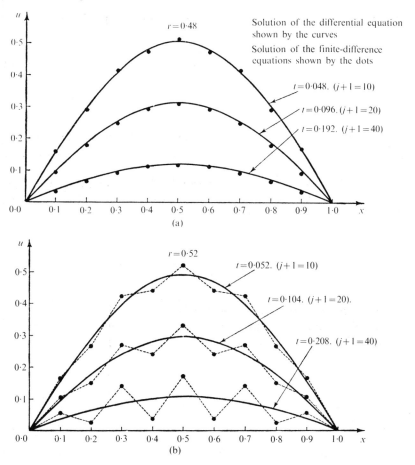

Fig. 2.3

because the process is valid only for $0 < k/h^2 \leqslant \frac{1}{2}$, i.e., $k \leqslant \frac{1}{2}h^2$, and $h = \delta x$ must be kept small in order to attain reasonable accuracy. Crank and Nicolson (1947) proposed, and used, a method that reduces the total volume of calculation and is valid (i.e., convergent and stable) for all finite values of r. They replaced $\partial^2 u/\partial x^2$ by the mean of its finite-difference representations on the $(j+1)$th and jth time rows and approximated the equation

$$\frac{\partial u}{\partial t} = \frac{\partial^2 u}{\partial x^2}$$

by

$$\frac{u_{i,j+1} - u_{i,j}}{k} = \frac{1}{2} \left\{ \frac{u_{i+1,j+1} - 2u_{i,j+1} + u_{i-1,j+1}}{h^2} + \frac{u_{i+1,j} - 2u_{i,j} + u_{i-1,j}}{h^2} \right\},$$

giving

$$-ru_{i-1,j+1} + (2+2r)u_{i,j+1} - ru_{i+1,j+1} = ru_{i-1,j} + (2-2r)u_{i,j} + ru_{i+1,j},$$
$$(2.10)$$

where $r = k/h^2$.

In general, the left side of equation (2.10) contains three un-known and the right side three known, pivotal values of u (Fig. 2.4).

If there are N internal mesh points along each time row then for $j = 0$ and $i = 1, 2, \ldots, N$, equation (2.10) gives N simultaneous equations for the N unknown pivotal values along the first time row in terms of known initial and boundary values. Similarly, $j = 1$ expresses N unknown values of u along the second time row in terms of the calculated values along the first, etc. A method such as this where the calculation of an unknown pivotal value necessitates the solution of a set of simultaneous equations is described as an *implicit* one.

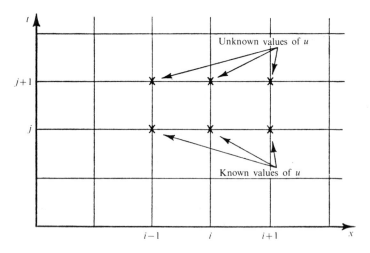

Fig. 2.4

Example 2.2

Use the Crank–Nicolson method to calculate a numerical solution of the previous worked example, namely,

$$\frac{\partial u}{\partial t} = \frac{\partial^2 u}{\partial x^2} \quad (0 < x < 1),$$

where (i) $u = 0$, $x = 0$ and 1, $t \geqslant 0$,

(ii) $u = 2x$, $0 \leqslant x \leqslant \frac{1}{2}$, $t = 0$,

(iii) $u = 2(1 - x)$, $\frac{1}{2} \leqslant x \leqslant 1$, $t = 0$.

Take $h = \frac{1}{10}$. Although the method is valid for all finite values of $r = k/h^2$, a large value will yield an inaccurate approximation for $\partial u/\partial t$. A suitable value is $r = 1$ and has the advantage of making the coefficient of $u_{i,j}$ zero in (2.10). Then $k = \frac{1}{100}$ and (2.10) reads as

$$-u_{i-1,j+1} + 4u_{i,j+1} - u_{i+1,j+1} = u_{i-1,j} + u_{i+1,j}. \tag{2.11}$$

The computational molecule corresponding to equation (2.11) is shown in Fig. 2.5. Denote $u_{i,j+1}$ by u_i ($i = 1, 2, \ldots, 9$). For this problem, because of symmetry, $u_6 = u_4$, $u_7 = u_3$, etc. (Fig. 2.6). The values of u for the first time step then satisfy

$$-0 + 4u_1 - u_2 = 0 \quad + 0 \cdot 4,$$
$$-u_1 + 4u_2 - u_3 = 0 \cdot 2 + 0 \cdot 6,$$
$$-u_2 + 4u_3 - u_4 = 0 \cdot 4 + 0 \cdot 8,$$
$$-u_3 + 4u_4 - u_5 = 0 \cdot 6 + 1 \cdot 0,$$
$$-2u_4 + 4u_5 \quad\quad = 0 \cdot 8 + 0 \cdot 8.$$

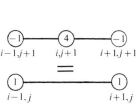

Fig. 2.5

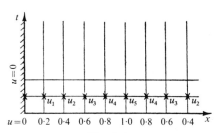

Fig. 2.6

As indicated in the next section these are easily solved by systematic eliminations to give

$$u_1 = 0.1989, \quad u_2 = 0.3956, \quad u_3 = 0.5834, \quad u_4 = 0.7381, \quad u_5 = 0.7691.$$

Hence the equations for the pivotal values of u along the next time row are

$$-0 \quad + 4u_1 - u_2 = 0 + 0.3956,$$
$$-u_1 + 4u_2 - u_3 = 0.1989 + 0.5834,$$
$$-u_2 + 4u_3 - u_4 = 0.3956 + 0.7381,$$
$$-u_3 + 4u_4 - u_5 = 0.5834 + 0.7691,$$
$$-2u_4 + 4u_5 \quad = 2 \times 0.7381$$

The solution of these equations is given in Table 2.8 together with figures comparing the finite-difference solution at $t = 0.1$ with the solution of the partial differential equation. The numerical solution is clearly a good one.

TABLE 2.8

	$x = 0$	0·1	0·2	0·3	0·4	0·5
$t = 0.00$	0	0·2	0·4	0·6	0·8	1·0
$t = 0.01$	0	0·1989	0·3956	0·5834	0·7381	0·7691
$t = 0.02$	0	0·1936	0·3789	0·5400	0·6461	0·6921
·	0	·				
·		·				
·		·				
$t = 0.10$	0	0·0948	0·1803	0·2482	0·2918	0·3069
Analytical solution $t = 0.10$	0	0·0934	0·1776	0·2444	0·2873	0·3021

Table 2.9 below displays both solutions at $x = 0.5$ for various values of t. A glance at Table 2.4 shows that in this example the accuracy of this implicit method over the time-range taken is about the same as for the explicit method which uses ten times as many time-steps.

TABLE 2.9

	Finite-difference solution ($x = 0.5$)	Analytical solution ($x = 0.5$)	Difference	Percentage error
$t = 0.01$	0·7691	0·7743	−0·0052	−0·7
$t = 0.02$	0·6921	0·6809	+0·0112	+1·6
$t = 0.10$	0·3069	0·3021	0·0048	1·6

As mentioned previously the greatest difference between the two solutions occurs at $x = 0\cdot5$ because of the ordinary discontinuity in the initial value of $\partial u/\partial x$ at this point. A glance at Table 2.25 in exercise 3 at the end of the chapter shows that this difference is less for an initial function that is continuous together with its derivatives.

Solution of the equations by Gauss's elimination method (Without pivoting)

When there are N-1 internal mesh points along each time row the Crank–Nicolson equations (2.10) can be written very generally as

$$+b_1 u_1 - c_1 u_2 \qquad\qquad\qquad\qquad = d_1,$$
$$-a_2 u_1 + b_2 u_2 - c_2 u_3 \qquad\qquad\qquad = d_2,$$
$$\cdot \qquad\qquad \cdot$$
$$\cdot \qquad\qquad \cdot$$
$$-a_i u_{i-1} + b_i u_i - c_i u_{i+1} \qquad\qquad = d_i,$$
$$\cdot \qquad\qquad \cdot$$
$$-a_{N-1} u_{N-2} + b_{N-1} u_{N-1} = d_{N-1},$$

where the a's, b's, c's and d's are known. The first equation can be used to eliminate u_1 from the second equation, the new second equation used to eliminate u_2 from the third equation and so on, until finally, the new last but one equation can be used to eliminate u_{N-2} from the last equation, giving one equation with only one unknown, u_{N-1}. The unknowns $u_{N-2}, u_{N-3}, \ldots u_2, u_1$ can then be found in turn by back-substitution. Noting that the coefficient c in each new equation is the same as in the corresponding old equation, assume that the following stage of the eliminations has been reached,

$$\alpha_{i-1} u_{i-1} - c_{i-1} u_i = S_{i-1},$$
$$-a_i u_{i-1} + b_i u_i - c_i u_{i+1} = d_i,$$

where $\alpha_1 = b_1$, $S_1 = d_1$.

Eliminating u_{i-1} leads to

$$\left(b_i - \frac{a_i c_{i-1}}{\alpha_{i-1}} \right) u_i - c_i u_{i+1} = d_i + \frac{a_i S_{i-1}}{\alpha_{i-1}},$$

i.e.,

$$\alpha_i u_i - c_i u_{i+1} = S_i, \qquad\qquad (2.12)$$

where $\alpha_i = b_i - \dfrac{a_i c_{i-1}}{\alpha_{i-1}}$ and $S_i = d_i + \dfrac{a_i S_{i-1}}{\alpha_{i-1}}$ $(i = 2, 3, \ldots)$.

The last pair of simultaneous equations are

$$\alpha_{N-2}u_{N-2} - c_{N-2}u_{N-1} = S_{N-2}$$

and

$$-a_{N-1}u_{N-2} + b_{N-1}u_{N-1} = d_{N-1}.$$

Elimination of u_{N-2} gives

$$\left(b_{N-1} - \frac{a_{N-1}c_{N-2}}{\alpha_{N-2}}\right)u_{N-1} = d_{N-1} + \frac{a_{N-1}S_{N-2}}{\alpha_{N-2}},$$

i.e.,

$$\alpha_{N-1}u_{N-1} = S_{N-1}, \qquad (2.13)$$

Equations (2.12) and (2.13) show that the solution can be calculated from

$$u_{N-1} = \frac{S_{N-1}}{\alpha_{N-1}},$$

$$u_i = \frac{1}{\alpha_i}(S_i + c_i u_{i+1}) \quad (i = N-2, N-3, \ldots, 1)$$

where the α's and S's are given recursively by

$$\alpha_i = b_1; \quad \alpha_i = b_i - \frac{a_i}{\alpha_{i-1}}c_{i-1},$$

$$S_1 = d_1; \quad S_i = d_i + \frac{a_i}{\alpha_{i-1}}S_{i-1} \quad (i = 2, 3, \ldots, N-1).$$

In many problems α_i and a_i/α_{i-1} are independent of time and need only be calculated once, irrespective of the number of time-steps.

As an illustration consider the last worked example for which the equations were

$$4u_1 - u_2 = 0.4,$$
$$-u_1 + 4u_2 - u_3 = 0.8,$$
$$-u_2 + 4u_3 - u_4 = 1.2,$$
$$-u_3 + 4u_4 - u_5 = 1.6,$$
$$-2u_4 + 4u_5 = 1.6.$$

Hence

$$a_2 = a_3 = a_4 = 1, \quad a_5 = 2; \quad b_1 = b_2 = b_3 = b_4 = b_5 = 4;$$
$$c_1 = c_2 = c_3 = c_4 = 1; \quad d_1 = 0.4, \quad d_2 = 0.8, \quad d_3 = 1.2, \quad d_4 = d_5 = 1.6,$$

so

$$\alpha_1 = b_1 = 4; \quad \alpha_i = b_i - \frac{a_i}{\alpha_{i-1}} c_{i-1} = 4 - \frac{a_i}{\alpha_{i-1}} \quad (i = 2, 3, 4, 5),$$

giving the following coefficients which are invariant for every time-step.

$$\alpha_1 = 4,$$

$$\frac{a_2}{\alpha_1} = \frac{1}{4} = 0.25, \qquad \alpha_2 = 4 - \frac{a_2}{\alpha_1} = 3.75,$$

$$\frac{a_3}{\alpha_2} = \frac{1}{3.75} = 0.2667, \qquad \alpha_3 = 4 - \frac{a_3}{\alpha_2} = 3.7333,$$

$$\frac{a_4}{\alpha_3} = \frac{1}{3.7333} = 0.2679, \quad \alpha_4 = 4 - \frac{a_4}{\alpha_3} = 3.7321,$$

$$\frac{a_5}{\alpha_4} = \frac{2}{3.7321} = 0.5359, \quad \alpha_5 = 4 - \frac{a_5}{\alpha_4} = 3.4641.$$

As

$$S_1 = d_1 = 0.4 \text{ and } S_i = d_i + \frac{a_i}{\alpha_{i-1}} S_{i-1} \quad (i = 2, 3, 4, 5),$$

$$S_1 = 0.4,$$

$$S_2 = 0.8 + \frac{a_2}{\alpha_1} S_1 = 0.8 + (0.25)(0.4) = 0.9,$$

$$S_3 = 1.2 + \frac{a_3}{\alpha_2} S_2 = 1.4400,$$

$$S_4 = 1.6 + \frac{a_4}{\alpha_3} S_3 = 1.9858,$$

$$S_5 = 1.6 + \frac{a_5}{\alpha_4} S_4 = 2.6642,$$

and the solution for the first time-step is

$$u_5 = \frac{S_5}{\alpha_5} = 0.7691,$$

$$u_4 = \frac{1}{\alpha_4}(S_4 + c_4 u_5) = 0.7381,$$

$$u_3 = \frac{1}{\alpha_3}(S_3 + c_3 u_4) = 0.5834,$$

$$u_2 = \frac{1}{\alpha_2}(S_2 + c_2 u_3) = 0.3956,$$

$$u_1 = \frac{1}{\alpha_1}(S_1 + c_1 u_2) = 0.1989.$$

A comment on the stability of the elimination method

The non-pivoting elimination method previously described for solving the set of linear equations $\mathbf{Au} = \mathbf{d}$, with a tridiagonal matrix $\mathbf{A}$, is always stable, that is, with no growth of rounding errors, if
  (i) $a_i > 0$, $b_i > 0$ and $c_i > 0$,
  (ii) $b_i > a_{i+1} + c_{i-1}$ for $i = 1, 2, \ldots, N-1$, defining $c_0 = a_N = 0$, and
  (iii) $b_i > a_i + c_i$ for $i = 1, 2, \ldots, N-1$, defining $a_1 = c_{N-1} = 0$.
Conditions (i) and (ii), which ensure that the forward elimination is stable, state that the diagonal element must exceed the sum of the moduli of the other elements in the same column of the matrix $\mathbf{A}$ of coefficients. Conditions (i) and (iii), which ensure that the back substitution is stable, state that the diagonal element must exceed the sum of the moduli of the other elements in the same row. When these conditions are satisfied the algorithm is a very efficient one for programming on a digital computer, using a minimum of storage space.

Proof

To prove that the forward elimination procedure is stable it is necessary to show that the moduli of the multipliers $m_i = a_i/\alpha_{i-1}$ used to eliminate $u_1, u_2, \ldots$, are ≤ 1. By page 23,

$$\alpha_i = b_i - \frac{a_i c_{i-1}}{\alpha_{i-1}} = b_i - m_i c_{i-1}.$$

Therefore

$$m_{i+1} = \frac{a_{i+1}}{\alpha_i} = \frac{a_{i+1}}{b_i - m_i c_{i-1}}.$$

Hence

$$0 < m_2 < \frac{a_2}{b_1} < 1 \text{ since } b_1 > a_2 > 0 = c_0.$$

Similarly

$$0 < m_3 < \frac{a_3}{b_2 - m_2 c_1} \text{ since } a_3 > 0, \quad b_2 > c_1 \text{ and } 0 < m_2 < 1,$$

$$< \frac{a_3}{b_2 - c_1} \text{ since } c_1 > 0,$$

$$< \frac{a_3}{(a_3 + c_1) - c_1} = 1 \text{ since } b_2 > a_3 + c_1.$$

In this way, $0 < m_4, m_5, \ldots, m_{N-1} < 1$. The stability of the back substitution is proved in exercise 4, Chapter 2.

A weighted average approximation

A more general finite-difference approximation to $\partial u / \partial t = \partial^2 u / \partial x^2$ than those considered is given by

$$\frac{u_{i,j+1} - u_{i,j}}{\delta t} = \frac{1}{(\delta x)^2} \{ \theta(u_{i+1,j+1} - 2u_{i,j+1} + u_{i-1,j+1}) + (1 - \theta) \\ \times (u_{i+1,j} - 2u_{i,j} + u_{i-1,j}) \},$$

where, in practice, $0 \le \theta \le 1$. For readers familiar with finite-difference notation this replacement approximates the partial differential equation at the point $\{i\delta x, (j + \frac{1}{2})\delta t\}$ by the difference equation

$$\frac{1}{\delta t} \delta_t u_{i,j+\frac{1}{2}} = \frac{1}{(\delta x)^2} \{ \theta \delta_x^2 u_{i,j+1} + (1 - \theta) \delta_x^2 u_{i,j} \},$$

where the subscripts t and x denote differencing in the t and x directions respectively. $\theta = 0$ gives the explicit scheme, $\theta = \frac{1}{2}$ the Crank–Nicolson, and $\theta = 1$ a fully implicit backward time-difference method. The equations are unconditionally valid, i.e., stable and convergent for $\frac{1}{2} \le \theta \le 1$, but for $0 \le \theta < \frac{1}{2}$ we must have

$$r = \frac{\delta t}{(\delta x)^2} \le \frac{1}{2(1 - 2\theta)}. \quad \text{(See exercise 4, Chapter 3.)}$$

Derivative boundary conditions

Boundary conditions expressed in terms of derivatives occur very frequently in practice. When, for example, the surface of a heat-conducting material is thermally insulated, there is no heat flow normal to the surface and the corresponding boundary condition is $\partial u/\partial n = 0$ at every point of the insulated surface, where the differentiation of the temperature u is in the direction of the normal to the surface. Similarly, the rate at which heat is transferred by radiation from an external surface at temperature u into a surrounding medium at temperature v is often assumed to be proportional to $(u - v)$. As the fundamental assumption of heat-conduction theory is that the rate of flow across any surface is equal to $-K\partial u/\partial n$ units of heat per unit area per unit time in the direction of the outward normal, the corresponding boundary condition for surface radiation is

$$-K\frac{\partial u}{\partial n} = H(u - v).$$

The constant K is the thermal conductivity of the material and the constant H its coefficient of surface heat transfer. The negative sign indicates that heat is assumed to flow in the opposite direction to that in which u increases algebraically. This equation can be written as

$$\frac{\partial u}{\partial n} = -h(u - v),$$

where h is a positive constant.

Consider a thin rod that is thermally insulated along its length and which radiates heat from the end $x = 0$. The temperature at this end at time t is now unknown and its determination requires an extra equation. This equation can be the boundary condition itself when a forward difference is used for $\partial u/\partial x$, because the boundary condition at $x = 0$, the left-hand end, namely,

$$-\frac{\partial u}{\partial x} = -h(u - v),$$

will be represented by

$$\frac{u_{1,j} - u_{0,j}}{\delta x} = h(u_{0,j} - v),$$

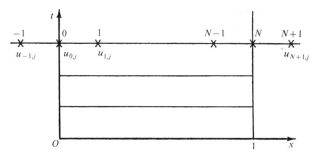

Fig. 2.7

giving one extra equation for the temperature $u_{0,j}$. A negative sign must be associated with $\partial u/\partial x$ because the outward normal to the rod at this end is in the negative direction of the x axis. Alternatively, the heat flow law, $-K\partial u/\partial n$, implies that when the positive direction of the x axis (and of u) is to the *right*, then the quantity of heat flowing from *right to left* across unit area per unit time is $+K\partial u/\partial x$, and this is proportional to the excess temperature at $x = 0$.

If we wish to represent $\partial u/\partial x$ more accurately at $x = 0$ by a central difference formula it is necessary to introduce the 'fictitious' temperature $u_{-1,j}$ at the external mesh point $(-\delta x, j\delta t)$ (Fig. 2.7), by imagining the rod to be extended very slightly. The boundary condition can then be represented by

$$\frac{u_{1,j} - u_{-1,j}}{2\delta x} = h(u_{0,j} - v).$$

The temperature $u_{-1,j}$ is unknown and necessitates another equation. This is obtained by assuming that the heat conduction equation is satisfied at the end of the rod. The unknown $u_{-1,j}$ can then be eliminated between these equations. Similar equations can be written down for radiation from the other end of the rod.

These methods are applied below to the problem of the cooling of a homogeneous rod by radiation from its ends into air at a constant temperature, the rod being at a different constant temperature initially.

Example 2.3

Solve the equation

$$\frac{\partial u}{\partial t} = \frac{\partial^2 u}{\partial x^2} \quad (0 < x < 1) \qquad (2.14)$$

satisfying the initial condition,

$$u = 1 \text{ for } 0 \leq x \leq 1 \text{ when } t = 0,$$

and the boundary conditions,

$$\frac{\partial u}{\partial x} = u \text{ at } x = 0, \text{ for all } t,$$

$$\frac{\partial u}{\partial x} = -u \text{ at } x = 1, \text{ for all } t,$$

using an explicit method and employing central-differences for the boundary conditions.

One explicit finite-difference representation of equation (2.14) is

$$\frac{u_{i,j+1} - u_{i,j}}{\delta t} = \frac{u_{i-1,j} - 2u_{i,j} + u_{i+1,j}}{(\delta x)^2},$$

i.e.,

$$u_{i,j+1} = u_{i,j} + r(u_{i-1,j} - 2u_{i,j} + u_{i+1,j}), \qquad (2.15)$$

where $r = \delta t / (\delta x)^2$.

At $x = 0$,

$$u_{0,j+1} = u_{0,j} + r(u_{-1,j} - 2u_{0,j} + u_{1,j}). \qquad (2.16)$$

The boundary condition at $x = 0$, in terms of central differences, can be written as

$$\frac{u_{1,j} - u_{-1,j}}{2\delta x} = u_{0,j}. \qquad (2.17)$$

Eliminating $u_{-1,j}$ between (2.16) and (2.17) gives

$$u_{0,j+1} = u_{0,j} + 2r\{u_{1,j} - (1 + \delta x)u_{0,j}\}. \qquad (2.18)$$

Let $\delta x = 0.1$. Then at $x = 1$, equation (2.15) becomes

$$u_{10,j+1} = u_{10,j} + r(u_{9,j} - 2u_{10,j} + u_{11,j}), \qquad (2.19)$$

and the boundary condition is

$$\frac{u_{11,j} - u_{9,j}}{2\delta x} = -u_{10,j}. \qquad (2.20)$$

Elimination of the 'fictitious' value $u_{11,j}$ between (2.19) and (2.20) yields

$$u_{10,j+1} = u_{10,j} + 2r\{u_{9,j} - (1 + \delta x)u_{10,j}\}. \qquad (2.21)$$

This result could have been deduced from the corresponding equation at $x = 0$ because of the symmetry with respect to $x = \frac{1}{2}$.

In Chapter 3 this scheme is proved to be valid for $r \leqslant 1/(2 + h\delta x)$, i.e., $r \leqslant 1/2 \cdot 1$ in this example.

Choose $r = \frac{1}{4}$. The difference equations (2.18), (2.15), then become

$$u_{0,j+1} = \tfrac{1}{2}(0 \cdot 9 u_{0,j} + u_{1,j}),$$

$$u_{i,j+1} = \tfrac{1}{4}(u_{i-1,j} + 2u_{i,j} + u_{i+1,j}) \quad (i = 1, 2, 3, 4),$$

and the use of symmetry rather than equation (2.21) gives

$$u_{5,j+1} = \tfrac{1}{4}(2u_{4,j} + 2u_{5,j}).$$

As the initial temperature is $u = 1$, the values of u at the end of the first time-step when $t = r(\delta x)^2 = \frac{1}{400}$, are

$$u_{0,1} = \tfrac{1}{2}(0 \cdot 9 + 1) = 0 \cdot 95,$$

$$u_{1,1} = \tfrac{1}{4}(1 + 2 + 1) = 1 = u_{2,1} = u_{3,1} = u_{4,1} = u_{5,1},$$

and the values at the end of the second time-step are

$$u_{0,2} = \tfrac{1}{2}(0 \cdot 9 \times 0 \cdot 95 + 1) = 0 \cdot 9275,$$

$$u_{1,2} = \tfrac{1}{4}(0 \cdot 95 + 2 + 1) = 0 \cdot 9875,$$

$$u_{2,2} = \tfrac{1}{4}(1 + 2 + 1) = 1 = u_{3,2} = u_{4,2} = u_{5,2}.$$

Similarly for subsequent time-steps. The values for several steps are recorded in Table 2.10.

TABLE 2.10

	$i = 0$ $x = 0$	1 0·1	2 0·2	3 0·3	4 0·4	5 0·5
$t = 0·0000$	1·0000	1·0000	1·0000	1·0000	1·0000	1·0000
0·0025	0·9500	1·0000	1·0000	1·0000	1·0000	1·0000
0·0050	0·9275	0·9875	1·0000	1·0000	1·0000	1·0000
0·0075	0·9111	0·9756	0·9969	1·0000	1·0000	1·0000
0·0100	0·8978	0·9648	0·9923	0·9992	1·0000	1·0000
0·0125	0·8864	0·9549	0·9872	0·9977	0·9998	1·0000
0·0150	0·8764	0·9459	0·9818	0·9956	0·9993	0·9999
0·0175	0·8673	0·9375	0·9762	0·9931	0·9985	0·9996
0·0200	0·8590	0·9296	0·9708	0·9902	0·9974	0·9991
····						
····						
0·1000	0·7175	0·7829	0·8345	0·8718	0·8942	0·9017
0·2500	0·5542	0·6048	0·6452	0·6745	0·6923	0·6983
0·5000	0·3612	0·3942	0·4205	0·4396	0·4512	0·4551
1·0000	0·1534	0·1674	0·1786	0·1867	0·1917	0·1933

The analytical solution of the partial differential equation satisfying these boundary and initial conditions is

$$u = 4 \sum_{n=1}^{\infty} \left\{ \frac{\sec \alpha_n}{(3 + 4\alpha_n^2)} e^{-4\alpha_n^2 t} \cos 2\alpha_n (x - \tfrac{1}{2}) \right\} \quad (0 < x < 1),$$

where α_n are the positive roots of

$$\alpha \tan \alpha = \tfrac{1}{2}.$$

Values of u calculated from this analytical solution are recorded in Table 2.11.

TABLE 2.11

$x =$	0	0·1	0·2	0·3	0·4	0·5
t						
0·0025	0·9460	0·9951	0·9999	1·0000	1·0000	1·0000
0·0050	0·9250	0·9841	0·9984	0·9999	1·0000	1·0000
0·0075	0·9093	0·9730	0·9950	0·9994	1·0000	1·0000
0·0100	0·8965	0·9627	0·9905	0·9984	0·9998	1·0000
0·0125	0·8854	0·9532	0·9855	0·9967	0·9994	0·9999

TABLE 2.11 *(continued)*

0·0150	0·8755	0·9444	0·9802	0·9945	0·9988	0·9996
0·0175	0·8666	0·9362	0·9748	0·9919	0·9979	0·9992
0·0200	0·8585	0·9286	0·9695	0·9891	0·9967	0·9985
....						
....						
0·1000	0·7176	0·7828	0·8342	0·8713	0·8936	0·9010
0·2500	0·5546	0·6052	0·6454	0·6747	0·6924	0·6984
0·5000	0·3619	0·3949	0·4212	0·4403	0·4519	0·4558
1·0000	0·1542	0·1682	0·1794	0·1875	0·1925	0·1941

The two solutions are compared at $x = 0\cdot2$ in Table 2.12.

TABLE 2.12

	Finite-difference solution ($x = 0\cdot2$)	Analytical solution ($x = 0\cdot2$)	Percentage error
$t = 0\cdot005$	1·0000	0·9984	0·16
0·050	0·9126	0·9120	0·07
0·100	0·8345	0·8342	0·04
0·250	0·6452	0·6454	−0·03
0·500	0·4205	0·4212	−0·16
1·000	0·1786	0·1794	−0·45

The finite-difference solution is clearly very accurate for this small value of r.

Because of the symmetry with respect to $x = \frac{1}{2}$ the solution above is the same for a rod of length $\frac{1}{2}$, thermally insulated along its length and at $x = \frac{1}{2}$, and which cools by radiation from $x = 0$ into a medium at zero temperature.

Example 2.4

Re-solve the worked example 2.3 using an explicit method and employing a forward-difference for the boundary condition at $x = 0$.

By equation (2.15), one explicit finite-difference representation of the partial differential equation is

$$u_{i,j+1} = u_{i,j} + r(u_{i-1,j} - 2u_{i,j} + u_{i+1,j}).$$

Hence, for $i = 1$,

$$u_{1,j+1} = u_{1,j} + r(u_{0,j} - 2u_{1,j} + u_{2,j}). \qquad (2.22)$$

The boundary condition at $x = 0$, namely $\partial u/\partial x = u$, in terms of a forward difference is

$$\frac{u_{1,j} - u_{0,j}}{\delta x} = u_{0,j},$$

so

$$u_{0,j} = \frac{u_{1,j}}{1 + \delta x}. \qquad (2.23)$$

Eliminating $u_{0,j}$ between (2.22) and (2.23) gives

$$u_{1,j+1} = \left(1 - 2r + \frac{r}{1 + \delta x}\right)u_{1,j} + ru_{2,j}. \qquad (2.24)$$

This scheme is valid for $0 < r \leqslant \frac{1}{2}$ (see exercise 7, Chapter 3), but in order to compare its solution with the previous one, put $r = \frac{1}{4}$ and $\delta x = 0 \cdot 1$. The relevant equations are then

$$u_{1,j+1} = \tfrac{8}{11}u_{1,j} + \tfrac{1}{4}u_{2,j},$$
$$u_{0,j+1} = \tfrac{10}{11}u_{1,j+1},$$
$$u_{i,j+1} = \tfrac{1}{4}(u_{i-1,j} + 2u_{i,j} + u_{i+1,j}) \quad (i = 2, 3, 4),$$

and

$$u_{5,j+1} = \tfrac{1}{4}(2u_{4,j} + 2u_{5,j}), \text{ because of symmetry.}$$

The solution of these equations for an initial value of $u = 1$ is shown in Table 2.13. A comparison with the analytical solution at $x = 0 \cdot 2$ is given in Table 2.14.

TABLE 2.13

$x =$	0	0·1	0·2	0·3	0·4	0·5
$t = 0 \cdot 0000$	1·0000	1·0000	1·0000	1·0000	1·0000	1·0000
0·0025	0·8884	0·9773	1·0000	1·0000	1·0000	1·0000
0·0050	0·8734	0·9607	0·9943	1·0000	1·0000	1·0000
0·0075	0·8612	0·9473	0·9873	0·9986	1·0000	1·0000
0·0100	0·8507	0·9358	0·9801	0·9961	0·9996	1·0000
0·0125	0·8415	0·9256	0·9730	0·9930	0·9989	0·9998
0·0150	0·8331	0·9164	0·9662	0·9895	0·9976	0·9993
0·0175	0·8255	0·9080	0·9596	0·9857	0·9960	0·9985

Table 2.13 (*continued*)

0·0200	0·8184	0·9003	0·9532	0·9817	0·9941	0·9973
....						
....						
0·1000	0·6869	0·7556	0·8102	0·8498	0·8738	0·8818
0·2500	0·5206	0·5727	0·6142	0·6444	0·6628	0·6689
0·5000	0·3283	0·3611	0·3873	0·4063	0·4179	0·4218
1·0000	0·1305	0·1435	0·1540	0·1615	0·1661	0·1677

Table 2.14

	Finite-difference solution ($x = 0 \cdot 2$)	Analytical solution ($x = 0 \cdot 2$)	Percentage error
$t = 0 \cdot 005$	0·9943	0·9984	−0·4
0·050	0·8912	0·9120	−2·3
0·100	0·8102	0·8342	−2·9
0·250	0·6142	0·6454	−4·8
0·500	0·3873	0·4212	−8·0
1·000	0·1540	0·1794	−14·2

Although this solution is not as good as the previous one it is still sufficiently accurate for many practical purposes.

Example 2.5

Solve example 2.3 by the Crank–Nicolson method.

This method represents $\partial u/\partial t = \partial^2 u/\partial x^2$ by

$$\frac{u_{i,j+1} - u_{i,j}}{\delta t} = \frac{1}{2}\left\{\frac{u_{i+1,j+1} - 2u_{i,j+1} + u_{i-1,j+1}}{(\delta x)^2} + \frac{u_{i+1,j} - 2u_{i,j} + u_{i-1,j}}{(\delta x)^2}\right\},$$

which can be written as

$$-ru_{i-1,j+1} + (2+2r)u_{i,j+1} - ru_{i+1,j+1} = ru_{i-1,j} + (2-2r)u_{i,j} + ru_{i+1,j}.$$
$$(2.25)$$

The central difference representation of the boundary condition at $x = 0$ is

$$\frac{u_{1,j} - u_{-1,j}}{2\delta x} = u_{0,j},$$

from which it follows that

$$u_{-1,j} = u_{1,j} - 2\delta x u_{0,j}$$

and

$$u_{-1,j+1} = u_{1,j+1} - 2\delta x u_{0,j+1}.$$

The last two equations enable us to eliminate $u_{-1,j}$ and $u_{-1,j+1}$ from the equation obtained by putting $i = 0$ in (2.25).

The boundary condition at $x = 1$ can be dealt with in the same way although in this problem it is easier to make use of the symmetry with respect to $x = \frac{1}{2}$, namely, $u_{6,j} = u_{4,j}$.

This scheme is formally valid for all finite values of r but we must keep it reasonably small if we want a close approximation to the solution of the partial differential equation. Choose $r = 1$ and $\delta x = 0 \cdot 1$. A small amount of algebra soon shows that the equations for the unknown pivotal values $u_{0,j+1}$, $u_{1,j+1}$, ..., $u_{5,j+1}$ are

$$2 \cdot 1 u_{0,j+1} - u_{1,j+1} = -0 \cdot 1 u_{0,j} + u_{1,j},$$

$$-u_{i-1,j+1} + 4 u_{i,j+1} - u_{i+1,j+1} = u_{i-1,j} + u_{i+1,j} \quad (i = 1, 2, 3, 4),$$

$$-u_{4,j+1} + 2 u_{5,j+1} = u_{4,j}.$$

For the first time-step these give

$$2 \cdot 1 u_0 - u_1 = 0 \cdot 9,$$

$$-u_0 + 4 u_1 - u_2 = 2 \cdot 0,$$

$$-u_1 + 4 u_2 - u_3 = 2 \cdot 0,$$

$$-u_2 + 4 u_3 - u_4 = 2 \cdot 0,$$

$$-u_3 + 4 u_4 - u_5 = 2 \cdot 0,$$

$$-u_4 + 2 u_5 = 1 \cdot 0.$$

These can be solved by the direct elimination method previously described. Table 2.15 records the solution for several time-steps and Table 2.16 compares it with the analytical solution at $x = 0 \cdot 2$.

TABLE 2.15

$i =$	0	1	2	3	4	5
$t = 0 \cdot 00$	1·0	1·0	1·0	1·0	1·0	1·0
0·01	0·8908	0·9707	0·9922	0·9979	0·9994	0·9997
0·02	0·8624	0·9293	0·9720	0·9900	0·9964	0·9979

TABLE 2.15 (*continued*)

. . .
. . .

0·10	0·7179	0·7834	0·8349	0·8720	0·8944	0·9018
0·25	0·5547	0·6054	0·6458	0·6751	0·6929	0·6989
0·50	0·3618	0·3949	0·4212	0·4404	0·4520	0·4559
1·00	0·1540	0·1680	0·1793	0·1874	0·1923	0·1940

TABLE 2.16

	Finite-difference solution ($x = 0·2$)	Analytical solution ($x = 0·2$)	Percentage error
$t = 0·01$	0·9922	0·9905	0·17
0·05	0·9131	0·9120	0·12
0·10	0·8349	0·8342	0·08
0·25	0·6458	0·6454	0·06
0·50	0·4212	0·4212	0·00
1·00	0·1793	0·1794	−0·06

Two-dimensional Parabolic equations. Alternating-direction implicit method

This section is concerned with the numerical solution of the equation

$$\frac{\partial u}{\partial t} = \kappa \left(\frac{\partial^2 u}{\partial x^2} + \frac{\partial^2 u}{\partial y^2} \right), \qquad (2.26)$$

over the rectangular region $0 < x < a$, $0 < y < b$, where u is known initially at all points within and on the boundary of the rectangle, and is known subsequently at all points on the boundary.

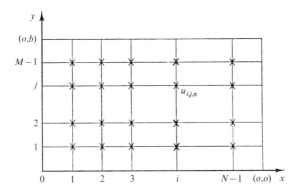

Fig. 2.8

Define the co-ordinates, (x, y, t), of the mesh points of the solution domain by

$$x = i\,\delta x, \quad y = j\,\delta y, \quad t = n\,\delta t,$$

where i, j, n are positive integers, and denote the values of u at these mesh points by

$$u(i\,\delta x, j\,\delta y, n\,\delta t) = u_{i,j,n}.$$

The explicit finite-difference representation of equation (2.26),

$$\frac{u_{i,j,n+1} - u_{i,j,n}}{\delta t} = \frac{\kappa}{(\delta x)^2}\left(u_{i-1,j,n} - 2u_{i,j,n} + u_{i+1,j,n}\right)$$

$$+ \frac{\kappa}{(\delta y)^2}\left(u_{i,j-1,n} - 2u_{i,j,n} + u_{i,j+1,n}\right),$$

appears attractively simple but is computationally laborious because the condition for its validity, which is

$$\kappa\left(\frac{1}{(\delta x)^2} + \frac{1}{(\delta y)^2}\right)\delta t \leq \tfrac{1}{2},$$

necessitates extremely small values for δt. For most problems it is an impractical method.

The Crank–Nicolson method, namely

$$\frac{u_{i,j,n+1} - u_{i,j,n}}{\delta t} = \frac{\kappa}{2}\left\{\left(\frac{\partial^2 u}{\partial x^2} + \frac{\partial^2 u}{\partial y^2}\right)_{i,j,n} + \left(\frac{\partial^2 u}{\partial x^2} + \frac{\partial^2 u}{\partial y^2}\right)_{i,j,n+1}\right\},$$

is valid for all values of δx, δy and δt, but requires the solution of $(M-1)(N-1)$ simultaneous algebraic equations for each step forward in time, where $N\,\delta x = a$, $M\,\delta y = b$. Unlike the one-dimensional case they cannot be solved by a simple recursive process. For large values of M and N they would often be solved iteratively.

Peaceman and Rachford, 1955, reference 30, put forward the following method and showed, for a typical problem with a rectangular region in the $x-y$ plane, that it involved about twenty-five times less work than the explicit method and about seven times less work than the Crank–Nicolson method.

Assume the solution is known for time $t = n\,\delta t$. Their method consists of replacing only one of the second-order derivatives, $\partial^2 u/\partial x^2$ say, by an implicit difference approximation in terms of unknown pivotal values of u from the $(n+1)$th time-level, the other second-order derivative, $\partial^2 u/\partial y^2$, being replaced by an explicit finite-difference approximation. Application of the corresponding finite-difference equation to each of the $(N-1)$ mesh points along a row parallel to Ox (Fig. 2.8), then gives $(N-1)$ equations for the $(N-1)$ unknown values of u at these mesh points for time $t = (n+1)\,\delta t$. When there are $(M-1)$ rows parallel to Ox the advancement of the solution over the whole rectangle to the $(n+1)$th time-step involves the solution of $(M-1)$ independent systems of equations, each system containing $(N-1)$ unknowns. The solution of these systems is much easier than the solution of the $(N-1)\times (M-1)$ equations associated with fully implicit methods.

The advancement of the solution to the $(n+2)$th time-level is then achieved by replacing $\partial^2 u/\partial y^2$ by an implicit finite-difference approximation and $\partial^2 u/\partial x^2$ by an explicit one, and writing down the finite-difference equation corresponding to each mesh point along columns parallel to Oy. This gives $(N-1)$ independent systems of equations, each system involving $(M-1)$ unknowns.

The time interval δt must be the same for each advancement.

Provided the solution for successive time-steps is derived by alternating between rows and columns as described above the method is valid for all ratios of $\delta t/(\delta x)^2$ and $\delta t/(\delta y)^2$. Each step on its own is unstable and unilateral repetition leads to an unacceptable growth of errors.

The detail is as follows. (Fig. 2.8.) The equation,

$$\frac{u_{i,j,n+1} - u_{i,j,n}}{\kappa\,\delta t} = \frac{u_{i-1,j,n+1} - 2u_{i,j,n+1} + u_{i+1,j,n+1}}{(\delta x)^2}$$
$$+ \frac{u_{i,j-1,n} - 2u_{i,j,n} + u_{i,j+1,n}}{(\delta y)^2},$$

is used to advance the solution from the nth to the $(n+1)$th time-step, and the equation

$$\frac{u_{i,j,n+2} - u_{i,j,n+1}}{\kappa\,\delta t} = \frac{u_{i-1,j,n+1} - 2u_{i,j,n+1} + u_{i+1,j,n+1}}{(\delta x)^2}$$
$$+ \frac{u_{i,j-1,n+2} - 2u_{i,j,n+2} + u_{i,j+1,n+2}}{(\delta y)^2},$$

for advancement from the $(n+1)$th to the $(n+2)$th time-step.

At present very little is known about the conditions under which this method is valid, and superior to others, for non-rectangular regions and more general equations. Numerical studies appear to indicate that it can be used with advantage under more general conditions but may not possess the same degree of superiority over alternative methods as it does with rectangular regions.

The parabolic equation in cylindrical and in spherical polar co-ordinates

The non-dimensional form of the equation for heat-conduction in three dimensions is $\partial u/\partial t = \nabla^2 u$, which, in cylindrical polar co-ordinates (r, θ, z) is

$$\frac{\partial u}{\partial t} = \frac{\partial^2 u}{\partial r^2} + \frac{1}{r}\frac{\partial u}{\partial r} + \frac{1}{r^2}\frac{\partial^2 u}{\partial \theta^2} + \frac{\partial^2 u}{\partial z^2}.$$

Assuming, for simplicity, that u is independent of z, this reduces to the two-dimensional equation

$$\frac{\partial u}{\partial t} = \frac{\partial^2 u}{\partial r^2} + \frac{1}{r}\frac{\partial u}{\partial r} + \frac{1}{r^2}\frac{\partial^2 u}{\partial \theta^2}. \tag{2.27}$$

For non-zero values of r there is no difficulty in expressing each derivative in terms of standard finite-difference approximations, as shown in Chapter 5, but at $r = 0$ the right side appears to contain singularities. This complication can be dealt with by replacing the polar co-ordinate form of $\nabla^2 u$ by its Cartesian equivalent which transforms equation (2.27) to the equation

$$\frac{\partial u}{\partial t} = \frac{\partial^2 u}{\partial x^2} + \frac{\partial^2 u}{\partial y^2}. \tag{2.28}$$

Now construct a circle of radius δr, centre the origin, and denote the four points in which Ox, Oy meet this circle by 1, 2, 3, 4. Denote the corresponding function values by u_1, u_2, u_3 and u_4 and the value at the origin by u_0. Then

$$\nabla^2 u = \frac{(u_1 + u_2 + u_3 + u_4 - 4u_0)}{(\delta r)^2} + O\{(\delta r)^2\}.$$

Rotation of the axes through a small angle clearly leads to a similar equation. Repetition of this rotation and the addition of all such equations then gives

$$\nabla^2 u = \frac{4(u_M - u_0)}{(\delta r)^2} + O\{(\delta r)^2\},$$

where u_M is a mean value of u round the circle. The best mean value available is given, of course, by adding all values and dividing by their number.

When a two-dimensional problem in cylindrical co-ordinates possesses circular symmetry, then $\partial^2 u/\partial \theta^2 = 0$, and equation (2.27) simplifies to

$$\frac{\partial u}{\partial t} = \frac{\partial^2 u}{\partial r^2} + \frac{1}{r}\frac{\partial u}{\partial r}. \tag{2.29}$$

Assuming $\partial u/\partial r = 0$ at $r = 0$, which it will be if the problem is symmetrical with respect to the origin, it is seen that $(1/r)\partial u/\partial r$ assumes the indeterminate form $0/0$ at this point.

By Maclaurin's expansion,

$$u'(r) = u'(0) + ru''(0) + \tfrac{1}{2}r^2 u'''(0) + \dots,$$

but $u'(0) = 0$, so the limiting value of $(1/r)\partial u/\partial r$ as r tends to zero is the value of $\partial^2 u/\partial r^2$ at $r = 0$. Hence equation (2.29) at $r = 0$ can be replaced by

$$\frac{\partial u}{\partial t} = 2\frac{\partial^2 u}{\partial r^2}. \tag{2.30}$$

This result can also be deduced from equation (2.28) because $\partial^2 u/\partial x^2 = \partial^2 u/\partial y^2$ from the circular symmetry, and we can make the x axis coincide with the direction of r. The finite-difference representation of (2.30) is further simplified by the condition $\partial u/\partial r = 0$ at $r = 0$ because this gives $u_{-1,j} = u_{1,j}$. For example, the explicit approximation

$$\frac{(u_{0,j+1} - u_{0,j})}{\delta t} = \frac{2(u_{1,j} - 2u_{0,j} + u_{-1,j})}{(\delta r)^2}$$

to equation (2.30) simplifies to

$$\frac{(u_{0,j+1} - u_{0,j})}{\delta t} = \frac{4(u_{1,j} - u_{0,j})}{(\delta r)^2} \quad \text{(See example 2.6)}$$

A complication identical to the one above also arises at $r = 0$ with the spherical polar form of $\nabla^2 u$, namely

$$\frac{\partial^2 u}{\partial r^2} + \frac{2}{r}\frac{\partial u}{\partial r} + \frac{\cot \theta}{r}\frac{\partial u}{\partial \theta} + \frac{1}{r^2}\frac{\partial^2 u}{\partial \theta^2} + \frac{1}{r^2 \sin^2 \theta}\frac{\partial^2 u}{\partial \phi^2}.$$

By the same argument as in the two dimensional case, this can be replaced at $r = 0$ by $\partial^2 u/\partial x^2 + \partial^2 u/\partial y^2 + \partial^2 u/\partial z^2$ and approximated by

$6(u_M - u_0)/(\delta r)^2$, where u_M is the mean value of u over the sphere of radius δr, centre the origin. The factor 6 occurs because Ox, Oy, Oz meet the sphere in six points. If, however the problem is symmetrical with respect to the origin, i.e., independent of θ and ϕ, $\nabla^2 u$ reduces to $\partial^2 u/\partial r^2 + (2/r)\partial u/\partial r$, with $\partial u/\partial r$ zero at $r = 0$. By either of the previous arguments it follows that the heat conduction equation at $r = 0$ becomes

$$\frac{\partial u}{\partial t} = 3\frac{\partial^2 u}{\partial r^2}.$$

It is of interest to note that symmetrical heat flow problems for hollow cylinders and spheres that *exclude* $r = 0$ can be solved by simpler equations than those considered because the change of independent variable defined by $R = \log r$ transforms the cylindrical equation

$$\frac{\partial u}{\partial t} = \frac{\partial^2 u}{\partial r^2} + \frac{1}{r}\frac{\partial u}{\partial r} \quad \text{to} \quad e^{2R}\frac{\partial u}{\partial t} = \frac{\partial^2 u}{\partial R^2},$$

and the change of dependent variable given by $u = w/r$ transforms the spherical equation

$$\frac{\partial u}{\partial t} = \frac{\partial^2 u}{\partial r^2} + \frac{2}{r}\frac{\partial u}{\partial r} \quad \text{to} \quad \frac{\partial w}{\partial t} = \frac{\partial^2 w}{\partial r^2}.$$

Example 2.6

The function U is a solution of the equation

$$\frac{\partial U}{\partial t} = \frac{\partial^2 U}{\partial x^2} + \frac{2}{x}\frac{\partial U}{\partial x}, \quad 0 < x < 1,$$

and satisfies the initial conditions

$$U = 1 - x^2 \text{ when } t = 0, \quad 0 \leqslant x \leqslant 1,$$

and the boundary conditions

$$\frac{\partial U}{\partial x} = 0 \text{ at } x = 0, \quad t > 0; \quad U = 0 \text{ at } x = 1, \quad t > 0.$$

Using a rectangular grid defined by $\delta x = 0.1$ and $\delta t = 0.001$, calculate a finite-difference solution to $4D$ by an explicit method at the points $(0, 0.001)$, $(0.1, 0.001)$ and $(0.9, 0.001)$ in the $x - t$ plane.

(See Chapter 3, exercise 6 for the stability of the difference scheme.)
At $x = 0$, $(2/x)(\partial U/\partial x)$ is indeterminate. As

$$\lim_{x \to 0} \frac{2}{x} \frac{\partial U}{\partial x} = \lim_{x \to 0} 2 \frac{\partial^2 U}{\partial x^2},$$

the equation can be replaced at $x = 0$ by

$$\frac{\partial U}{\partial t} = 3 \frac{\partial^2 U}{\partial x^2}.$$

This may be approximated by the difference equation

$$\frac{u_{0,j+1} - u_{0,j}}{\delta t} = \frac{3(u_{-1,j} - 2u_{0,j} + u_{1,j})}{(\delta x)^2}. \tag{2.31}$$

If $(\partial U/\partial x)_{i,j}$ is approximated by $(u_{i+1,j} - u_{i-1,j})/2(\delta x)$, it follows that
$u_{-1,j} = u_{1,j}$ since $(\partial U/\partial x)_{0,j} = 0$. Hence equation (2.31) reduces to

$$u_{0,j+1} = u_{0,j} + 3r(2u_{1,j} - 2u_{0,j}) = (1 - 6r)u_{0,j} + 6ru_{1,j},$$

where $r = \delta t/(\delta x)^2 = 0 \cdot 1$ in this example. Therefore

$$u_{0,j+1} = \tfrac{1}{5}(2u_{o,j} + 3u_{1,j}). \tag{2.32}$$

At $x \neq 0$ the differential equation can be approximated by

$$\frac{1}{\delta t}(u_{i,j+1} - u_{i,j}) = \frac{1}{(\delta x)^2}(u_{i-1,j} - 2u_{i,j} + u_{i+1,j}) + \frac{2}{2i(\delta x)^2}(u_{i+1,j} - u_{i-1,j})$$

giving

$$u_{i,j+1} = r\left(1 - \frac{1}{i}\right)u_{i-1,j} + (1 - 2r)u_{i,j} + r\left(1 + \frac{1}{i}\right)u_{i+1,j}.$$

Therefore

$$u_{1,j+1} = \tfrac{1}{5}(4u_{1,j} + u_{2,j}) \tag{2.33}$$

and

$$u_{9,j+1} = \tfrac{4}{5}(\tfrac{1}{9}u_{8,j} + u_{9,j}) \text{ since } u_{10,j} = 0. \tag{2.34}$$

By equations (2.32), (2.33) and (2.34) the solution values at the
points in question are as shown below

$x =$	0	0·1	0·2	$\cdots$	0·8	0·9
$t = 0$	1·0000	0·9900	0·9600	$\cdots$	0·3600	0·1900
$t = 0·001$	0·9940	0·9840				0·1840

Miscellaneous methods for improving accuracy

(i) *Reduction of the local truncation error* (Douglas' equations)

All derivatives can be expressed exactly in terms of infinite series of forward, backward, or central-differences. For example

$$\frac{\partial^2 U}{\partial x^2} = \frac{1}{h^2}(\delta_x^2 U - \tfrac{1}{12}\delta_x^4 U + \tfrac{1}{90}\delta_x^6 U + \cdots) \qquad (2.35)$$

where the subscript x denotes differencing in the x-direction and the central-differences are defined by

$$\delta_x U_{i,j} = U_{i+\frac{1}{2},j} - U_{i-\frac{1}{2},j}$$

and

$$\delta_x^2 U_{i,j} = \delta_x(\delta_x U_{i,j}) = U_{i+1,j} - 2U_{i,j} + U_{i-1,j}, \text{ etc.} \qquad (2.36)$$

In the approximation methods already considered the right-hand side of (2.35) has been truncated after the first term. If it is truncated after two or more terms the accuracy of the approximation method will always be improved but this normally increases the number of unknowns in an implicit method and complicates the boundary procedure. For equations involving second-order derivatives however it is possible to eliminate the fourth-order central-differences yet leave the number of unknowns unchanged. For example, if the equation

$$\frac{\partial U}{\partial t} = \frac{\partial^2 U}{\partial x^2}$$

is approximated at the point $(i, j+\frac{1}{2})$ by

$$\frac{1}{k}(u_{i,j+1} - u_{i,j}) = \frac{1}{2}\left\{\left(\frac{\partial^2 u}{\partial x^2}\right)_{i,j+1} + \left(\frac{\partial^2 u}{\partial x^2}\right)_{i,j}\right\}$$

$$= \frac{1}{2h^2}(\delta_x^2 - \tfrac{1}{12}\delta_x^4 + \tfrac{1}{90}\delta_x^6 + \ldots)(u_{i,j+1} + u_{i,j}),$$

then the terms involving δ_x^4 can be eliminated by operating on both sides with $(1+\tfrac{1}{12}\delta_x^2)$. This gives that

$$(1+\tfrac{1}{12}\delta_x^2)(u_{i,j+1} - u_{i,j}) = \tfrac{1}{2}r(\delta_x^2 u_{i,j+1} + \delta_x^2 u_{i,j}) + O(\delta_x^6)$$

which can be written as

$$\{1 + (\tfrac{1}{12} - \tfrac{1}{2}r)\delta_x^2\}u_{i,j+1} = \{1 + (\tfrac{1}{12} + \tfrac{1}{2}r)\delta_x^2\}u_{i,j}$$

when terms of order δ_x^6 are neglected. By (2.36) it follows that the differential equation at the point $(i, j+\frac{1}{2})$ can be approximated by the implicit algebraic equation

$$(1-6r)u_{i-1,j+1}+(10+12r)u_{i,j+1}+(1-6r)u_{i+1,j+1}$$
$$=(1+6r)u_{i-1,j}+(10-12r)u_{i,j}+(1+6r)u_{i+1,j}, \quad (2.37)$$

where $r = k/h^2$. The resulting tridiagonal system of equations can be solved by the algorithm on page 23 and requires exactly the same amount of arithmetic as the Crank–Nicolson method. Whereas, however, the local truncation error (see Chapter 3) of the Crank–Nicolson equation is $O(h^2)+O(k^2)$, it is $O(h^4)+O(k^2)$ for the Douglas equation. As proved in Chapter 3, exercise 11, the equations are stable and consistent for all positive r. The numerical solution of example 2.1 by equations (2.37) for $h = 0.1$ and $r = 1$ at $t = 0.1$ is compared with the Crank–Nicolson solution for $r = 1$ in Table 2.17.

TABLE 2.17

$x =$	0·1	0·2	0·3	0·4	0·5
Solution of P.D.E.	0·0934	0·1776	0·2444	0·2873	0·3021
Douglas solution	0·0941	0·1789	0·2463	0·2895	0·3044
C–N solution	0·0948	0·1803	0·2482	0·2918	0·3069

An explicit difference equation of h^4 accuracy is developed in exercise 12.

(ii) *Use of three time-level difference equations*

The finite-difference approximation of a parabolic equation needs only two time-levels. Three (or more) time-level schemes can be constructed but naturally this is done only to achieve some advantage over two-level schemes, such as a smaller local truncation error, greater stability, or the transformation of a non-linear problem to a linear one as is demonstrated further on in this chapter. For example, the three-level difference equation

$$\frac{3}{2}\frac{(u_{i,j+1}-u_{i,j})}{k}-\frac{1}{2}\frac{(u_{i,j}-u_{i,j-1})}{k}=\frac{(u_{i+1,j+1}-2u_{i,j+1}+u_{i-1,j+1})}{h^2}$$

approximating $\partial U/\partial t = \partial^2 U/\partial x^2$ has a truncation error of the same order as the Crank–Nicolson equation, namely $O(k^2)+O(h^2)$, but is the better one to use when the initial data is discontinuous or varies very rapidly with x. The Crank–Nicolson approximation should be used when the initial data and its derivatives are continuous (reference 31). In order to solve the first set of equations for $u_{i,2}$ it is necessary to calculate a solution along the first time-level by some other method, it being assumed that the initial data along $t = 0$ are known. This first time-level solution must be of the same accuracy as that given by the three-levels equation. A three-level variation of the Douglas equation is

$$\frac{1}{12k}\{\tfrac{3}{2}(u_{i+1,j+1}-u_{i+1,j})-\tfrac{1}{2}(u_{i+1,j}-u_{i+1,j-1})\}$$

$$+\frac{5}{6k}\{\tfrac{3}{2}(u_{i,j+1}-u_{i,j})-\tfrac{1}{2}(u_{i,j}-u_{i,j-1})\}$$

$$+\frac{1}{12k}\{\tfrac{3}{2}(u_{i-1,j+1}-u_{i-1,j})-\tfrac{1}{2}(u_{i-1,j}-u_{i-1,j-1})\}$$

$$=\frac{1}{h^2}(u_{i+1,j+1}-2u_{i,j+1}+u_{i-1,j+1}),$$

and like the Douglas equation its truncation error is $O(k^2)+O(h^4)$. A number of such schemes for constant and variable coefficients and for one and two space dimensions are discussed in references 23 and 31.

(iii) *Deferred correction method*

In this method the approximating difference equations are solved as usual. Their solution is then used to calculate a correction term, at each mesh point of the solution domain, which is added to the approximating difference equation at each mesh point. The corrected equations are then re-solved and the process repeated if necessary. The correction terms are numbers obtained by differencing the numerical solution in either the x-direction or the t-direction, or both directions. One method for deriving a correction term for the Crank–Nicolson equations is given in Chapter 3, exercise 16, but the following is better as it is based on a general result. Define the averaging operator μ by $\mu f_{j+\frac{1}{2}}=\tfrac{1}{2}(f_j+f_{j+1})$ and use the following results which are proved in most introductory books to numerical analysis, namely,

$$k\frac{\partial}{\partial t}\equiv 2\sinh^{-1}(\tfrac{1}{2}\delta_t)\quad\text{and}\quad \mu_t\equiv(1+\tfrac{1}{4}\delta_t^2)^{\frac{1}{2}}.$$

From these it follows that

$$\tfrac{1}{2}k\left\{\left(\frac{\partial U}{\partial t}\right)_{i,j}+\left(\frac{\partial U}{\partial t}\right)_{i,j+1}\right\}=k\mu_t\left(\frac{\partial U}{\partial t}\right)_{i,j+\frac{1}{2}}$$
$$=\{(1+\tfrac{1}{4}\delta_t^2)^{\frac{1}{2}}\,2\sinh^{-1}(\tfrac{1}{2}\delta_t)\}\,U_{i,j+\frac{1}{2}}.$$

The expansion of the right-hand side into positive powers of δ_t leads to

$$\tfrac{1}{2}k\left\{\left(\frac{\partial U}{\partial t}\right)_{i,j}+\left(\frac{\partial U}{\partial t}\right)_{i,j+1}\right\}=(\delta_t+\tfrac{1}{12}\delta_t^3-\tfrac{1}{120}\delta_t^5+\ldots)\,U_{i,j+\frac{1}{2}}$$

which can be rearranged as

$$\delta_t U_{i,j+\frac{1}{2}}=\tfrac{1}{2}k\left\{\left(\frac{\partial U}{\partial t}\right)_{i,j}+\left(\frac{\partial U}{\partial t}\right)_{i,j+1}\right\}+C_t U_{i,j+\frac{1}{2}},$$

giving

$$U_{i,j+1}-U_{i,j}=\tfrac{1}{2}k\frac{\partial}{\partial t}(U_{i,j}+U_{i,j+1})+C_t U_{i,j+\frac{1}{2}}, \qquad (2.38)$$

where

$$C_t\equiv-\tfrac{1}{12}\delta_t^3+\tfrac{1}{120}\delta_t^5+\ldots.$$

Equation (2.38) is a general result relating the value of a continuous function at the $(j+1)$th time-level to its value at the jth time-level in terms of first time-derivatives and central-differences in the t-direction. For the equation

$$\frac{\partial U}{\partial t}=\frac{\partial^2 U}{\partial x^2}$$

it follows that

$$\frac{\partial}{\partial t}\equiv\frac{\partial^2}{\partial x^2}.$$

Hence by (2.38),

$$U_{i,j+1}-U_{i,j}=\tfrac{1}{2}k\frac{\partial^2}{\partial x^2}(U_{i,j}+U_{i,j+1})+C_t U_{i,j+\frac{1}{2}}.$$

Using equation (2.35) it is seen that

$$U_{i,j+1}-U_{i,j}=\frac{1}{2}\frac{k}{h^2}(\delta_x^2-\tfrac{1}{12}\delta_x^4+\tfrac{1}{90}\delta_x^6+\ldots)(U_{i,j+1}+U_{i,j})+C_t U_{i,j+\frac{1}{2}}$$
$$=\tfrac{1}{2}r(\delta_x^2 U_{i,j+1}+\delta_x^2 U_{i,j})+C \qquad (2.39)$$

where

$$C = \tfrac{1}{2}r\{(-\tfrac{1}{12}\delta_x^4 U_{i,j+1} + \tfrac{1}{90}\delta_x^6 U_{i,j+1} + \ldots)$$
$$+ (-\tfrac{1}{12}\delta_x^4 U_{i,j} + \tfrac{1}{90}\delta_x^6 U_{i,j} + \ldots)\} + (-\tfrac{1}{12}\delta_t^3 U_{i,j+\frac{1}{2}} + \tfrac{1}{120}\delta_t^5 U_{i,j+\frac{1}{2}}).$$

The first approximation to the solution of (2.39) would be found by putting $C = 0$ and solving the Crank–Nicolson equations

$$u_{i,j+1} - u_{i,j} = \tfrac{1}{2}r(\delta_x^2 u_{i,j+1} + \delta_x^2 u_{i,j})$$

over the solution domain $[0 < x < 1] \times [0 < t \le T]$, say. The correction term at each mesh point would then be calculated from a truncated approximation to C such as

$$C' = -\frac{r}{24}(\delta_x^4 u_{i,j+1} + \delta_x^4 u_{i,j}) - \tfrac{1}{12}\delta_t^3 u_{i,j+\frac{1}{2}}.$$

This method, in effect, includes higher-order difference terms in the approximations to the derivatives but keeps the matrix of coefficients of the approximation equations tridiagonal which allows the algorithm on page 23 to be used.

(iv) *Richardson's deferred approach to the limit*

For this method two or more solutions approximating the problem must be known for two or more different mesh sizes and the difference between the solution of the partial differential equation and the solution of the approximating equations must be known as a function of the mesh lengths.

Let U represent the solution of the differential equation and $u_{r,s}$ represent the solution of the finite-difference equations for a mesh of size (h_r, k_s). Now assume, for example, that the discretization error

$$U - u(h, k) = Ak + Bh^2 + Ck^2 + Dh^4 + \ldots,$$

as it is for the classical explicit method for finite t.

If two solutions $u_{1,1}$ and $u_{2,2}$ are known then

$$U - u_{1,1} = Ak_1 + Bh_1^2 + Ck_1^2 + Dh_1^4 + \ldots$$

and

$$U - u_{2,2} = Ak_2 + Bh_2^2 + Ck_2^2 + Dh_2^4 + \ldots.$$

Hence B (or A, but not both) can be eliminated to give that

$$U = \frac{1}{h_2^2 - h_1^2}(h_2^2 u_{1,1} - h_1^2 u_{2,2}) + A\frac{k_1 h_2^2 - k_2 h_1^2}{h_2^2 - h_1^2} + \ldots$$

If the term involving A is negligible then U is an improvement on $u_{1,1}$ and $u_{2,2}$. For the special case $h_2 = 2h_1$, $k_2 = 2k_1$,

$$U = \tfrac{1}{3}(4u_{1,1} - u_{2,2}) + \tfrac{2}{3}kA + \ldots.$$

If three different solutions are known then A and B can be eliminated. For the Crank–Nicolson equations,

$$U - u(h, k) = Ah^2 + Bk^2 + Ch^4 + \ldots.$$

If $h_2 = \lambda h_1$ and $k_2 = \lambda k_1$ it is easily shown that

$$U = u_{2,2} + \frac{\lambda^2}{1 - \lambda^2}(u_{2,2} - u_{1,1}) + O(h^4).$$

Solution of non-linear parabolic equations

There is no difficulty in formally applying finite-difference methods to non-linear parabolic equations. The difficulties are associated with the difference equations themselves. If they are linear they can usually be solved quite easily, although we still have the problem of determining the conditions that must be satisfied for stability and convergence because the coefficients of the unknowns will be functions of the solution at earlier time-levels. If they are non-linear we have also the problem of their solution. Direct methods, in general, are difficult, so they are usually solved iteratively after being linearized in some way. Taylor's expansion provides a standard way of doing this and the method is usually referred to as Newton's method. As will be seen later this is probably not the best method for parabolic equations but is included for completeness.

Linearization by Newton's method

Let

$$f_i(u_1, u_2, \ldots, u_N) = 0, \quad i = 1(1)N, \qquad (2.40)$$

represent N equations in the N dependent variables $u_1, u_2, \ldots, u_N$.

Let V_i be a known approximation to the exact solution value u_i, $i = 1(1)N$.

Put $u_i = V_i + \epsilon_i$ and substitute into equation (2.40). Then by Taylor's expansion to first-order terms in ϵ_i, $i = 1(1)N$,

$$f_i(V_1, V_2, \ldots, V_N) + \left[\frac{\partial f_i}{\partial u_1} \epsilon_1 + \frac{\partial f_i}{\partial u_2} \epsilon_2 + \ldots + \frac{\partial f_i}{\partial u_N} \epsilon_N \right]_{u_i = V_i} = 0,$$
$$i = 1(1)N.$$

(2.41)

The subscript notation on the second bracket indicates that the dependent variables $u_1, u_2, \ldots, u_N$ appearing in the coefficients of $\epsilon_1, \epsilon_2, \ldots, \epsilon_N$ are replaced by $V_1, V_2, \ldots, V_N$ respectively after the differentiations. Equation (2.41) represents N linear equations for the N unknowns $\epsilon_1, \epsilon_2, \ldots, \epsilon_N$ because $V_1, V_2, \ldots, V_N$ are known. When the ϵ's have been calculated the process is repeated, the starting values of the dependent variables for the next iteration being $(V_i + \epsilon_i)$, $i = 1(1)N$. This process of successive approximations is continued until the u_i's have been found to the required degree of accuracy, such as $|\epsilon_i| < 10^{-8}$, $i = 1(1)N$. Some numerical results for a particular problem are given on page 55.

Example 2.7

The function U satisfies the non-linear equation

$$\frac{\partial U}{\partial t} = \frac{\partial^2 U^2}{\partial x^2}, \quad 0 < x < 1,$$

the initial condition $U = 4x(1-x)$, $0 < x < 1$, $t = 0$, and the boundary conditions $U = 0$ at $x = 0$ and 1, $t \geq 0$.

If the equation is approximated at the point $\{ih, (j + \frac{1}{2})k\}$ in the $x - t$ plane by the difference scheme

$$\frac{1}{k} \delta_t u_{i,j+\frac{1}{2}} = \frac{1}{2h^2} (\delta_x^2 u_{i,j+1}^2 + \delta_x^2 u_{i,j}^2),$$

use Newton's method to derive a set of linear equations giving an improved value $(V_i + \epsilon_i)$ to the approximate value V_i at the mesh points defined by $x_i = \frac{1}{6}i$, $i = 1(1)5$, $t = k = \frac{1}{36}$.

If the V_i are taken equal to the initial values at $x_i = \frac{1}{6}i$, $i = 1(1)5$, $t = 0$, show that these equations reduce to

$$-19\epsilon_1 + 8\epsilon_2 + \tfrac{14}{9} = 0,$$
$$5\epsilon_1 - 25\epsilon_2 + 9\epsilon_3 - \tfrac{22}{9} = 0,$$

and

$$16\epsilon_2 - 27\epsilon_3 - \tfrac{34}{9} = 0.$$

The approximation equation is

$$\frac{1}{k}(u_{i,j+1} - u_{i,j}) = \frac{1}{2h^2}\{(u_{i-1,j+1}^2 - 2u_{i,j+1}^2 + u_{i+1,j+1}^2)$$

$$+ (u_{i-1,j}^2 - 2u_{i,j}^2 + u_{i+1,j}^2)\}.$$

Put $p = h^2/k$ and denote $u_{i,j+1}$ by u_i. The equation can then be written as

$$u_{i-1}^2 - 2(u_i^2 + pu_i) + u_{i+1}^2 + \{u_{i-1,j}^2 - 2(u_{i,j}^2 - pu_{i,j}) + u_{i+1,j}^2\}$$

$$= 0 \equiv f_i(u_{i-1}, u_i, u_{i+1}).$$

By equation (2.41),

$$\left[\frac{\partial f_i}{\partial u_{i-1}}\epsilon_{i-1} + \frac{\partial f_i}{\partial u_i}\epsilon_i + \frac{\partial f_i}{\partial u_{i+1}}\epsilon_{i+1}\right]_{u_i = V_i} + f_i(V_{i-1}, V_i, V_{i+1}) = 0,$$

hence

$$2V_{i-1}\epsilon_{i-1} - 2(2V_i + p)\epsilon_i + 2V_{i+1}\epsilon_{i+1} + [\{V_{i-1}^2 - 2(V_i^2 + pV_i) + V_{i+1}^2\}$$

$$+ \{u_{i-1,j}^2 - 2(u_{i,j}^2 - pu_{i,j}) + u_{i+1,j}^2\}] = 0, \quad (2.42)$$

where V_i is an approximation to $u_{i,j+1}$.

The problem is symmetric with respect to $x = \tfrac{1}{2}$. When the V_i are taken equal to $u_{i,0}$, equation (2.42) for $j = 0$ reduces to

$$2u_{i-1,0}\epsilon_{i-1} - 2(2u_{i,0} + p)\epsilon_i + 2u_{i+1,0}\epsilon_{i+1} + \{2u_{i-1,0}^2 - 4u_{i,0}^2 + 2u_{i+1,0}^2\} = 0.$$

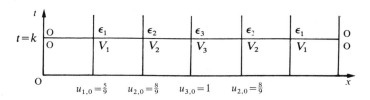

Fig. 2.9

This gives the equations quoted for $p = 1$ and the initial values indicated in Fig. 2.9 since $\epsilon_0 = 0$.

Richtmyer's linearization method

Richtmyer, reference 31, considers the equation

$$\frac{\partial U}{\partial t} = \frac{\partial^2 U^m}{\partial x^2}, \quad m \text{ a positive integer} \geq 2,$$

which he approximates by the implicit weighted average difference scheme

$$\frac{1}{k}(u_{i,j+1} - u_{i,j}) = \frac{1}{h^2}[\theta \, \delta_x^2(u_{i,j+1}^m) + (1 - \theta) \, \delta_x^2(u_{i,j}^m)]. \quad (2.43)$$

By Taylor's expansion about the point (i, j),

$$u_{i,j+1}^m = u_{i,j}^m + k \frac{\partial u_{i,j}^m}{\partial t} + \dots$$

$$= u_{i,j}^m + k \frac{\partial u_{i,j}^m}{\partial u_{i,j}} \frac{\partial u_{i,j}}{\partial t} + \dots.$$

Hence to terms of order k,

$$u_{i,j+1}^m = u_{i,j}^m + m u_{i,j}^{m-1}(u_{i,j+1} - u_{i,j}),$$

a result which replaces the non-linear unknown $u_{i,j+1}^m$ by an approximation linear in $u_{i,j+1}$.

Putting $\omega_i = u_{i,j+1} - u_{i,j}$ in equation (2.43) leads to

$$\frac{1}{k}\omega_i = \frac{1}{h^2}[\theta \, \delta_x^2(u_{i,j}^m + m u_{i,j}^{m-1} \, \omega_i) + (1 - \theta) \, \delta_x^2 u_{i,j}^m]$$

$$= \frac{1}{h^2}[m\theta \, \delta_x^2 u_{i,j}^{m-1} \, \omega_i + \delta_x^2 u_{i,j}^m]$$

$$= \frac{1}{h^2}[m\theta(u_{i-1,j}^{m-1} \, \omega_{i-1} - 2u_{i,j}^{m-1} \, \omega_i + u_{i+1,j}^{m-1} \, \omega_{i+1})$$

$$+ (u_{i-1,j}^m - 2u_{i,j}^m + u_{i+1,j}^m)],$$

which gives a set of linear equations for the ω_i. The solution at the $(j+1)$th time-level is obtained from $u_{i,j+1} = \omega_i + u_{i,j}$. For known boundary values at $x = 0$ and 1, where $Nh = 1$ and $r = k/h^2$, the equations in matrix form for $m = 2$ are

$$\begin{bmatrix} (1+4r\theta u_{1,j}) & -2r\theta u_{2,j} & & & \\ -2r\theta u_{1,j} & (1+4r\theta u_{2,j}) & -2r\theta u_{3,j} & & \\ & \cdot & & & \\ & & \cdot & & \\ & & -2r\theta u_{N-3,j}(1+4r\theta u_{N-2,j}) & -2r\theta u_{N1,j} \\ & & -2r\theta u_{N-2,j}(1+4r\theta u_{N-1,j}) \end{bmatrix} \begin{bmatrix} \omega_1 \\ \omega_2 \\ \\ \\ \omega_{N-2} \\ \omega_{N-1} \end{bmatrix}$$

$$= \begin{bmatrix} -2ru_{1,j} & ru_{2,j} & & & \\ ru_{1,j} & -2ru_{2,j} & ru_{3,j} & & \\ & \cdot & \cdot & & \\ & & \cdot & & \\ & & & ru_{N-2,j} & -2ru_{N-1,j} \end{bmatrix} \begin{bmatrix} u_{1,j} \\ u_{2,j} \\ \\ \\ u_{N-1,j} \end{bmatrix} + \begin{bmatrix} ru_{0,j}^2 + 2r\theta u_{0,j}(u_{0,j+1} - u_{0,j}) \\ 0 \\ 0 \\ \cdot \\ \cdot \\ 0 \\ ru_{N,j}^2 + 2r\theta u_{N,j}(u_{N,j+1} - u_{N,j}) \end{bmatrix}$$

$$(2.44)$$

These are easily solved by the algorithm on page 23. The stability of this scheme is considered in Chapter 3, exercise 12, and numerical results for a particular problem are given on page 55.

A three time-level method

Lees, reference 19, considered the non-linear equation

$$b(U)\frac{\partial U}{\partial t} = \frac{\partial}{\partial x}\left\{a(U)\frac{\partial U}{\partial x}\right\}, \quad a(U) > 0, \quad b(U) > 0, \quad (2.45)$$

and investigated a difference scheme that
(i) achieved linearity in the unknowns $u_{i,j+1}$ by evaluating all coefficients of $u_{i,j+1}$ at a time-level of known solution values,
(ii) preserved stability by averaging $u_{i,j}$ over three time-levels, and
(iii) maintained accuracy by using central-difference approximations.

As

$$\left(\frac{\partial U}{\partial x}\right)_{i,j} \simeq \frac{1}{h}\left(U_{i+\frac{1}{2},j} - U_{i-\frac{1}{2},j}\right) = \frac{1}{h}\,\delta_x U_{i,j},$$

an obvious central-difference approximation to (2.45) is

$$b(u_{i,j})\frac{1}{2k}\left(u_{i,j+1} - u_{i,j-1}\right) = \frac{1}{h}\,\delta_x\left\{a(u_{i,j})\frac{1}{h}\,\delta_x u_{i,j}\right\}$$

$$= \frac{1}{h^2}\,\delta_x\{a(u_{i,j})(u_{i+\frac{1}{2},j} - u_{i-\frac{1}{2},j})\}$$

$$= \frac{1}{h^2}\{a(u_{i+\frac{1}{2},j})(u_{i+1,j} - u_{i,j}) - a(u_{i-\frac{1}{2},j})(u_{i,j} - u_{i-1,j})\},$$

but this is certainly unstable for $a = b = 1$ (reference Chapter 3, exercise 9(b)). If, however, $u_{i+1,j}$, $u_{i,j}$, and $u_{i-1,j}$ are replaced by $\frac{1}{3}(u_{i+1,j+1} + u_{i+1,j} + u_{i+1,j-1})$, $\frac{1}{3}(u_{i,j+1} + u_{i,j} + u_{i,j-1})$, and $\frac{1}{3}(u_{i-1,j+1} + u_{i-1,j} + u_{i-1,j-1})$ respectively, and the coefficients $a(u_{i+\frac{1}{2},j})$ and $a(u_{i-\frac{1}{2},j})$ are replaced by $a\{\frac{1}{2}(u_{i+1,j} + u_{i,j})\}$ and $a\{\frac{1}{2}(u_{i,j} + u_{i-1,j})\}$ respectively in order to avoid mid-point values of u, Lees proved that for sufficiently small values of h and k,

$$\max_{i,j}|U_{i,j} - u_{i,j}| \leq A(h^2 + k^2),$$

where A is a constant.

For this method the equation considered in the preceding section needs to be written as

$$\frac{\partial U}{\partial t} = \frac{\partial^2 U^m}{\partial x^2} = \frac{\partial}{\partial x}\left(mU^{m-1}\frac{\partial U}{\partial x}\right).$$

A comparison of results for a particular problem

If it is assumed that $U = U(x - vt)$, v constant, is a solution of the equation

$$\frac{\partial U}{\partial t} = \frac{\partial^2 U^2}{\partial x^2}, \quad 0 < x < 1, \tag{2.46}$$

substitution into (2.46) and integration with respect to $(x - vt)$ gives that

$$\frac{A}{v}\log\left(U - \frac{A}{v}\right) + U = B - \tfrac{1}{2}v(x - vt),$$

where *A* and *B* are constants. Choosing $A = 1$ and $v = 2$, then $B = U(0,0) = 1 \cdot 5$ leads to the particular solution

$$(2U - 3) + \log(U - \tfrac{1}{2}) = 2(2t - x). \qquad (2.47)$$

For any given *x* and *t* equation (2.47) can be solved iteratively for *U* by, for example, the Newton-Raphson method. In a similar way initial values for this particular solution can be found by putting $t = 0$ in (2.47), and $x = 0(0 \cdot 1)1$, say. Boundary values at $x = 0$ and 1 for known values of *t* can be found in the same manner. These initial values and boundary values can then be used as the boundary data for approximation methods whose accuracy can be checked against the analytical solution of the differential equation.

The solution of equations (2.43) for $\theta = \tfrac{1}{2}$, $m = 2$, $h = 0 \cdot 1$, $r = k/h^2 = \tfrac{1}{2}$. and $t = 0 \cdot 5$, which corresponds to 100 time-steps, by Newton's, Richtmyer's, and the three time-levels methods are given below. All agree with the analytical solution of the differential equation to five decimal places after rounding, and with each other to 6D.

x	Solution of the P.D.E.	Newton's, Richtmyer's, and the three time-levels methods
0·1	2·149703	2·149701
0·3	1·997951	1·997948
0·5	1·849962	1·849958
0·7	1·706244	1·706240
0·9	1·567391	1·567389

EXERCISES AND SOLUTIONS
F.D.S. *means finite difference solution.*
A.S. *means analytical solution of the partial differential equation.*

1. Calculate a finite-difference solution of the equation

$$\frac{\partial u}{\partial t} = \frac{\partial^2 u}{\partial x^2} \quad (0 < x < 1),$$

satisfying the initial condition

$$u = \sin \pi x \text{ when } t = 0 \text{ for } 0 \leqslant x \leqslant 1,$$

and the boundary conditions

$$u = 0 \text{ at } x = 0 \text{ and } 1 \text{ for } t > 0,$$

using an explicit method with $\delta x = 0.1$ and $r = 0.1$.

Show by the method of separation of the variables, or merely verify, that the analytical solution is $u = e^{-\pi^2 t} \sin \pi x$. Hence check the accuracy of the numerical solution for $t = 0.005$.

Solution

Solutions for $r = 0.1$ and 0.5 and comparisons with the analytical solution at $x = 0.5$ are given below. They show clearly that when the initial function and all its derivatives are continuous, and the boundary values at $(0, 0)$ and $(1,0)$ remain equal to the initial values at these points, then the finite-difference solution can be very accurate indeed. Only one-half of the solution is shown because the problem is symmetric with respect to $x = 0.5$.

TABLE 2.18

	$x = 0$	0·1	0·2	0·3	0·4	0·5
t						
(F.D.S.) 0·005	0	0·2942	0·5596	0·7702	0·9054	0·9520
(A.S.) 0·005	0	0·2941	0·5595	0·7701	0·9053	0·9519
· · ·	·					
· · ·	·					
(F.D.S.) 0·01	0	0·2801	0·5327	0·7332	0·8602	0·9063
(A.S.) 0·01	0	0·2800	0·5325	0·7330	0·8617	0·9060
· · ·	·					
· · ·	·					
(F.D.S.) 0·02	0	0·2538	0·4828	0·6645	0·7812	0·8214
(A.S.) 0·02	0	0·2537	0·4825	0·6641	0·7807	0·8209
· · ·	·					
· · ·	·					
(F.D.S.) 0·10	0	0·1156	0·2198	0·3025	0·3556	0·3739
(A.S.) 0·10	0	0·1152	0·2191	0·3015	0·3545	0·3727

TABLE 2.19

	Finite-difference solution ($x = 0.5$)	*Analytical solution* ($x = 0.5$)	*Percentage error*
$t = 0.005$	0·9520	0·9519	negligible
$t = 0.01$	0·9063	0·9060	negligible

TABLE 2.19 (*Continued*)

$t = 0.02$	0·8214	0·8209	negligible
$t = 0.10$	0·3739	0·3727	0·3

$\mathbf{r = 0.5}$ $\qquad u_{i,j+1} = \tfrac{1}{2}(u_{i-1,j} + u_{i+1,j}).$

TABLE 2.20

	$x = 0$	0·1	0·2	0·3	0·4	0·5
t						
(F.D.S.) 0·005	0	0·2939	0·5590	0·7694	0·9045	0·9511
(F.D.S.) 0·010	0	0·2795	0·5317	0·7318	0·8602	0·9045
	0					
(F.D.S.) 0·02	0	0·2528	0·4809	0·6619	0·7781	0·8181
	.					
	.					
	.					
(F.D.S.) 0·10	0	0·1133	0·2154	0·2965	0·3486	0·3665
(A.S.) 0·10	0	0·1152	0·2191	0·3015	0·3545	0·3727

TABLE 2.21

	Finite-difference solution ($x = 0.5$)	Analytical solution ($x = 0.5$)	Percentage error
$t = 0.005$	0·9511	0·9519	−0·08
$t = 0.01$	0·9045	0·9060	−0·17
$t = 0.02$	0·8181	0·8209	−0·34
$t = 0.10$	0·3665	0·3727	−1·66

2. Calculate a numerical solution of the equation $\partial u/\partial t = \partial^2 u/\partial x^2$, $0 < x < 1$, satisfying the initial condition $u = 1$ when $t = 0$, $0 < x < 1$, and the boundary condition $u = 0$ at $x = 0$ and 1, $t \geqslant 0$.

Show by the method of separation of the variables that the analytical solution is

$$u = \frac{4}{\pi} \sum_{n=0}^{\infty} \frac{1}{(2n+1)} e^{-(2n+1)^2\pi^2 t} \sin(2n+1)\pi x.$$

Compare these solutions at $x = 0.1$ for small values of t. (See the comments below the tables in the solution on page 58.)

$\mathbf{r = 0.1}$ $\qquad u_{i,j+1} = \tfrac{1}{10}(u_{i-1,j} + 8u_{i,j} + u_{i+1,j}).$

This problem concerns the temperature changes in a uniform heat-insulated rod that is initially at a constant temperature and which is cooled by having its ends reduced to zero temperature at zero time and subsequently kept at zero.

Solution

The finite-difference solution given by equation (2.6) is shown in Table 2.22, and the analytical solution in Table 2.23. Only one-half the solution is given because the problem is symmetrical with respect to $x = \frac{1}{2}$.

TABLE 2.22 (Finite-difference solution)

$x = 0$	0·1	0·2	0·3	0·4	0·5
$t = 0{\cdot}000$ 0	1·0000	1·0000	1·0000	1·0000	1·0000
0·001 0	0·9000	1·0000	1·0000	1·0000	1·0000
0·002 0	0·8200	0·9900	1·0000	1·0000	1·0000
0·005 0	0·6566	0·9335	0·9927	0·9996	1·0000
0·010 0	0·5113	0·8283	0·9566	0·9919	0·9979
0·050 0	0·2429	0·4589	0·6263	0·7313	0·7669
0·100 0	0·1460	0·2776	0·3820	0·4490	0·4721
0·200 0	0·0546	0·1038	0·1428	0·1679	0·1766

TABLE 2.23 (Analytical solution)

$x = 0$	0·1	0·2	0·3	0·4	0·5
$t = 0{\cdot}000$ 0	1·0000	1·0000	1·0000	1·0000	1·0000
0·001 0	0·9747	1·0000	1·0000	1·0000	1·0000
0·002 0	0·8862	0·9984	1·0000	1·0000	1·0000
0·005 0	0·6827	0·9545	0·9973	0·9999	1·0000
0·010 0	0·5205	0·8427	0·9661	0·9953	0·9992
0·050 0	0·2442	0·4616	0·6304	0·7363	0·7723
0·100 0	0·1467	0·2790	0·3839	0·4513	0·4745
0·200 0	0·0547	0·1040	0·1431	0·1682	0·1769

Comments

An obvious difficulty arises in the solution domain at $(0, 0)$ because the limiting value of the initial temperature is unity as x tends to zero, whereas the limiting value of the boundary temperatures is zero as t tends to zero. In other words the temperature is discontinuous at $(0, 0)$ and its value could equally well have been chosen as 1 or $\frac{1}{2}$ instead of zero. Because of this discontinuity the finite-difference solution is a poor one near $x = 0$ for small values of *t*. It will be noticed, however, on comparing Tables 2.22 and 2.23, that the accuracy of the finite-difference solution near $x = 0$ improves as t increases. This is characteristic of parabolic equations and indicates that an implicit method would give a better solution in the neighbourhood of $(0, 0)$ than an explicit one because it would not draw its information exclusively from the first row and column. In general, however, we cannot calculate an accurate solution near a point of discontinuity by finite-difference methods unless we remove the discontinuity by a suitable transformation such as below. An alternative approach is to calculate an analytical solution that is continuous near the discontinuity.

For the problem above it is possible to remove the discontinuity at $(0, 0)$ by changing the independent variables from (x, t) to (X, T) by means of the equations

$$X = xt^{-\frac{1}{2}}; \quad T = t^{\frac{1}{2}}.$$

This expands the origin $x = t = 0$ into the positive half of the X axis and collapses the positive half of the x axis into the point at infinity along the x axis. Consquently the discontinuity in u at $x = t = 0$ is transformed into a smooth change along the positive half of the X axis. Further details are given in reference 12.

3. Derive the Crank–Nicolson equations for the problem in exercise 1 and solve them directly for at least two time-steps.

Evaluate the corresponding analytical solution and calculate the percentage error in the numerical solution.

Solution

The solution for $x = 0(0 \cdot 1)0 \cdot 5$ and $r = 1$ is given in Table 2.24. $\{x = 0(0 \cdot 1)0 \cdot 5$ means that x varies from zero to $0 \cdot 5$ by intervals of $0 \cdot 1.\}$

TABLE 2.24

	$x = 0$	$0 \cdot 1$	$0 \cdot 2$	$0 \cdot 3$	$0 \cdot 4$	$0 \cdot 5$
t						
(F.D.S.) $0 \cdot 01$	0	$0 \cdot 2802$	$0 \cdot 5329$	$0 \cdot 7335$	$0 \cdot 8623$	$0 \cdot 9067$
(A.S.) $\quad 0 \cdot 01$	0	$0 \cdot 2800$	$0 \cdot 5325$	$0 \cdot 7330$	$0 \cdot 8617$	$0 \cdot 9060$
(F.D.S.) $0 \cdot 02$	0	$0 \cdot 2540$	$0 \cdot 4832$	$0 \cdot 6651$	$0 \cdot 7818$	$0 \cdot 8221$
(A.S.) $\quad 0 \cdot 02$	0	$0 \cdot 2537$	$0 \cdot 4825$	$0 \cdot 6641$	$0 \cdot 7807$	$0 \cdot 8209$
(F.D.S.) $0 \cdot 10$	0	$0 \cdot 1160$	$0 \cdot 2207$	$0 \cdot 3037$	$0 \cdot 3571$	$0 \cdot 3754$
(A.S.) $\quad 0 \cdot 10$	0	$0 \cdot 1152$	$0 \cdot 2191$	$0 \cdot 3015$	$0 \cdot 3545$	$0 \cdot 3727$

TABLE 2.25

	Finite-difference solution $(x = 0.5)$	Analytical solution $(x = 0.5)$	Percentage error
$t = 0.01$	0.9067	0.9060	0.08
0.02	0.8221	0.8209	0.15
0.10	0.3754	0.3727	0.72

4. Prove that the back-substitution procedure of the non-pivoting elimination algorithm given on page 23 for solving a tridiagonal system of linear equations is stable when $a_i > 0$, $b_i > 0$, $c_i > 0$, and $b_i > a_i + c_i$, $i = 1, 2, \ldots, (N-1)$, where $a_1 = c_{N-1} = 0$.

Solution

$$u_i = \frac{1}{\alpha_i}(S_i + c_i u_{i+1}) = p_{i+1} u_{i+1} + \frac{S_i}{\alpha_i}, \quad \text{(say)},$$

where

$$\alpha_1 = b_1, \quad \alpha_i = b_i - \frac{a_i c_{i-1}}{\alpha_{i-1}}, \quad S_1 = d_1, \quad S_i = d_i + \frac{a_i S_{i-1}}{\alpha_{i-1}},$$

$$a_1 = 0 = c_{N-1}, \quad i = 1, 2, \ldots, N-1.$$

There will be no build-up of errors in the back-substitution process if $|p_{i+1}| < 1$,

where

$$p_{i+1} = \frac{c_i}{\alpha_i} = \frac{c_i}{b_i - a_i p_i}, \quad i = 1(1)(N-1).$$

Now $p_2 = (c_1/b_1)$, since $a_1 = 0$. Also $b_1 > c_1$ by hypothesis.
    Therefore $0 < p_2 < 1$.

$$p_3 = \frac{c_2}{b_2 - a_2 p_2}.$$

As $c_2 > 0$, $0 < p_2 < 1$, and $b_2 > a_2 > 0$, it follows that

$$0 < p_3 < \frac{c_2}{b_2 - a_2}.$$

By hypothesis, $b_2 > a_2 + c_2$. Hence

$$0 < p_3 < \frac{c_2}{(a_2 + c_2) - a_2} = 1.$$

Similarly $0 < p_4, p_5, \ldots, p_{N-1} < 1$.

5. Use the Gaussian elimination algorithm on page 23 to calculate a finite-difference approximation to worked example 2.1 for one time-step taking $\delta x = 0\cdot 1$ and $r = 1$ using either
(i) the fully implicit backward time-difference equation

$$u_{i,j+1} - u_{i,j} = r(u_{i-1,j+1} - 2u_{i,j+1} + u_{i+1,j+1}),$$

or (ii) the Douglas equation (2.37).

Solution

(i) $\qquad -ru_{i-1,j+1} + (1+2r)u_{i,j+1} - ru_{i+1,j+1} = u_{i,j}.$

The equations for the first time-step are $3u_1 - u_2 = 0\cdot 2$, $-u_1 + 3u_2 - u_3 = 0\cdot 4$, $-u_2 + 3u_3 - u_4 = 0\cdot 6$, $-u_3 + 3u_4 - u_5 = 0\cdot 8$, and $-2u_4 + 3u_5 = 1\cdot 0$.

$\alpha_1 = 3$, $\alpha_2 = 2\cdot 66667$, $\alpha_3 = 2\cdot 62500$, $\alpha_4 = 2\cdot 61905$, $\alpha_5 = 2\cdot 23636$.
$S_1 = 0\cdot 2$, $S_2 = 0\cdot 46667$, $S_3 = 0\cdot 77500$, $S_4 = 1\cdot 09524$, $S_5 = 1\cdot 83636$.
$u_5 = S_5/\alpha_5$ and $u_i = (S_i + c_i u_{i+1})/\alpha_i$ giving $u_5 = 0.8211$, $u_4 = 0.7317$, $u_3 = 0.5740$, $u_2 = 0.3902$, $u_1 = 0.1967$.

(ii) $(1-6r)u_{i-1,j+1} + (10+12r)u_{i,j+1} + (1-6r)u_{i+1,j+1}$
$$= (1+6r)u_{i-1,j} + (10-12r)u_{i,j} + (1-6r)u_{i+1,j}.$$

The equations for the first time-step are $22u_1 - 5u_2 = 2\cdot 4$, $-5u_1 + 22u_2 - 5u_3 = 4\cdot 8$, $-5u_2 + 22u_3 - 5u_4 = 7\cdot 2$, $-5u_3 + 22u_4 - 5u_5 = 9\cdot 6$, and $-10u_4 + 22u_5 = 9\cdot 2$.

$\alpha_1 = 22$, $\alpha_2 = 20\cdot 86364$, $\alpha_3 = 20\cdot 80174$, $\alpha_4 = 20\cdot 79818$, $\alpha_5 = 19\cdot 59594$.
$S_1 = 2\cdot 4$, $S_2 = 5\cdot 34545$, $S_3 = 8\cdot 48104$, $S_4 = 11\cdot 63855$, $S_5 = 14\cdot 79594$.
$u_5 = 0\cdot 7551$, $u_4 = 0\cdot 7411$, $u_3 = 0\cdot 5858$, $u_2 = 0\cdot 3966$, and $u_1 = 0\cdot 1992$.

6. The function u satisfies the equation

$$\frac{\partial u}{\partial t} = \frac{\partial^2 u}{\partial x^2} \quad (0 < x < 1),$$

and the boundary conditions

$$\frac{\partial u}{\partial x} = h_1(u - v_1) \text{ at } x = 0, \quad \frac{\partial u}{\partial x} = -h_2(u - v_2) \text{ at } x = 1,$$

where h_1, h_2, v_1, v_2 are positive constants.

(*a*) When the boundary conditions are approximated by central differences (see worked example 2.3) show that one explicit difference scheme is

$$u_{0,j+1} = \{1 - 2r(1 + h_1\,\delta x)\}u_{0,j} + 2ru_{1,j} + 2rh_1v_1\,\delta x,$$

$$u_{i,j+1} = ru_{i-1,j} + (1 - 2r)u_{i,j} + ru_{i+1,j}, \quad (i = 1, 2 \ldots, N-1),$$

$$u_{N,j+1} = 2ru_{N-1,j} + \{1 - 2r(1 + h_2\,\delta x)\}u_{N,j} + 2rh_2v_2\,\delta x,$$

where $N\,\delta x = 1$ and $r = \delta t/(\delta x)^2$.

(*b*) when the boundary conditions are approximated by forward-differences at $x = 0$ and backward-differences at $x = 1$ (see worked example 2.4), show that another explicit difference scheme is

$$u_{1,j+1} = \{1 - 2r + r/(1 + h_1\,\delta x)\}u_{1,j} + ru_{2,j} + rh_1v_1\,\delta x/(1 + h_1\,\delta x),$$

$$u_{0,j+1} = (u_{1,j+1} + h_1v_1\,\delta x)/(1 + h_1\,\delta x),$$

$$u_{i,j+1} = ru_{i-1,j} + (1 - 2r)u_{i,j} + ru_{i+1,j}, \quad (i = 2, 3, \ldots, N-2),$$

$$u_{N-1,j+1} = \{1 - 2r + r/(1 + h_2\,\delta x)\}u_{N-1,j} + ru_{N-2,j} + rh_2v_2\,\delta x/(1 + h_2\,\delta x),$$

$$u_{N,j+1} = (u_{N-1,j+1} + h_2v_2\,\delta x)/(1 + h_2\,\delta x).$$

7. A uniform solid rod of one-half a unit of length is thermally insulated along its length and its initial temperature at zero time is $0°C$. One end is thermally insulated and the other supplied with heat at a steady rate. Show that the subsequent temperatures at points within the rod are given, in non-dimensional form, by the solution of the equation

$$\frac{\partial u}{\partial t} = \frac{\partial^2 u}{\partial x^2} \quad 0 < x < \tfrac{1}{2}),$$

satisfying the initial condition

$$u = 0 \text{ when } t = 0 \quad (0 \le x \le \tfrac{1}{2}),$$

and the boundary conditions

$$\frac{\partial u}{\partial x} = 0 \text{ at } x = 0, \quad t > 0, \quad \frac{\partial u}{\partial x} = f \text{ at } x = \tfrac{1}{2}, \quad t > 0$$

where f is a constant.

Solve this problem numerically for $f = 1$ using
(*a*) an explicit method with $\delta x = 0\cdot 1$ and $r = \tfrac{1}{4}$,
(*b*) an implicit method with $\delta x = 0\cdot 1$ and $r = 1$.

Solution

The solution given by the explicit method of worked example 2.3, for which the equations are

$$u_{0,j+1} = \tfrac{1}{2}(u_{0,j} + u_{1,j}),$$

$$u_{i,j+1} = \tfrac{1}{4}(u_{i-1,j} + 2u_{i,j} + u_{i+1,j}) \quad (i = 1, 2, 3, 4),$$

$$u_{5,j+1} = \tfrac{1}{2}(u_{4,j} + u_{5,j} + 0\cdot1),$$

is recorded in Table 2.26. The Crank–Nicolson solution, as in the worked example 2.3, is shown in Table 2.27. The analytical solution of the differential equation is

$$u = 2t + \frac{1}{2}\left\{ \frac{12x^2 - 1}{6} - \frac{2}{\pi^2} \sum_{n=1}^{\infty} \frac{(-1)^n}{n^2} e^{-4\pi^2 n^2 t} \cos 2n\pi x \right\}$$

$$= 2\sqrt{t} \sum_{n=0}^{\infty} \left\{ i \operatorname{erfc} \frac{(2n+1-2x)}{4\sqrt{t}} + i \operatorname{erfc} \frac{(2n+1+2x)}{4\sqrt{t}} \right\},$$

and is evaluated in Table 2.28. Comparisons are made in Table 2.29.

The finite-difference solutions are clearly very accurate except for small values of t.

A simple calculation shows that the effect of the exponential component of the analytical solution is negligible for values of t in excess of $0\cdot1$.

A point of interest in this example is that the difference between the analytical solution and both finite-difference solutions for values of t in excess of $0\cdot1$ is $0\cdot0017 = 0\cdot01/6 = (\delta x)^2/6$. It is proved analytically at the end of Chapter 3 that the transient component of the solution of any explicit or implicit finite-difference scheme for a parabolic equation satisfying the boundary conditions above does not tend to zero as t increases, as does the transient, i.e., exponential component of the solution of the differential equation, but tends to a value of $k(\delta x)^2$, k constant. In this example $k = \tfrac{1}{6}$.

TABLE 2.26 (Explicit method)

$x =$	0	0·1	0·2	0·3	0·4	0·5
$t = 0\cdot005$	0·0000	0·0000	0·0000	0·0000	0·0125	0·0750
0·0075	0·0000	0·0000	0·0000	0·0031	0·0250	0·0938
0·01	0·0000	0·0000	0·0008	0·0078	0·0367	0·1094

TABLE 2.26 (*continued*)

0·02	0·0009	0·0027	0·0103	0·0313	0·0767	0·1571
0·03	0·0062	0·0104	0·0248	0·0554	0·1095	0·1934
0·05	0·0291	0·0364	0·0594	0·1007	0·1636	0·2509
0·10	0·1169	0·1265	0·1556	0·2044	0·2735	0·3631
0·20	0·3150	0·3250	0·3550	0·4050	0·4750	0·5650
0·50	0·9150	0·9250	0·9550	1·0050	1·0750	1·1650
1·00	1·9150	1·9250	1·9550	2·0050	2·0750	2·1650

TABLE 2.27 (Crank–Nicolson method)

$x =$	0	0·1	0·2	0·3	0·4	0·5
$t = 0·01$	0·0003	0·0006	0·0022	0·0083	0·0309	0·1155
0·02	0·0023	0·0039	0·0108	0·0302	0·0770	0·1540
0·03	0·0077	0·0115	0·0252	0·0552	0·1080	0·1925
0·05	0·0301	0·0373	0·0597	0·1004	0·1627	0·2499
0·10	0·1172	0·1268	0·1557	0·2043	0·2732	0·3628
0·20	0·3150	0·3250	0·3550	0·4050	0·4750	0·5650
0·50	0·9150	0·9250	0·9550	1·0050	1·0750	1·1650
100	1·9150	1·9250	1·9550	2·0050	2·0750	2·1650

TABLE 2.28 (Analytical solution)

$x =$	0	0·1	0·2	0·3	0·4	0·5
$t = 0·0025$	0·0000	0·0000	0·0000	0·0001	0·0050	0·0564
0·0050	0·0000	0·0000	0·0001	0·0017	0·0167	0·0798
0·0075	0·0000	0·0000	0·0006	0·0053	0·0286	0·0977
0·01	0·0000	0·0002	0·0017	0·0101	0·0399	0·1128
0·02	0·0016	0·0035	0·0117	0·0333	0·0791	0·1596
0·03	0·0074	0·0117	0·0264	0·0573	0·1115	0·1954
0·05	0·0307	0·0381	0·0610	0·1023	0·1653	0·2526
0·10	0·1186	0·1282	0·1573	0·2061	0·2751	0·3647
0·20	0·3167	0·3267	0·3567	0·4067	0·4766	0·5666
0·50	0·9167	0·9267	0·9567	1·0067	1·0767	1·1667
1·00	1·9167	1·9267	1·9567	2·0067	2·0767	2·1667

TABLE 2.29

t	Analytical solution $x = 0.3$	Explicit solution $x = 0.3$	Percentage error	Crank–Nicolson solution $x = 0.3$	Percentage error
0.0075	0.0053	0.0031	−41.5		
0.01	0.0101	0.0078	−22.8	0.0083	−17.8
0.05	0.1023	0.1007	− 1.56	0.1004	− 1.85
0.10	0.2061	0.2044	− 0.82	0.2043	− 0.87
0.50	1.0067	1.0050	− 0.17	1.0050	− 0.17
1.00	2.0067	2.0050	− 0.08	2.0050	− 0.08

8. The function U satisfies the equation

$$\frac{\partial U}{\partial t} = x \frac{\partial^2 U}{\partial x^2}, \quad 0 < x < \tfrac{1}{2},$$

the boundary conditions

$$U = 0 \text{ at } x = 0, \quad t > 0, \quad \frac{\partial U}{\partial x} = -\tfrac{1}{2}U \text{ at } x = \tfrac{1}{2}, \quad t > 0,$$

and the initial condition $U = x(1 - x)$ when $t = 0$, $0 \leqslant x \leqslant \tfrac{1}{2}$.

When all the derivatives with respect to x are expressed in terms of central-difference formulae show that the simplest explicit difference equations approximating this problem at the point (ih, jk) in the $x - t$ plane can be written as

$$u_{i,j+1} = irh u_{i-1,j} + (1 - 2irh)u_{i,j} + irh u_{i+1,j}, \quad i = 1(1)(N-1),$$

and

$$u_{N,j+1} = 2Nrh u_{N-1,j} + (1 - 2Nrh - Nrh^2)u_{N,j},$$

where $r = k/h^2$ and $Nh = \tfrac{1}{2}$.

Take $h = 0.1$ and $r = 0.5$ and calculate a numerical solution to 4D at the points along the first time-level corresponding to $i = 3$ and 5. Will the solution at points near the point $(\tfrac{1}{2}, 0)$ be very accurate? Give a reason for your answer. (The stability of these equations is considered in Chapter 3, exercise 8.)

Solution

The simplest explicit approximation is

$$\frac{u_{i,j+1} - u_{i,j}}{k} = \frac{ih(u_{i-1,j} - 2u_{i,j} + u_{i+1,j})}{h^2},$$

giving

$$u_{i,j+1} = irhu_{i-1,j} + (1-2irhu)u_{i,j} + irhu_{i+1,j}, \quad i = 1(1)(N-1).$$

Mentally extend the interval of integration $0 \le x \le \frac{1}{2}$ to $0 \le x \le \frac{1}{2} + h$. Then the previous equation holds for $i = N$. Eliminate $u_{N+1,j}$ by means of the approximating boundary condition equation

$$(u_{N+1,j} - u_{N-1,j})/2h = -\tfrac{1}{2}u_{N,j}$$

to give

$$u_{N,j+1} = 2Nrhu_{N-1,j} + (1 - 2Nrh - Nrh^2)u_{N,j}.$$

For $h = 0\cdot1$ and $r = \frac{1}{2}$,

$$u_{i,j+1} = 0\cdot05iu_{i-1,j} + (1 - 0\cdot1i)u_{i,j} + 0\cdot05iu_{i+1,j}, \quad i = 1(1)4,$$

and

$$u_{5,j+1} = 0\cdot5u_{4,j} + 0\cdot475u_{5,j}.$$

$i =$	0	1	2	3	4	5
$t = 0$	0	0·0900	0·1600	0·2100	0·2400	0·2500
$t = 0\cdot005$	0			0·2070		0·2388

As $(x, 0) \to (\frac{1}{2}, 0)$, $\partial U/\partial x = 1 - 2x \to 0$. At $(\frac{1}{2}, 0)$, $U = \frac{1}{2}(1 - \frac{1}{2}) = \frac{1}{4}$. As $(\frac{1}{2}, t) \to (\frac{1}{2}, 0)$, $\partial U/\partial x = -\frac{1}{2}U \to -\frac{1}{2} \cdot \frac{1}{4} = -\frac{1}{8}$. Therefore $\partial U/\partial x$ is discontinuous at $(\frac{1}{2}, 0)$ so the finite-difference solution will not be very accurate at mesh points near $(\frac{1}{2}, 0)$. The solution of the differential equation is continuous at $(\frac{1}{2}, 0)$ and the effect of this discontinuous derivative on the difference solution is very small a few mesh lengths from $(\frac{1}{2}, 0)$.

9. The function V satisfies the non-linear equation

$$\frac{\partial V}{\partial t} = \frac{\partial^2 V}{\partial x^2} + \left(\frac{\partial V}{\partial x}\right)^2, \quad 0 < x < 1,$$

the initial condition $V = 0$ when $t = 0$, $0 \le x \le 1$, and the boundary conditions

$$\frac{\partial V}{\partial x} = 1 \text{ at } x = 0, t > 0; \quad V = 0 \text{ at } x = 1, t > 0.$$

Show that the change of dependent variable defined by $V = \log U$, $U \neq 0$, transforms the problem to the solution of the equation

$$\frac{\partial U}{\partial t} = \frac{\partial^2 U}{\partial x^2}, \quad 0 < x < 1,$$

where U satisfies the conditions

$$U = 1 \text{ when } t = 0, \quad 0 \leqslant x \leqslant 1,$$

$$\frac{\partial U}{\partial x} = U \text{ at } x = 0, \ t > 0; \quad U = 1 \text{ at } x = 1, \ t > 0.$$

Using a rectangular mesh defined by $\delta x = 0 \cdot 1$ and $\delta t = 0 \cdot 0025$, approximate the equation for U by the classical explicit scheme, the derivative boundary condition being approximated by a central-difference formula.

Hence calculate a numerical solution for V at the points $(0, 0 \cdot 005)$ and $(0, 0 \cdot 0075)$ in the $x - t$ plane.

Solution

$$\frac{\partial V}{\partial t} = \frac{1}{U}\frac{\partial U}{\partial t}; \quad \frac{\partial V}{\partial x} = \frac{1}{U}\frac{\partial U}{\partial x}; \quad \frac{\partial^2 V}{\partial x^2} = -\frac{1}{U^2}\left(\frac{\partial U}{\partial x}\right)^2 + \frac{1}{U}\frac{\partial^2 U}{\partial x^2}.$$

By $V = \log U$, $U = 1$ when $V = 0$. As $\partial V/\partial x = (1/U)(\partial U/\partial x)$ and $\partial V/\partial x = 1$ at $x = 0$, $t > 0$, therefore $\partial U/\partial x = U$ at $x = 0$, $t > 0$. The approximation equation is

$$u_{i,j+1} = \tfrac{1}{4}(u_{j-i,j} + 2u_{i,j} + u_{i+1,j}), \quad i = 1(1)9.$$

Putting $i = 0$ and eliminating $u_{-1,j}$ by means of the derivative boundary equation $(u_{i,j} - u_{-1,j})/0 \cdot 2 = u_{0,j}$, leads to $u_{0,j+1} = 0 \cdot 45 u_{0,j} + 0 \cdot 5 u_{1,j}$.

$x =$	0	0·1	0·2
$t = 0$	1·0000	1·0000	1·0000
$t = 0 \cdot 0025$	0·9500	1·0000	1·0000
$t = 0 \cdot 0050$	0·9275	0·9875	1·0000
$t = 0 \cdot 0075$	0·9111		

$V(0, 0 \cdot 005) = \log_e 0 \cdot 9275 = -0 \cdot 0753$. $V(0, 0 \cdot 0075) = \log_e 0 \cdot 9111 = -0 \cdot 0931$.

10. The function U is a solution of the equation

$$\frac{\partial U}{\partial t} = \frac{\partial^2 U}{\partial r^2} + \frac{1}{r}\frac{\partial U}{\partial r} + \frac{1}{r^2}\frac{\partial^2 U}{\partial \theta^2}, \quad 0 < r < 1,$$

at every point $P(r, \theta, t)$ of the open bounded domain $0 < r < 1$, $t > 0$, and satisfies the initial condition $U = r \sin \frac{1}{2}\theta$, $0 \le r \le 1$, $t = 0$, and the boundary condition $\partial U/\partial r = -U$ at $r = 1$, $t > 0$, where (r, θ, t) are the cylindrical polar co-ordinates of P.

Take $\delta r = 0 \cdot 1$, $\delta\theta = \pi/16$, and $\delta t = 0 \cdot 01(\delta r)^2$, and use an explicit method to calculate the numerical values of $u(o, \pi, \delta t)$ and $u(1, \pi, \delta t)$, where u is an approximation to U.

Solution

As shown on page 40 the equation may be approximated at $(0, \theta_j, t_n)$ by $(u_{o,j,n+1} - u_{o,j,n})/\delta t = 4(u_M - u_{0,j,n})$, where u_M is the mean value of u round the circle $r = 0 \cdot 1$ at time $t_n = n\,\delta t$. Initially $u(r, \theta, 0) = U(r, \theta, 0)$. Hence

$$u_M = \frac{1}{2\pi}\int_0^{2\pi} r \sin\frac{\theta}{2}\,d\theta = \frac{0 \cdot 1}{2\pi}\int_0^{2\pi}\sin\frac{\theta}{2}\,d\theta = 0 \cdot 0637.$$

As $u_{0,j,0} = 0$ for all j,

$$u(0, \pi, \delta t) = u_{0,16,1} = 4\,\delta t u_M/(\delta r)^2 = 0 \cdot 04 u_M = 0 \cdot 0025.$$

As shown on page 214 the equation can be approximated at the point (r_i, θ_j, t_n) by

$$\frac{u_{i,j,n+1} - u_{i,j,n}}{\delta t} = \frac{1}{(\delta r)^2}\left\{\left(1 - \frac{1}{2i}\right)u_{i-1,j,n} + \left(1 + \frac{1}{2i}\right)u_{i+1,j,n}\right.$$
$$\left. -2\left(1 + \frac{1}{i^2(\delta\theta)^2}\right)u_{i,j,n} + \frac{1}{i^2(\delta\theta)^2}(u_{i,j-1,n} + u_{i,j+1,n})\right\}.$$

By the boundary condition, $(u_{11,j,n} - u_{9,j,n})/2(\delta r) = -u_{10,j,n}$. Hence $u_{11,j,n} = u_{9,j,n} - 0 \cdot 2u_{10,j,n}$. Putting $i = 10$, $j = 16$, and $n = 0$ in the difference equation leads to

$$u_{10,16,1} = u_{10,16,0} + 0 \cdot 01\left\{\tfrac{19}{20}u_{9,16,0} + \tfrac{21}{20}(u_{9,16,0} - 0 \cdot 2u_{10,16,0})\right.$$
$$\left. -2\left(1 + \frac{16^2}{100\pi^2}\right)u_{10,16,0} + \frac{16^2}{100\pi^2}(u_{10,15,0} + u_{10,17,0})\right\},$$

where $u_{10,16,0} = \sin(\pi/2) = 1$, $u_{9,16,0} = 0 \cdot 9$, and $u_{10,15,0} = u_{10,17,0} = 0 \cdot 9952$. Therefore $u_{10,16,1} = 1 - 0 \cdot 01882 = 0 \cdot 9812$ to 4D.

11. The cooling temperature u of a nylon thread being wound onto a spinning bobbin is given by the solution of the equation

$$a \frac{\partial u}{\partial t} = \frac{\partial^2 u}{\partial r^2} + \frac{1}{r} \frac{\partial u}{\partial r}, \quad (0 < r < 1),$$

satisfying the initial condition

$$u = 1 \text{ when } t = 0 \text{ for all } r,$$

and the boundary conditions

$$\frac{\partial u}{\partial r} = 0 \text{ at } r = 0, \quad \frac{1}{u} \frac{\partial u}{\partial r} = -F(t) \text{ at } r = 1,$$

where a is a constant and $F(t)$ an empirical function of t.

When $r_i = i \, \delta r$, $n \, \delta r = 1$, $t_j = j \, \delta t$, and the boundary conditions are represented by central-difference approximations, show that the Crank–Nicolson method gives the following equations for a numerical solution of this problem.

$$a \frac{(u_{0,j+1} - u_{0,j})}{\delta t} = \frac{2}{(\delta r)^2} (u_{1,j+1} - u_{0,j+1} + u_{1,j} - u_{0,j}),$$

$$a \frac{(u_{i,j+1} - u_{i,j})}{\delta t} = \frac{1}{2(\delta r)^2} \left\{ \left(1 + \frac{1}{2i}\right) u_{i+1,j+1} - 2u_{i,j+1} \right.$$

$$\left. + \left(1 - \frac{1}{2i}\right) u_{i-1,j+1} + \left(1 + \frac{1}{2i}\right) u_{i+1,j} - 2u_{i,j} + \left(1 - \frac{1}{2i}\right) u_{i-1,j} \right\},$$

$(i = 1, 2, \ldots, n-1)$, and

$$a \frac{(u_{n,j+1} - u_{n,j})}{\delta t} = \frac{1}{(\delta r)^2} \left[-\left\{ 1 + \left(1 + \frac{1}{2n}\right) \delta r . F_{j+1} \right\} u_{n,j+1} \right.$$

$$\left. + u_{n-1,j+1} - \left\{ 1 + \left(1 + \frac{1}{2n}\right) \delta r . F_j \right\} u_{n,j} + u_{n-1,j} \right],$$

where the terms suffixed $(j + 1)$ represent unknowns. (Reference 1.)

12. Show that $U_{i,j+1} = \{\exp(k(\partial/\partial t))\} U_{i,j}$, where $U_{i,j} = U(x_i, t_j)$, $x_i = ih$, $i = 0, \pm 1, \pm 2, \ldots$, and $t_j = jk$, $j = 0, 1, 2, \ldots$. Using the result

$$D_x \equiv \frac{2}{h} \sinh^{-1}(\tfrac{1}{2}\delta_x) = \frac{1}{h}(\delta_x - \tfrac{1}{24}\delta_x^3 + \tfrac{3}{640}\delta_x^5 + \ldots),$$

where $D_x \equiv (\partial/\partial x)$ and $\delta_x U_{i,j} = U_{i+\frac{1}{2},j} - U_{i-\frac{1}{2},j}$, show that the *exact* difference replacement of the equation $(\partial U/\partial t) = (\partial^2 U/\partial x^2)$ is given

by

$$U_{i,j+1} = \{1 + r\delta_x^2 + \tfrac{1}{2}r(r-\tfrac{1}{6})\,\delta_x^4 + \tfrac{1}{6}r(r^2 - \tfrac{1}{2}r + \tfrac{1}{15})\,\delta_x^6 + \ldots\}U_{i,j},$$

where $r = k/h^2$.

If only second-order central-differences are retained show that the function $u_{i,j}$ approximating $U_{i,j}$ is the solution of the classical explicit equation

$$u_{i,j+1} = ru_{i-1,j} + (1-2r)u_{i,j} + ru_{i+1,j}.$$

If only second and fourth-order differences are retained show that $u_{i,j}$ is the solution of the explicit equation

$$u_{i,j+1} = \tfrac{1}{2}(2 - 5r + 6r^2)u_{i,j} + \tfrac{2}{3}r(2 - 3r)(u_{i+1,j} + u_{i-1,j}) \\ - \tfrac{1}{12}r(1 - 6r)(u_{i+2,j} + u_{i-2,j}).$$

(This equation is stable for $r \leqslant \tfrac{2}{3}$.)

Solution

By Taylor's series,

$$U_{i,j+1} = U(x_i, t_j + k) = U_{i,j} + k\frac{\partial U_{i,j}}{\partial t} + \frac{1}{2}k^2\frac{\partial^2 U_{i,j}}{\partial x^2} + \ldots$$

$$= \left(1 + k\frac{\partial}{\partial t} + \frac{1}{2!}k^2\frac{\partial^2}{\partial t^2} + \frac{1}{3!}k^3\frac{\partial^3}{\partial t^3} + \ldots\right)U_{i,j} = \left\{\exp\left(k\frac{\partial}{\partial t}\right)\right\}U_{i,j}.$$

For the equation

$$\frac{\partial U}{\partial t} = \frac{\partial^2 U}{\partial x^2}, \quad \frac{\partial}{\partial t} \equiv \frac{\partial^2}{\partial x^2}.$$

Hence

$$U_{i,j+1} = \left\{\exp\left(k\frac{\partial^2}{\partial x^2}\right)\right\}U_{i,j} = \left\{\exp(kD_x^2)\right\}U_{i,j}$$

$$= (1 + kD_x^2 + \tfrac{1}{2}k^2D_x^4 + \tfrac{1}{6}k^3D_x^6 + \ldots)U_{i,j}$$

where

$$D_x \equiv \frac{1}{h}(\delta_x - \tfrac{1}{24}\delta_x^3 + \tfrac{3}{640}\delta_x^5 + \ldots)$$

Evaluate D_x^2, D_x^4, and D_x^6 to terms in δ_x^6 and substitute into the expression for $u_{i,j+1}$ etc. The results stated follow from

$$\delta_x^2 u_{i,j} = u_{i-1,j} - 2u_{i,j} + u_{i,1,j}$$

and

$$\delta_x^4 u_{i,j} = \delta_x^2(u_{i-1,j} - 2u_{i,j} + u_{i+1,j}) = u_{i-2,j} - 4u_{i-1,j} + 6u_{i,j} - 4u_{i+1,j} + u_{i+2,j}.$$

13(a). The equation

$$\frac{\partial U}{\partial t} = \frac{\partial}{\partial x}\left\{a(x)\frac{\partial U}{\partial x}\right\}, \quad a(x) > 0,$$

is approximated at the point (ih, jk) by the difference equation

$$\frac{1}{k}\Delta_t u_{i,j} = \frac{1}{h}\delta_x\left(a_i\frac{1}{h}\delta_x u_{i,j}\right).$$

Show that this gives the explicit difference equation

$$u_{i,j+1} = ra_{i-\frac{1}{2}}u_{i-1,j} + \{1 - r(a_{i-\frac{1}{2}} + a_{i+\frac{1}{2}})\}u_{i,j} + ra_{i+\frac{1}{2}}u_{i+1,j},$$

where $r = k/h^2$.

(b) Using the result

$$U(x, t+k) = \left\{\exp\left(k\frac{\partial}{\partial t}\right)\right\}U(x, t),$$

show that the equation

$$\frac{\partial U}{\partial t} = \frac{\partial^2 U}{\partial x^2}$$

may be replaced exactly by the equation

$$\exp(-\tfrac{1}{2}kD_x^2)U_{i,j+1} = \exp(\tfrac{1}{2}kD_x^2)U_{i,j},$$

where

$$D_x^2 U_{i,j} = \frac{\partial^2 U_{i,j}}{\partial x^2} = \frac{1}{h^2}(\delta_x^2 - \tfrac{1}{12}\delta_x^4 + \tfrac{1}{90}\delta_x^6 + \ldots)U_{i,j}.$$

Hence show that to second-order differences the differential equation may be approximated by the (Crank–Nicolson) equation

$$-ru_{i-1,j+1} + (2+2r)u_{i,j+1} - ru_{i+1,j+1} = ru_{i-1,j} + (2-2r)u_{i,j} + ru_{i+1,j}.$$

(c) Eliminate the fourth-order term in

$$D_x^2 \equiv \frac{1}{h^2}(\delta_x^2 - \tfrac{1}{12}\delta_x^4 + \tfrac{1}{90}\delta_x^6 + \ldots).$$

Hence use the first part of (b) to show that $(\partial U/\partial t) = (\partial^2 U/\partial x^2)$ may be approximated by the Douglas difference equation

$$(1-6r)u_{i+1,j+1}+(10+12r)u_{i,j+1}+(1-6r)u_{i+1,j+1}$$
$$=(1+6r)u_{i-1,j}+(10-12r)u_{i,j}+(1+6r)u_{i+1,j},$$

where $r=k/h^2$. (As mentioned previously this uses the same six grid points as the Crank–Nicolson equation but gives a more accurate solution.)

Solution

(a) As $\delta_x u_{i,j}=u_{i+\frac{1}{2},j}-u_{i-\frac{1}{2},j}$, the approximation leads to

$$\frac{1}{k}(u_{i,j+1}-u_{i,j})=\frac{1}{h^2}\delta_x\{a_i(u_{i+\frac{1}{2},j}-u_{i-\frac{1}{2},j})\}$$

$$=\frac{1}{h^2}\{a_{i+\frac{1}{2}}(u_{i+1,j}-u_{i,j})-a_{i-\frac{1}{2},j}(u_{i,j}-u_{i-1,j})\}\text{ etc.}$$

(b) $U_{i,j+1}=\{\exp(k(\partial/\partial t))\}U_{i,j}$. Operate on both sides with $\exp(-\frac{1}{2}k(\partial/\partial t))$. Hence $\{\exp(-\frac{1}{2}k(\partial/\partial t))\}U_{i,j+1}=\{\exp(k(\partial/\partial t))\}U_{i,j}$. By the differential equation $(\partial/\partial t)\equiv D_x^2$, therefore $\{\exp(-\frac{1}{2}kD_x^2)\}U_{i,j+1}=\{\exp(\frac{1}{2}kD_x^2)\}U_{i,j}$. Substituting for D_x^2 in terms of δ_x^2 and expanding to terms in δ_x^2 shows that

$$(1-\tfrac{1}{2}r\delta_x^2)u_{i,j+1}=(1+\tfrac{1}{2}r\delta_x^2)u_{i,j}$$

approximates the partial differential equation to this order of accuracy. Hence the result.

(c) Operate on both sides of

$$D_x^2\equiv\frac{1}{h^2}(\delta_x^2-\tfrac{1}{12}\delta_x^4+\ldots)$$

with $(1+\tfrac{1}{12}\delta_x^2)$ to give that

$$(1+\tfrac{1}{12}\delta_x^2)D_x^2\equiv\frac{1}{h_2}\delta_x^2+O(\delta_x^6).$$

By part (b), $(1-\frac{1}{2}kD_x^2+\ldots)U_{i,j+1}=(1+\frac{1}{2}kD_x^2+\ldots)U_{i,j}$. Operate on both sides with $(1+\tfrac{1}{12}\delta_x^2)$ and use the previous identity. Neglecting terms of order δ^6 and above yields the approximation equation

$$(1+\tfrac{1}{12}\delta_x^2-\tfrac{1}{2}r\delta_x^2)u_{i,j+1}=(1+\tfrac{1}{12}\delta_x^2+\tfrac{1}{2}r\delta_x^2)u_{i,j}.$$

Hence the result.

14. The function U satisfies the non-linear equation

$$\frac{\partial U}{\partial t} = \frac{\partial^2 U^2}{\partial x^2}, \quad 0 < x < 1,$$

the initial condition $U = 1$, $0 \le x \le 1$, $t = 0$, and the boundary conditions

$$\frac{\partial U}{\partial x} = 0 \text{ at } x = 0, \; t < 0, \quad \frac{\partial U}{\partial x} = -U \text{ at } x = 1, \; t > 0.$$

If the equation is approximated by the difference equation

$$\frac{1}{k} \, \delta_t u_{i,j+\frac{1}{2}} = \frac{1}{2h^2} \left(\delta_x^2 u_{i,j+1}^2 + \delta_x^2 u_{i,j}^2 \right),$$

and the derivative boundary conditions are approximated by the usual central-difference formulae, show that the corresponding non-linear approximation equations are

$$u_0^2 + p u_0 - u_1^2 + \{ u_{0,j}^2 - p u_{0,j} - u_{1,j}^2 \} = 0,$$

$$u_{i-1}^2 - 2(u_i^2 + p u_i) + u_{i+1}^2 + \{ u_{i-1,j}^2 - 2(u_{i,j}^2 - p u_{i,j}) + u_{i+1,j}^2 \} = 0,$$

$$i = 1(1)(N-1),$$

and

$$u_{N-1}^2 - 2 h u_{N-1} u_N - (1 - 2h^2) u_N^2 - p u_N$$
$$+ \{ u_{N-1,j}^2 - 2 h u_{N-1,j} u_{N,j} - (1 - 2h^2) u_{N,j}^2 + p u_{N,j} \} = 0,$$

where $p = h^2/k$, $Nh = 1$ and u_i denotes the unknown $u_{i,j+1}$, $i = 0(1)N$.

Briefly describe Newton's method for deriving a set of linear equations giving an improved value $(V_i + \varepsilon_i)$ to the approximate solution values V_i, $i = 1(1)N$, of the N non-linear equations

$$f_i(u_1, u_2, \ldots, u_N) = 0, \quad i = 1(1)N.$$

Apply this method to the approximation equations previously obtained and write out the linear equations giving the first iteration values at mesh points along the time-level $t = k$, taking V_i, $i = 0(1)N$, equal to the initial value 1.

Solution

$$\frac{1}{k} (u_{i,j+1} - u_{i,j}) = \frac{1}{2h^2} (u_{i-1,j+1}^2 - 2 u_{i,j+1}^2 + u_{i+1,j+1}^2 + u_{i-1,j}^2 - 2 u_{i,j}^2 + u_{i+1,j}^2).$$

Put $p = h^2/k$ and denote $u_{i,j+1}$ by u_i. Then

$$u_{i-1}^2 - 2(u_i^2 + pu_i) + u_{i+1}^2 + \{u_{i-1,j}^2 - 2(u_{i,j}^2 - pu_{i,j}) + u_{i+1,j}^2\} = 0.$$

By the B.C.,

$$\left(\frac{\partial U}{\partial x}\right)_{0,j} \simeq \frac{u_{1,j} - u_{-1,j}}{2h} = 0 \Rightarrow u_{-1,j} = u_{1,j}.$$

Put $i = 0$ in the approximation equation and eliminate $u_{-1,j}$ to give

$$u_0^2 + pu_0 - u_1^2 + (u_{0,j}^2 - pu_{0,j} - u_{1,j}^2) = 0.$$

By the B.C. at

$$i = N, \quad \frac{1}{2h}(u_{N+1,j+1} - u_{N-1,j+1}) = -u_{N,j} \text{ etc.}$$

By Newton's method,

$$\left(\frac{\partial f_i}{\partial u_1}\epsilon_1 + \ldots + \frac{\partial f_i}{\partial u_N}\epsilon_N\right)_{u_i = V_i} + f_i(V_1, V_2, \ldots, V_N) = 0.$$

Hence

$$(2V_0 + p)\epsilon_o - 2V_1\epsilon_1 + \{V_0^2 + pV_0 - V_1^2 + (u_{0,j}^2 - pu_{0,j} - u_{1,j}^2)\} = 0,$$

$$2V_{i-1}\epsilon_{i-1} - 2(2V_i + p)\epsilon_i + 2V_{i+1}\epsilon_{i+1}$$
$$+ \{V_{i-1}^2 - 2(V_i^2 + pV_i) + V_{i-1}^2 + (u_{i-1,j}^2 - 2[u_{i,j}^2 - pu_{i,j}] + u_{i+1,j}^2)\} = 0,$$

and

$$(2V_{N-1} - 2hV_N)\epsilon_{N-1} + \{-2hV_{N-1} - 2(1 - 2h^2)V_N - p\}\epsilon_N$$
$$+ \{V_{N-1}^2 - 2hV_{N-1}V_N - (1 - 2h^2)V_N^2 - pV_N$$
$$+ u_{N-1,j}^2 - 2hu_{N-1,j}u_{N,j} - (1 - 2h^2)u_{N,j}^2 + pu_{N,j}\} = 0.$$

Put $V_i = 1$, $i = 0(1)N$, $j = 0$, and $u_{i,0} = 1$, $i = 0(1)N$ etc.

3 Convergence, stability, and consistency

THIS chapter is concerned with the conditions that must be satisfied if the solution of the finite-difference equations is to be a reasonably accurate approximation to the solution of the corresponding parabolic or hyperbolic partial differential equation.

These conditions are associated with two different but inter-related problems. The first concerns the convergence of the exact solution of the approximating difference equations to the solution of the differential equation; the second concerns the unbounded growth or controlled decay of any errors associated with the solution of the finite-difference equations.

Descriptive treatment of convergence

Let U represent the *exact* solution of a partial differential equation with independent variables x and t, and u the *exact* solution of the difference equations used to approximate the partial differential equation. Then the finite-difference equation is said to be convergent when u tends to U at a fixed point or along a fixed t-level as δx and δt both tend to zero.

Although the conditions under which u converges to U have been established for linear elliptic, parabolic and hyperbolic second-order partial differential equations with solutions satisfying fairly general boundary and initial conditions, they are not yet known for non-linear equations except in a few particular cases. (The equation,

$$a\frac{\partial^2 U}{\partial x^2} + b\frac{\partial^2 U}{\partial x \partial t} + c\frac{\partial^2 U}{\partial t^2} + d\frac{\partial U}{\partial x} + e\frac{\partial U}{\partial t} + fU + g = 0,$$

is linear when the coefficients a, b, ..., g, are constants or functions of x and t only. Otherwise it is non-linear. If the coefficients of the second-order derivatives are functions of x, t, U, $\partial U/\partial x$ and $\partial U/\partial t$ but not of second-order derivatives the equation is described as quasi-linear even though it is non-linear. The important feature of linear equations is that the sum of separate solutions is also a solution.)

The difference $(U-u)$ is called the *discretization error*. Some texts call it the truncation error but in this book the latter term will be reserved for the difference between the differential equation and its approximating difference equation. The magnitude of the discretization error at any mesh point depends on the finite-sizes of the mesh lengths δx and δt, i.e., on the distances between consecutive, discrete grid-points, and on the number of terms in the truncated series of differences used to approximate the derivatives. Readers familiar with the calculus of finite-differences will have recognised the approximation used earlier for $\partial U/\partial t$ as the first term in either the series

$$(\delta t)\left(\frac{\partial U}{\partial t}\right)_{i,j} = (\Delta_t - \tfrac{1}{2}\Delta_t^2 + \tfrac{1}{3}\Delta_t^3 - \ldots)U_{i,j}$$

or the series

$$(\delta t)\left(\frac{\partial U}{\partial t}\right)_{i,j+\frac{1}{2}} = (\delta_t - \tfrac{1}{24}\delta_t^3 + \tfrac{3}{640}\delta_t^5 + \ldots)U_{i,j+\frac{1}{2}}$$

and the approximation for $\partial^2 U/\partial x^2$ as the first term in the series

$$(\delta x)^2 \frac{\partial^2 U_{i,j}}{\partial x^2} = (\delta_x^2 - \tfrac{1}{12}\delta_x^4 + \tfrac{1}{90}\delta_x^6 - \ldots)U_{i,j},$$

where the subscripts t and x denote the directions in which the differences are calculated. The symbols Δ and δ are the forward and central difference operators defined by $\Delta_t u_{i,j} = u_{i,j+1} - u_{i,j}$ and $\delta_t u_{i,j+\frac{1}{2}} = u_{i,j+1} - u_{i,j}$, so that $\delta_x^2 u_{i,j} = \delta_x(\delta_x u_{i,j}) = \delta_x(u_{i+\frac{1}{2},j} - u_{i-\frac{1}{2},j}) = u_{i+1,j} - 2u_{i,j} + u_{i-1,j}$. Better approximations can be obtained by truncating the series after two or more terms but have the disadvantage of involving more pivotal values of u. It will be shown later that the discretization error can be analysed in terms of preceding local truncation errors. (See page 104.)

The discretization error can usually be diminished by decreasing δx and δt, subject invariably to some relationship between them, but as this leads to an increase in the number of equations to be solved, this method of improvement is limited by such factors as cost of computation and computer storage requirements, etc.

In general the problem of convergence is a difficult one to investigate usefully because the final expression for the discretization error is usually in terms of unknown derivatives for which no bounds can be estimated. Fortunately however, the convergence of difference equations approximating *linear* parabolic and hyperbolic differential equations can be investigated in terms of stability and consistency which are much easier to deal with.

Local truncation error and consistency

Let $F_{i,j}(u) = 0$ represent the difference equation at the (i, j)th mesh point. If u is replaced by U at the mesh points of the difference equation then the value of $F_{i,j}(U)$ is called the local truncation error at the (i, j)th mesh point. If this tends to zero as the mesh lengths tend to zero the difference equation is said to be *consistent* with the partial differential equation. (See page 96.)

Descriptive treatment of stability

The equations that are actually solved are, of course, the finite-difference equations and if it were possible to carry out all calculations to an infinite number of decimal places we would obtain their exact solution u. In practice however, each calculation is carried out to a finite number of decimal places or significant figures, a procedure that introduces a rounding error every time it is used, and the solution actually computed is not u but N, (say). N will be called the numerical solution and $(u - N)$ the global rounding error R. The total error at the (i, j)th mesh point is

$$U_{i,j} - N_{i,j} = (U_{i,j} - u_{i,j}) + (u_{i,j} - N_{i,j})$$
$$= \text{discretization error} + \text{global rounding error}.$$

Assuming for the moment that the discretization error can be controlled it would at first sight seem reasonable to say that the difference equations are stable if there is a limit to the magnification of the global rounding error $R_{i,j}$ as j tends to infinity, i.e., if the growth of $R_{i,j}$ is bounded for all i as j tends to infinity. *But this is not possible* because the global rounding error depends not only on the difference equations themselves but also on the manner in which a particular computer rounds off numbers and carries out its arithmetic. In other words it is arbitrary in the sense that different computing machines introduce different rounding errors and it may not be possible to deduce a *useful* conclusion from its mathematical analysis. (See page 95.)This error also differs from the discretization error in that it cannot be made to converge to zero as the mesh lengths tend to zero. The opposite in fact would happen because smaller mesh sizes imply an increase in the number of arithmetic operations. In practice, fortunately, the individual rounding errors introduced by large modern computing machines are extremely small and numerical studies indicate that the global rounding error is invariably smaller than the discretization error.

The most fruitful way to define stability is in terms of the boundedness of the exact solution of the difference equations because it is amenable to mathematical analysis, involves only known coefficients and boundary values, and has several useful consequences. First, if $u_{i,j}$ is bounded as j increases it follows that the magnification of each rounding error is also bounded because the same arithmetic operations are applied to both numbers when propagating the solution forward in time, assuming t represents time. (In this process $R_{i,j}$ is the global rounding error at the (i, j)th mesh point in relation to the arithmetic needed to propagate the solution forward from the initial line to the jth time-level, but is the local rounding error at this mesh point in relation to subsequent time-level calculations.) Secondly, it is obvious that the discretization error $e_{i,j} = U_{i,j} - u_{i,j}$ is bounded when $u_{i,j}$ is bounded because $U_{i,j}$ is a fixed number for a given equation with known boundary and initial conditions. (As was remarked earlier, the discretization error can be expressed in terms of local truncation errors. Because of this the difference equations are sometimes defined as stable when local rounding errors and local truncation errors do not increase unboundedly with increasing time-levels of calculations.) Finally, for linear differential equations, this definition of stability is related to convergence through the concept of consistency in that stability and consistency together guarantee convergence.

The following point should, however, be noted.

Convergence is concerned with the conditions that must be satisfied in order for the discretization error to tend to zero *at a fixed point or fixed time-level* $T = j\delta t$ as δt (and δx) tends to zero, i.e., as the number of time-levels of calculations tends to infinity. But in an actual computation δt and δx are kept constant as the solution is propagated forward time-level by time-level from $t = 0$ to $t = j\delta t$. In both processes however, for the same values of j, essentially the same sets of difference equations are solved, the only difference between, say, the nth equation of each set, being different constant terms arising from different boundary conditions. As a consequence the conditions necessary for the boundedness of $u_{i,j}$ (and of any form of local error) as j tends to infinity when δt and δx are kept constant, together with consistency, ensure convergence.

Analytical treatment of convergence (Direct proof)

Convergence is more difficult to investigate than stability and only one example will be considered in detail at this stage. A second example involving derivative boundary conditions and employing a standard method for solving the finite-difference equations is given at the end of the chapter. The method below derives a difference equation for the discretization error e which fortunately, in our example, can be dealt with fairly easily.

Denote the exact solution of the partial differential equation by U and the exact solution of the finite-difference equation by u. Then $e = U - u$.

Consider the equation

$$\frac{\partial U}{\partial t} = \frac{\partial^2 U}{\partial x^2}, \quad 0 < x < 1, \tag{3.1}$$

where U is known for $0 < x < 1$ when $t = 0$, and at $x = 0$ and 1 when $t \geq 0$.

The simplest explicit finite-difference approximation to (3.1) is

$$\frac{u_{i,j+1} - u_{i,j}}{k} = \frac{u_{i-1,j} - 2u_{i,j} + u_{i+1,j}}{h^2}. \tag{3.2}$$

At the mesh points,

$$u_{i,j} = U_{i,j} - e_{i,j}, \quad u_{i,j+1} = U_{i,j+1} - e_{i,j+1}, \text{ etc.}$$

Substitution into (3.2) leads to

$$e_{i,j+1} = re_{i-1,j} + (1 - 2r)e_{i,j} + re_{i+1,j} + U_{i,j+1} - U_{i,j} + \\ + r(2U_{i,j} - U_{i-1,j} - U_{i+1,j}). \tag{3.3}$$

By Taylor's theorem,

$$U_{i+1,j} = U(x_i + h, t_j) = U_{i,j} + h\left(\frac{\partial U}{\partial x}\right)_{i,j} + \frac{h^2}{2!}\frac{\partial^2 U}{\partial x^2}(x_i + \theta_1 h, t_j),$$

$$U_{i-1,j} = U(x_i - h, t_j) = U_{i,j} - h\left(\frac{\partial U}{\partial x}\right)_{i,j} + \frac{h^2}{2!}\frac{\partial^2 U}{\partial x^2}(x_i - \theta_2 h, t_j),$$

$$U_{i,j+1} = U(x_i, t_j + k) = U_{i,j} + k\frac{\partial U(x_i, t_j + \theta_3 k)}{\partial t},$$

where $0 < \theta_1 < 1, 0 < \theta_2 < 1$ and $0 < \theta_3 < 1$. Substitution into equation (3.3) gives

$$e_{i,j+1} = r e_{i-1,j} + (1 - 2r) e_{i,j} + r e_{i+1,j} +$$

$$+ k \left\{ \frac{\partial U(x_i, t_j + \theta_3 k)}{\partial t} - \frac{\partial^2 U(x_i + \theta_4 h, t_j)}{\partial x^2} \right\}, \quad (3.4)$$

where $-1 < \theta_4 < 1$.

This is a difference equation for $e_{i,j}$ which fortunately we need not solve.

Let E_j denote the maximum value of $|e_{i,j}|$ along the jth time-row and M the maximum modulus of the expression in the braces for all i and j. When $r \leq \frac{1}{2}$, all the coefficients of e in equation (3.4) are positive or zero, so

$$\begin{aligned} |e_{i,j+1}| &\leq r|e_{i-1,j}| + (1 - 2r)|e_{i,j}| + r|e_{i+1,j}| + kM \\ &\leq rE_j + (1 - 2r)E_j + rE_j + kM \\ &= E_j + kM. \end{aligned}$$

As this is true for all values of i it is true for $\max_i |e_{i,j+1}|$. Hence

$$E_{j+1} \leq E_j \times kM \leq (E_{j-1} + kM) + kM = E_{j-1} + 2kM,$$

etc., from which it follows that

$$E_j \leq E_0 + jkM = tM,$$

because the initial values for u and U are the same, i.e., $E_0 = 0$. When h tends to zero, $k = rh^2$ also tends to zero and M tends to

$$\left(\frac{\partial U}{\partial t} - \frac{\partial^2 U}{\partial x^2} \right)_{i,j}.$$

Since U is a solution of equation (3.1) the limiting value of M and therefore of E_j is zero. As $|U_{i,j} - u_{i,j}| \leq E_j$, this proves that u converges to U as h tends to zero when $r \leq \frac{1}{2}$ and t is finite.

When $r > \frac{1}{2}$ it can be shown that the complementary function of the difference equation (3.4) tends to infinity as h tends to zero. There is no need however to do this when our main purpose is to find the conditions necessary for a useful numerical solution because we shall prove later that this finite-difference scheme is stable for $r \leq \frac{1}{2}$ but unstable for $r > \frac{1}{2}$.

The proof above implies that $\partial U/\partial t$ and $\partial^2 U/\partial x^2$ are uniformly continuous and bounded throughout the solution domain. This was so in the worked example 2.1 in which $\partial^2 U/\partial x^2$ was initially zero in spite of the discontinuity in $\partial U/\partial x$. If it is assumed that U possesses continuous bounded derivatives up to order three in t and order six in x, exercise 1 shows that the discretization error is of order h^2, except when $r = \frac{1}{6}$ in which case it is of order h^4.

Analytical treatment of stability

There are two standard ways of investigating the boundedness of the solution of the finite-difference equations. In one we express the equations in matrix form and examine the eigenvalues of an associated matrix; in the other we use a finite Fourier series. The Fourier method is the easier of the two in that it requires no knowledge of matrix algebra but is the less rigorous because it neglects the boundary conditions.

In order to eliminate constants arising from non-zero boundary conditions the subsequent analysis is in terms of the difference of two solutions, **u** and **u***, where **u** is the solution of the difference equations corresponding to known initial values and **u*** is the solution of the same equations but with perturbed initial values. This difference has been called the 'error' **e**. (See exercise 2.) The 'errors' introduced at mesh points along the initial line could of course correspond to local rounding errors, it then being assumed that all subsequent arithmetic is exact.

Matrix method
Explicit finite-difference scheme

Consider the equation

$$\frac{\partial U}{\partial t} = \frac{\partial^2 U}{\partial x^2}, \quad 0 < x < 1,$$

where $U = 0$ at $x = 0$ and 1, $t > 0$, and U is known when $t = 0$.

The explicit finite-difference approximation

$$u_{i,j+1} = ru_{i-1,j} + (1 - 2r)u_{i,j} + ru_{i+1,j} \tag{3.5}$$

leads to the equations

$$u_{1,j+1} = 0 + (1-2r)u_{1,j} + ru_{2,j},$$
$$u_{2,j+1} = \quad\quad ru_{1,j} + (1-2r)u_{2,j} + ru_{3,j},$$

$$\cdot\;\cdot\;\cdot\;\cdot\;\cdot\;\cdot\;\cdot\;\cdot\;\cdot\;\cdot\;\cdot\;\cdot\;\cdot\;\cdot\;\cdot\;\cdot$$

$$u_{N-1,j+1} = \quad\quad\quad\quad\quad\quad\quad ru_{N-2,j} + (1-2r)u_{N-1,j} + 0,$$

where $N\delta x = 1$, and $u_{0,j} = u_{N,j} = 0$. These can be written as

$$
\begin{bmatrix}
u_{1,j+1} \\
u_{2,j+1} \\
u_{3,j+1} \\
\cdot \\
\cdot \\
\cdot \\
u_{N-1,j+1}
\end{bmatrix}
=
\begin{bmatrix}
(1-2r) & r & & & & \\
r & (1-2r) & r & & & \\
& r & (1-2r) & r & & \\
& & & \cdot & & \\
& & & & \cdot & \\
& & & & & \cdot \\
& & & & r & (1-2r)
\end{bmatrix}
\begin{bmatrix}
u_{1,j} \\
u_{2,j} \\
u_{3,j} \\
\cdot \\
\cdot \\
\cdot \\
u_{N-1,j}
\end{bmatrix}
$$

or as

$$\mathbf{u}_{j+1} = \mathbf{A}\mathbf{u}_j,$$

where the small roman bold letter $\mathbf{u}_j$ denotes the column vector

$$
\begin{bmatrix}
u_{1,j} \\
u_{2,j} \\
\cdot \\
\cdot \\
\cdot \\
u_{N-1,j}
\end{bmatrix}
$$

and the capital roman bold letter $\mathbf{A}$ the $(N-1)$ square matrix

$$
\begin{bmatrix}
(1-2r) & r & & & \\
r & (1-2r) & r & & \\
& & \cdot & & \\
& & & \cdot & \\
& & & & \cdot \\
& & & r & (1-2r)
\end{bmatrix}.
$$

Hence

$$\mathbf{u}_j = \mathbf{A}\mathbf{u}_{j-1} = \mathbf{A}(\mathbf{A}\mathbf{u}_{j-2}) = \ldots = \mathbf{A}^j\mathbf{u}_0,$$

where $\mathbf{u}_0$ is the vector of initial values. Now suppose we introduce errors at every pivotal point along $t = 0$ and start the computation with the vector of values $\mathbf{u}_0^{\star}$ instead of $\mathbf{u}_0$. We shall then calculate

$$\mathbf{u}_1^{\star} = \mathbf{A}\mathbf{u}_0^{\star}, \qquad \mathbf{u}_2^{\star} = \mathbf{A}\mathbf{u}_1^{\star} = \mathbf{A}^2\mathbf{u}_0^{\star},$$

and finally,

$$\mathbf{u}_j^{\star} = \mathbf{A}^j\mathbf{u}_0^{\star},$$

assuming we introduce no further errors.

Define the error vector $\mathbf{e}$ by

$$\mathbf{e} = \mathbf{u} - \mathbf{u}^{\star}.$$

Then

$$\mathbf{e}_j = \mathbf{u}_j - \mathbf{u}_j^{\star} = \mathbf{A}^j(\mathbf{u}_0 - \mathbf{u}_0^{\star}) = \mathbf{A}^j\mathbf{e}_0,$$

showing that the formula for the propagation of the errors is the same as that for the calculation of $\mathbf{u}$. This also shows, incidentally, that when the finite-difference equations are linear we need consider the propagation of only one line of errors because the overall effect of several lines will be given by the addition of the effect produced by each line considered separately.

The finite difference scheme will be stable when $\mathbf{e}_j$ remains bounded as j increases indefinitely. This can always be investigated by expressing the initial error vector in terms of the eigenvectors of $\mathbf{A}$.

Assume that the matrix $\mathbf{A}$ is non-deficient, i.e., has $(N-1)$ linearly independent eigenvectors $\mathbf{v}_s$, which will always be so if the eigenvalues λ_s of $\mathbf{A}$ are all distinct or $\mathbf{A}$ is real and symmetric. Then these eigenvectors can be used as a basis for our $(N-1)$-dimensional vector space and the error vector $\mathbf{e}_0$, with its $(N-1)$ components, can be expressed uniquely as a linear combination of them, namely,

$$\mathbf{e}_0 = \sum_{s=1}^{N-1} c_s\mathbf{v}_s,$$

where the c_s, $s = 1(1)(N-1)$, are known scalars.

The errors along the time-level $t = k$, resulting from the initial errors $\mathbf{e}_0$, will be given by

$$\mathbf{e}_1 = \mathbf{A}\mathbf{e}_0 = \mathbf{A}\sum c_s\mathbf{v}_s = \sum c_s\mathbf{A}\mathbf{v}_s.$$

But $\mathbf{A}\mathbf{v}_s = \lambda_s\mathbf{v}_s$ by the definition of an eigenvalue.
Therefore

$$\mathbf{e}_1 = \sum c_s\lambda_s\mathbf{v}_s.$$

Similarly,

$$\mathbf{e}_j = \sum c_s \lambda_s^j \mathbf{v}_s. \tag{3.6}$$

This shows that the errors will not increase exponentially with j provided the eigenvalue with the largest modulus has a modulus less than or equal to unity.

The matrix $\mathbf{A}$ can be written as

$$
\begin{bmatrix}
(1-2r) & r & & & & \\
r & (1-2r) & r & & & \\
 & \cdot & \cdot & \cdot & & \\
 & & r & (1-2r) & r & \\
 & & & r & (1-2r)
\end{bmatrix}
=
\begin{bmatrix}
1 & 0 & 0 & & & \\
0 & 1 & 0 & & & \\
 & \cdot & \cdot & \cdot & & \\
 & & & 0 & 1 & 0 \\
 & & & & 0 & 1
\end{bmatrix}
$$

$$
+ \quad r
\begin{bmatrix}
-2 & 1 & & & \\
1 & -2 & 1 & & \\
 & \cdot & \cdot & \cdot & \\
 & & 1 & -2 & 1 \\
 & & & 1 & -2
\end{bmatrix}
$$

i.e., as

$$\mathbf{A} = \mathbf{I} + r\mathbf{T}_{N-1},$$

where $\mathbf{T}_{N-1}$ is an $(N-1) \times (N-1)$ matrix whose eigenvalues λ_s and eigenvectors $\mathbf{v}_s$ are given by

$$\lambda_s = -4 \sin^2\left(\frac{s\pi}{2N}\right), \quad (s = 1, 2, \ldots, N-1),$$

$$\mathbf{v}_s = \left(\sin\frac{s\pi}{N}, \sin\frac{2s\pi}{N}, \ldots, \sin\frac{(N-1)s\pi}{N}\right). \quad \text{(See page 86.)}$$

These values can be verified by substitution into $\mathbf{T}_{N-1}\mathbf{v}_s = \lambda_s \mathbf{v}_s$.

Hence the eigenvalues of $\mathbf{A}$, as shown later in '*A note on eigenvalues and eigenvectors*', are

$$1 + r\left\{-4 \sin^2\frac{s\pi}{2N}\right\}.$$

Therefore the condition for stability of the explicit scheme is

$$\left|1 - 4r \sin^2 \frac{s\pi}{2N}\right| \le 1.$$

The only useful inequality is

$$-1 \le 1 - 4r \sin^2 \frac{s\pi}{2N},$$

giving

$$r \le 1/2 \sin^2 \frac{s\pi}{2N} > \tfrac{1}{2},$$

proving that the scheme is stable for $r \le \tfrac{1}{2}$.

A note on eigenvalues and eigenvectors

Let $\mathbf{x}$ be an eigenvector of the matrix $\mathbf{A}$ corresponding to the eigenvalue λ. Then $\mathbf{Ax} = \lambda\mathbf{x}$. Hence $\mathbf{A}(\mathbf{Ax}) = \mathbf{A}^2\mathbf{x} = \lambda\mathbf{Ax} = \lambda^2\mathbf{x}$, showing that the matrix $\mathbf{A}^2$ has an eigenvalue λ^2 corresponding to the eigenvector $\mathbf{x}$. Similarly $\mathbf{A}^p\mathbf{x} = \lambda^p\mathbf{x}$, $p = 3, 4, \ldots$.

(i) If $f(\mathbf{A}) = a_p\mathbf{A}^p + a_{p-1}\mathbf{A}^{p-1} + \ldots + a_0\mathbf{I}$ is a polynomial in $\mathbf{A}$ with scalar coefficients $a_p, \ldots, a_0$, then $f(\mathbf{A})\mathbf{x} = (a_p\lambda^p + \ldots + a_0)\mathbf{x} = f(\lambda)\mathbf{x}$, showing that $f(\mathbf{A})$ has an eigenvalue $f(\lambda)$ corresponding to the eigenvector $\mathbf{x}$.

(ii) The eigenvalue of $[f_1(\mathbf{A})]^{-1}f_2(\mathbf{A})$ corresponding to the eigenvector $\mathbf{x}$ is $f_2(\lambda)/f_1(\lambda)$, where $f_1(\mathbf{A})$ and $f_2(\mathbf{A})$ are polynomials in $\mathbf{A}$. The proof is as follows. By (i),

$$f_1(\mathbf{A})\mathbf{x} = f_1(\lambda)\mathbf{x} \quad \text{and} \quad f_2(\mathbf{A})\mathbf{x} = f_2(\lambda)\mathbf{x}.$$

Premultiply both equations by $[f_1(\mathbf{A})]^{-1}$ and write as

$$[f_1(\mathbf{A})]^{-1}\mathbf{x} = \mathbf{x}/f_1(\lambda) \quad \text{and} \quad [f_1(\mathbf{A})]^{-1}f_2(\mathbf{A})\mathbf{x} = f_2(\lambda)[f_1(\mathbf{A})]^{-1}\mathbf{x}.$$

Then the elimination of $[f_1(\mathbf{A})]^{-1}\mathbf{x}$ between these two equations shows that

$$[f_1(\mathbf{A})]^{-1}f_2(\mathbf{A})\mathbf{x} = \{f_2(\lambda)/f_1(\lambda)\}\mathbf{x},$$

which states, by the definition of an eigenvalue, that $f_2(\lambda)/f_1(\lambda)$ is an eigenvalue of $[f_1(\mathbf{A})]^{-1}f_2(\mathbf{A})$ corresponding to the eigenvector $\mathbf{x}$. In a similar manner the eigenvalue of $f_2(\mathbf{A})[f_1(\mathbf{A})]^{-1}$ corresponding to the eigenvector $\mathbf{x}$ is $f_2(\lambda)/f_1(\lambda)$.

In particular, the eigenvalue of $[f_1(\mathbf{A})]^{-1}$ corresponding to the eigenvector $\mathbf{x}$ is $1/f_1(\lambda)$.

The eigenvalues of a common tridiagonal matrix

The eigenvalues of the $N \times N$ matrix

$$
\begin{bmatrix}
a & b & & & & \\
c & a & b & & & \\
  & c & a & b & & \\
  &   &   & \cdot & \cdot & \cdot \\
  &   &   & c & a & b \\
  &   &   &   & c & a
\end{bmatrix}
$$

are

$$
\lambda_s = a + 2\{\sqrt{(bc)}\}\cos \frac{s\pi}{N+1}, \quad s = 1(1)N,
$$

where a, b, and c may be real or complex. A proof is given on page 113.

Crank–Nicolson implicit scheme

The Crank–Nicholson equations (2.10) can be written as

$$
\begin{bmatrix}
(2+2r) & -r & & & \\
-r & (2+2r) & -r & & \\
 & \cdot & \cdot & \cdot & \\
 & & -r & (2+2r) & -r \\
 & & & -r & (2+2r)
\end{bmatrix}
\begin{bmatrix}
u_{1,j+1} \\
u_{2,j+1} \\
\cdot \\
\cdot \\
u_{N-1,j+1}
\end{bmatrix}
$$

$$
=
\begin{bmatrix}
(2-2r) & r & & & \\
r & (2-2r) & r & & \\
 & \cdot & \cdot & \cdot & \\
 & & r & (2-2r) & r \\
 & & & r & (2-2r)
\end{bmatrix}
\begin{bmatrix}
u_{1,j} \\
u_{2,j} \\
\cdot \\
\cdot \\
u_{N-1,j}
\end{bmatrix}
$$

i.e., as

$$
(2\mathbf{I} - r\mathbf{T}_{N-1})\mathbf{u}_{j+1} = (2\mathbf{I} + r\mathbf{T}_{N-1})\mathbf{u}_j,
$$

Hence

$$
\mathbf{u}_{j+1} = (2\mathbf{I} - r\mathbf{T}_{N-1})^{-1}(2\mathbf{I} + r\mathbf{T}_{N-1})\mathbf{u}_j.
$$

By the previous argument these finite-difference equations will be stable when the moduli of the eigenvalues of

$$(2\mathbf{I} - r\mathbf{T}_{N-1})^{-1}(2\mathbf{I} + r\mathbf{T}_{N-1}) = \mathbf{A}$$

are each less than one. As the eigenvalues of $\mathbf{T}_{N-1}$ are $-4 \sin^2(s\pi/2N)$, the eigenvalues of $\mathbf{A}$ are

$$\frac{2 - 4r \sin^2\left(\dfrac{s\pi}{2N}\right)}{2 + 4r \sin^2\left(\dfrac{2\pi}{2N}\right)}, \quad (s = 1, 2, \ldots, N-1),$$

and these are clearly less than one in modulus for all positive values of r, proving that the Crank–Nicholson equations have unrestricted stability.

Useful theorems on bounds for eigenvalues

Gerschgorin's first theorem

The largest of the moduli of the eigenvalues of the square matrix **A** cannot exceed the largest sum of the moduli of the elements along any row or any column.

Proof

Let λ_i be an eigenvalue of the $N \times N$ matrix **A**, and $\mathbf{x}_i$ the corresponding eigenvector with components $v_1, v_2 \ldots v_n$. Then the equation

$$\mathbf{A}\mathbf{x}_i = \lambda_i \mathbf{x}_i$$

in detail, is

$$a_{1,1}v_1 + a_{1,2}v_2 + \ldots + a_{1,n}v_n = \lambda_i v_1,$$
$$a_{2,1}v_1 + a_{2,2}v_2 + \ldots + a_{2,n}v_n = \lambda_i v_2,$$
$$\begin{matrix} \cdot & \cdot & \cdot & \cdot \\ \cdot & \cdot & \cdot & \cdot \end{matrix}$$
$$a_{s,1}v_1 + a_{s,2}v_2 + \ldots + a_{s,n}v_n = \lambda_i v_s,$$
$$\begin{matrix} \cdot & \cdot & \cdot & \cdot \\ \cdot & \cdot & \cdot & \cdot \end{matrix}$$

Let v_s be the largest in modulus of v_1, v_2, ..., v_n. Select the sth equation and divide by v_s, giving

$$\lambda_i = a_{s,1}\left(\frac{v_1}{v_2}\right) + a_{s,2}\left(\frac{v_2}{v_s}\right) + \ldots + a_{s,n}\left(\frac{v_n}{v_s}\right).$$

Therefore

$$|\lambda_i| \leqslant |a_{s,1}| + |a_{s,2}| + \ldots + |_{s,n}|,$$

because

$$\left|\frac{v_i}{v_s}\right| \leqslant 1, \quad i = 1, 2, \ldots, n.$$

In particular this holds for $|\lambda_i| = \max|\lambda_s|$, $s = 1(1)N$.

Since the eigenvalues of the transpose of $\mathbf{A}$ are the same as those of $\mathbf{A}$ the theorem is also true for columns.

Gerschgorin's circle theorem or Brauer's theorem

Let P_s be the sum of the moduli of the elements along the sth row excluding the diagonal element $a_{s,s}$. Then each eigenvalue of $\mathbf{A}$ lies inside or on the boundary of at least one of the circles $|\lambda - a_{s,s}| = P_s$.

Proof

By the previous proof,

$$\lambda_i = a_{s,1}\left(\frac{v_1}{v_s}\right) + a_{s,2}\left(\frac{v_2}{v_s}\right) + \ldots + a_{s,s} + \ldots + a_{s,n}\left(\frac{v_n}{v_s}\right).$$

Hence

$$|\lambda_i - a_{s,s}| = \left|a_{s,1}\left(\frac{v_1}{v_s}\right) + \ldots + 0 + \ldots + a_{s,n}\left(\frac{v_n}{v_s}\right)\right|$$

$$\leqslant |a_{s,1}| + |a_{s,2}| + \ldots + 0 + \ldots + |a_{s,n}|$$

$$= P_s,$$

which completes the proof.

As an illustrative example consider the Crank–Nicolson equations

$$(2\mathbf{I} - r\mathbf{T}_{N-1})\mathbf{u}_{j+1} = (2\mathbf{I} + r\mathbf{T}_{N-1})\mathbf{u}_j = \{4\mathbf{I} - (2\mathbf{I} - r\mathbf{T}_{N-1})\}\mathbf{u}_j,$$

which can be written as

$$\mathbf{B}\mathbf{u}_{j+1} = (4\mathbf{I} - \mathbf{B})\mathbf{u}_j,$$

giving

$$\mathbf{u}_{j+1} = (4\mathbf{B}^{-1} - \mathbf{I})\mathbf{u}_j,$$

where

$$\mathbf{B} = \begin{bmatrix} (2+2r) & -r & & & \\ -r & (2+2r) & -r & & \\ & \cdot & \cdot & & \\ & & -r & (2+2r) & -r \\ & & & -r & (2+2r) \end{bmatrix}.$$

The finite-difference equations will be stable when the modulus of every eigenvalue of $(4\mathbf{B}^{-1} - \mathbf{I})$ does not exceed one, that is, when

$$\left| \frac{4}{\lambda} - 1 \right| \leq 1,$$

where λ is an eigenvalue of $\mathbf{B}$. This is equivalent to $\lambda \geq 2$.

For the matrix $\mathbf{B}$, $a_{s,s} = 2 + 2r$, $\max P_s = 2r$, so Brauer's theorem leads to

$$|\lambda - 2 - 2r| \leq 2r,$$

from which it follows that

$$-2r \leq \lambda - 2 - 2r \leq 2r,$$

or

$$2 \leq \lambda \leq 2 + 4r,$$

proving that the equations are unconditionally stable as $\lambda \geq 2$ for all values of r.

Stability criteria for derivative boundary conditions

Consider the equation

$$\frac{\partial u}{\partial t} = \frac{\partial^2 u}{\partial x^2}, \quad 0 < x < 1,$$

and the conditions,

$$\frac{\partial u}{\partial x} = h_1(u - v_1) \quad \text{at } x = 0, \quad t \geq 0,$$

$$\frac{\partial u}{\partial x} = -h_2(u - v_2) \quad \text{at } x = 1, \quad t \geq 0,$$

where h_1, h_2, v_1, v_2 are constants, $h_1 \geq 0$, $h_2 \geq 0$.

When the boundary conditions are approximated by the central-difference equations

$$(u_{1,j} - u_{-1,j})/2\delta x = h_1(u_{0,j} - v_1),$$

$$(u_{N+1,j} - u_{N-1,j})/2\delta x = -h_2(u_{N,j} - v_2), \quad (N\delta x = 1),$$

and the differential equation by the explicit scheme

$$u_{i,j+1} = ru_{i-1,j} + (1 - 2r)u_{i,j} + ru_{i+1,j},$$

elimination of $u_{-1,j}, u_{N+1,j}$, leads to the equations

$$
\begin{bmatrix} u_{0,j+1} \\ u_{1,j+1} \\ \\ u_{N-1,j+1} \\ u_{N,j+1} \end{bmatrix}
$$

$$
=
\begin{bmatrix}
\{1 - 2r(1 + h_1\delta x)\} & 2r & & \\
r & (1-2r) & r & \\
 & \cdot & \cdot & \cdot & \\
 & & r & (1-2r) & & r \\
 & & & 2r & \{1 - 2r(1 + h_2\delta x)\}
\end{bmatrix}
$$

$$
\times
\begin{bmatrix} u_{0,j} \\ u_{1,j} \\ \\ u_{N-1,j} \\ u_{N,j} \end{bmatrix}
+
\begin{bmatrix} 2rh_1v_1\delta x \\ 0 \\ \\ 0 \\ 2rh_2v_2\delta x \end{bmatrix}
$$

As each component of the last column vector is a constant the matrix determining the propagation of the error is

$$
\begin{bmatrix}
\{1 - 2r(1 + h_1\delta x)\} & 2r & & \\
r & (1-2r) & r & \\
 & & r & (1-2r) & & r \\
 & & & 2r & \{1 - 2r(1 + h_2\delta x)\}
\end{bmatrix}
$$

Application of Brauer's theorem to this matrix, with

$$a_{ss} = 1 - 2r(1 + h_1 \delta x) \quad \text{and} \quad P_s = 2r,$$

shows that some of its eigenvalues λ may lie on or within the circle

$$|\lambda - \{1 - 2r(1 + h_1 \delta x)\}| \leq 2r.$$

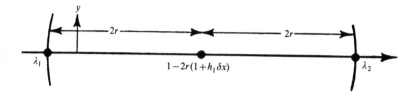

Fig. 3.1

Using Fig. 3.1,

$$\lambda_1 = 1 - 2r(2 + h_1 \delta x), \quad \lambda_2 = 1 - 2rh_1 \delta x,$$

and for stability,

$$|\lambda_1| \leq 1, \quad |\lambda_2| \leq 1.$$

Hence

$$-1 \leq 1 - 2r(2 + h_1 \delta x) \leq 1, \quad \text{giving } r \leq 1/(2 + h_1 \delta x),$$

and

$$-1 \leq 1 - 2rh_1 \delta x, \quad \text{giving } r \leq 1/h_1 \delta x.$$

The least of these is $r \leq 1/(2 + h_1 \delta x)$.

Similarly we require $r \leq 1/(2 + h_2 \delta x)$.

For rows $2(1)N$, $a_{s,s} = 1 - 2r$, $P_s = 2r$, giving $r \leq \frac{1}{2}$.

For overall stability, $r \leq \min\{1/(2 + h_1 \delta x); 1/(2 + h_2 \delta x)\}$.

Crank–Nicolson equations

It is easily shown that the Crank–Nicolson equations for the problem just considered propagate the error by the recursion formula

$$\mathbf{e}_{j+1} = (4\mathbf{B}^{-1} - \mathbf{I})\mathbf{e}_j,$$

where

$$\mathbf{B} = \begin{bmatrix} \{2+2r(1+h_1\delta x)\} & -2r & & & \\ -r & (2+2r) & -r & & \\ & & \cdot & \cdot & & \cdot \\ & & -r & (2+2r) & -r \\ & & & -2r & \{2+2r(1+h_2\delta x)\} \end{bmatrix}$$

As previously proved, this scheme is stable when the eigenvalues of $\mathbf{B} \geq 2$. By Gerschgorin's circle theorem,

$$|\lambda - \{2 + 2r(1 + h_1\delta x)\}| \leq 2r,$$

giving

$$2 + 2rh_1\delta x \leq \lambda.$$

Similarly, the remaining rows give that $2 \leq \lambda$ and $2 + 2rh_2\delta x \leq \lambda$.
Hence the equations are unconditionally stable.

Stability by the Fourier series method (von Neumann's method)

This method, developed by von Neumann during World War II, was first discussed in detail by O'Brien, Hyman and Kaplan in a paper published in 1951, reference 27. It expresses an initial line of errors in terms of a finite Fourier series, and considers the growth of a function that reduces to this series for $t = 0$ by a 'variables separable' method identical with that commonly used for deriving analytical solutions of partial differential equations. The Fourier series can be formulated in terms of sines or cosines but the algebra is easier if the complex exponential form is used, i.e., with $\sum a_n \cos(n\pi x/l)$ or $\sum b_n \sin(n\pi x/l)$ replaced by the equivalent $\sum A_n e^{in\pi x/l}$, where $i = \sqrt{-1}$ and l is the interval throughout which the function is defined. Clearly, we need to change our usual notation $u_{i,j}$ to $u(ph,qk) = u_{p,q}$, say. In terms of this notation

$$A_n e^{in\pi x/l} = A_n e^{in\pi ph/Nh} = A_n e^{i\beta_n ph},$$

where $\beta_n = n\pi/Nh$ and $Nh = l$.

Denote the errors at the pivotal points along $t = 0$, between $x = 0$ and Nh, by $E(ph) = E_p$, $p = 0, 1, \ldots, N$. Then the $(N+1)$ equations

$$E_p = \sum_{n=0}^{N} A_n e^{i\beta_n ph}, \qquad (p = 0, 1, \ldots, N),$$

are sufficient to determine the $(N+1)$ unknowns A_0, A_1, ..., A_n uniquely, showing that an arbitrary distribution of initial errors can be expressed in this complex exponential form. As our finite-difference equations will always be linear, and therefore separate solutions additive, we need only consider the propagation of the error due to a single term, such as $e^{i\beta ph}$. The coefficient A_n is a constant and can be neglected.

To investigate the propagation of this error as t increases we need to find a solution of the finite-difference equation which reduces to $e^{i\beta ph}$ when $t = qk = 0$. Assume

$$E_{p,q} = e^{i\beta x} e^{\alpha t} = e^{i\beta ph} e^{\alpha qk} = e^{i\beta ph} \xi^q, \qquad (3.7)$$

where $\xi = e^{\alpha k}$, and α, in general, is a complex constant. This obviously reduces to $e^{i\beta ph}$ when $q = 0$. The error will not increase as t increases provided

$$|\xi| \leq 1.$$

It should be noted that this method applies only to linear difference equations with constant coefficients, and strictly speaking only to initial value problems with periodic initial data. The criterion $|\xi| \leq 1$ is necessary and sufficient for two time-level difference equations but is not always sufficient for three or more level equations although it is always necessary. In practice the method often gives useful results even when its application is not fully justified (reference 31.)

Example 3.1

Investigate the stability of the fully-implicit finite-difference equation

$$(u_{p,q+1} - u_{p,q})/k = (u_{p-1,q+1} - 2u_{p,q+1} + u_{p+1,q+1})/h^2 \qquad (3.8)$$

approximating the parabolic equation $\partial u/\partial t = \partial^2 u/\partial x^2$.

As the error function $E_{p,q}$ satisfies the same difference equations as $u_{p,q}$, substitution of $E_{p,q}$ from equation (3.7) into equation (3.8) gives

$$e^{i\beta ph}\xi^{q+1} - e^{i\beta ph}\xi^q = r\{e^{i\beta(p-1)h}\xi^{q+1} - 2e^{i\beta ph}\xi^{q+1} + e^{i\beta(p+1)h}\xi^{q+1}\},$$

where $r = k/h^2$. Division by $e^{i\beta ph}\xi^q$ leads to

$$\xi - 1 = r\xi(e^{-i\beta h} - 2 + e^{i\beta h})$$
$$= r\xi(2\cos\beta h - 2)$$
$$= -4r\xi\sin^2(\beta h/2).$$

Hence

$$\xi = \frac{1}{1 + 4r\sin^2(\beta h/2)}.$$

For stability $|\xi| \leq 1$, which is so for all positive values of r.

Example 3.2

The hyperbolic equation $\partial^2 u/\partial t^2 = \partial^2 u/\partial x^2$ is approximated by the explicit scheme

$$(u_{p,q+1} - 2u_{p,q} + u_{p,q-1})/k^2 = (u_{p+1,q} - 2u_{p,q} + u_{p-1,q})/h^2. \quad (3.9)$$

Investigate its stability.

It is easily shown by the procedure of example 3.1 that the equation for ξ is

$$\xi^2 - 2A\xi + 1 = 0,$$

where

$$A = 1 - 2r^2\sin^2(\beta h/2), \quad r = k/h. \quad (3.10)$$

Hence the values of ξ are

$$\xi_1 = A + (A^2 - 1)^{\frac{1}{2}} \quad \text{and} \quad \xi_2 = A - (A^2 - 1)^{\frac{1}{2}}.$$

For stability

$$|\xi| \leq 1.$$

As r, k, β are real, $A \leq 1$ by (3.10).

When $A < -1$, $|\xi_2| > 1$, giving instability.

When

$$-1 \leq A \leq 1, A^2 \leq 1, \xi_1 = A + i(1 - A^2)^{\frac{1}{2}}, \xi_2 = A - i(1 - A^2)^{\frac{1}{2}},$$

hence

$$|\xi_1| = |\xi_2| = \{A^2 + (1 - A^2)\}^{\frac{1}{2}} = 1,$$

proving that equation (3.9) is stable for $-1 \leqslant A \leqslant 1$. By (3.10), we then have

$$-1 \leqslant 1 - 2r^2 \sin^2(\beta h/2) \leqslant 1.$$

The only useful inequality is

$$-1 \leqslant 1 - 2r^2 \sin^2(\beta h/2),$$

giving $r \leqslant 1$.

The global rounding error

For simplicity, assume that all boundary values are zero so that the finite-difference equations approximating the initial-value differential equation in the solution domain $0 < x < 1$, $t > 0$, can be written as

$$\mathbf{u}_j = \mathbf{A}\mathbf{u}_{j-1},$$

where $\mathbf{u}_0 = \mathbf{U}_0$ is the vector of known initial values and $\mathbf{A}$ is a square matrix of known elements of order $(N-1)$.

In general, the computer will not store the initial value $u_{i,0}$ exactly, but a numerical approximation $N_{i,0}$, so that

$$N_{i,0} = u_{i,0} - r_{i,0}, \quad \text{i.e.,} \quad \mathbf{N}_0 = \mathbf{u}_0 - \mathbf{r}_0,$$

where $\mathbf{r}_0$ is the vector of initial rounding errors. As rounding errors will be introduced at every stage of the calculations the numerical solution values calculated by the computer at the first time-level will be

$$\mathbf{N}_1 = \mathbf{A}\mathbf{N}_0 - \mathbf{r}_1 = \mathbf{A}\mathbf{u}_0 - \mathbf{A}\mathbf{r}_0 - \mathbf{r}_1.$$

Finally, at the jth time-level, the computed solution will be

$$\mathbf{N}_j = \mathbf{A}^j\mathbf{u}_0 - \mathbf{A}^j\mathbf{r}_0 - \mathbf{A}^{j-1}\mathbf{r}_1 - \ldots - \mathbf{r}_j.$$

If there were no rounding errors the exact solution of the difference equations would be

$$\mathbf{u}_j = \mathbf{A}^j\mathbf{u}_0.$$

Hence the difference between the exact solution and the computed solution, i.e., the global rounding error $\mathbf{R}_j$, at the jth time-level is

$$\mathbf{u}_j - \mathbf{N}_j = \mathbf{A}^j\mathbf{r}_0 + \mathbf{A}^{j-1}\mathbf{r}_1 + \ldots + \mathbf{r}_j.$$

This shows that the local rounding error vector at each time-level propagates forward in the same way as the exact solution vector at that time-level. As proved earlier, the effect of each local rounding error will diminish with increasing j if $\max_i |\lambda_i| < 1$, where λ_i, $i = 1(1)(N-1)$, are the eigenvalues of $\mathbf{A}$, but the global rounding error cannot possibly tend to zero because of the terms $\mathbf{r}_j$, $\mathbf{Ar}_{j-1}, \ldots$.

The local truncation error

Consider the differential equation

$$L(U_{i,j}) = \left(\frac{\partial U}{\partial t} - \frac{\partial^2 U}{\partial x^2}\right)_{i,j} = 0.$$

As mentioned on page 76 the derivatives in this equation may be replaced exactly at the point (ih, jk) by the forward-difference series

$$\left(\frac{\partial U}{\partial t}\right)_{i,j} = \frac{1}{k}\left(\Delta_t U_{i,j} - \tfrac{1}{2}\Delta^2 U_{i,j} + \tfrac{1}{3}\Delta_t^3 U_{i,j} - \ldots\right)$$

and the central-difference series

$$\left(\frac{\partial^2 U}{\partial x^2}\right)_{i,j} = \frac{1}{h^2}\left(\delta_x^2 U_{i,j} - \tfrac{1}{12}\delta_x^4 U_{i,j} + \tfrac{1}{90}\delta^6 U_{i,j} - \ldots\right).$$

To derive a finite-difference equation approximating the differential equation each series could be truncated after a certain number of terms. Say, for example, each series is truncated after the first term. The approximating difference equation is then the classical explicit equation

$$F(u_{i,j}) = 0 = \frac{1}{k}\Delta_t u_{i,j} - \frac{1}{h^2}\,\delta_x^2 u_{i,j} = \frac{u_{i,j+1} - u_{i,j}}{k} - \frac{u_{i-1,j} - 2u_{i,j} + u_{i+1,j}}{h^2}.$$

$$(3.11)$$

It is obvious from this that the difference between the exact solution U of the differential equation and the exact solution u of the difference equation at the point (ih, jk), i.e., the discretization error $U_{i,j} - u_{i,j}$, depends not only on h and k but also on the number of terms in the truncated series used to approximate each derivative.

The amount by which the exact solution U of the partial differential equation does *not* satisfy the difference equation at the point (ih, jk) is called the *local truncation error* $T_{i,j}$. For the scheme defined by equation (3.11),

$$T_{i,j} = F(U_{i,j}) = \frac{U_{i,j+1} - U_{i,j}}{k} - \frac{U_{i-1,j} - 2U_{i,j} + U_{i+1,j}}{h^2}. \quad (3.12)$$

This can be written as

$$U_{i,j+1} = kT_{i,j} + rU_{i-1,j} + (1-2r)U_{i,j} + rU_{i+1,j}. \quad (3.13)$$

Assume now that the solution U of the partial differential equation is known at all mesh points up to and including those at the jth time-level. We could then calculate a *local* approximation $u_{i,j+1}$ to $U_{i,j+1}$ by means of equation (3.11) and get

$$u_{i,j+1} = rU_{i-1,j} + (1-2r)U_{i,j} + rU_{i+1,j}. \quad (3.14)$$

By equations (3.13) and (3.14) it follows that

$$U_{i,j+1} - u_{i,j+1} = \text{(local) discretization error at } (i,j+1) = kT_{i,j}. \quad (3.15)$$

This shows that the local truncation error at the point (i, j) is a measure of the (local) discretization error at the point $(i, j+1)$ *when the finite-difference scheme is applied once only to the exact solution values of the partial differential equation*, all arithmetic being exact, i.e., without rounding errors. [The exact solution value $u_{i,j+1}$ of equation (3.11) obtained by applying it recursively without rounding errors to the initial values and boundary values will, in general, be different from that given by (3.14).] An expression for the (total) discretization error, as defined on page 75, in terms of preceding local truncation errors is derived on page 104.

By means of Taylor's expansion it is easy to express $T_{i,j}$ in terms of powers of h and k and partial derivatives of U at (ih, jk). Although U and its derivatives at this point are generally unknown the analysis is worth while because it provides a method for comparing the local accuracies of different difference schemes approximating the same partial differential equation.

Example 3.3

Calculate the order of the local truncation error of the classical explicit difference approximation to

$$\frac{\partial U}{\partial t} - \frac{\partial^2 U}{\partial x^2} = 0.$$

By Taylor's expansion

$$U_{i+1,j} = U\{(i+1)h, jk\} = U(x_i + h, t_j)$$

$$= U_{i,j} + h\left(\frac{\partial U}{\partial x}\right)_{i,j} + \tfrac{1}{2}h^2\left(\frac{\partial^2 U}{\partial x^2}\right)_{i,j} + \tfrac{1}{6}h^3\left(\frac{\partial^3 U}{\partial x^3}\right)_{i,j} + \ldots$$

$$U_{i-1,j} = U\{(i-1)h, jk\} = U(x_i - h, t_j)$$

$$= U_{i,j} - h\left(\frac{\partial U}{\partial x}\right)_{i,j} + \tfrac{1}{2}h^2\left(\frac{\partial^2 U}{\partial x^2}\right)_{i,j} - \tfrac{1}{6}h^3\left(\frac{\partial^3 U}{\partial x^3}\right)_{i,j} + \ldots$$

$$U_{i,j+1} = U(x_i, t_j + k)$$

$$= U_{i,j} + k\left(\frac{\partial U}{\partial t}\right)_{i,j} + \tfrac{1}{2}k^2\left(\frac{\partial^2 U}{\partial t^2}\right)_{i,j} + \tfrac{1}{6}k^3\left(\frac{\partial^3 U}{\partial t^3}\right)_{i,j} + \ldots.$$

Hence

$$T_{i,j} = \frac{U_{i,j+1} - U_{i,j}}{k} - \frac{U_{i-1,j} - 2U_{i,j} + U_{i+1,j}}{h^2}$$

$$= \left(\frac{\partial U}{\partial t} - \frac{\partial^2 U}{\partial x^2}\right)_{i,j} + \tfrac{1}{2}k\left(\frac{\partial^2 U}{\partial t^2}\right)_{i,j} - \tfrac{1}{12}h^2\left(\frac{\partial^4 U}{\partial x^4}\right)_{i,j}$$

$$+ \tfrac{1}{6}k^2\left(\frac{\partial^3 U}{\partial t^3}\right)_{i,j} - \tfrac{1}{360}h^4\frac{\partial^6 U}{\partial x^6} + \ldots. \quad (3.16)$$

But U is the solution of the differential equation so

$$\left(\frac{\partial U}{\partial t} - \frac{\partial^2 U}{\partial x^2}\right)_{i,j} = 0.$$

Therefore the principal part of the local truncation error is

$$\left(\tfrac{1}{2}k\frac{\partial^2 U}{\partial t^2} - \tfrac{1}{12}h^2\frac{\partial^4 U}{\partial x^4}\right)_{i,j}.$$

Hence

$$T_{i,j} = O(k) + O(h^2).$$

When $k = rh^2$, $0 < r \le \frac{1}{2}$, $T_{i,j}$ is $O(k)$ or $O(h^2)$, as one would expect by equations (1.8) and (1.10).

This error may be further reduced by choosing a special value for k/h^2 because equation (3.16) can be written as

$$T_{i,j} = \tfrac{1}{12}h^2 \left(6\frac{k}{h^2} \frac{\partial^2 U}{\partial t^2} - \frac{\partial^4 U}{\partial x^4} \right)_{i,j} + O(k^2) + O(h^4).$$

By the differential equation,

$$\frac{\partial}{\partial t} \equiv \frac{\partial^2}{\partial x^2},$$

so

$$\frac{\partial}{\partial t} \left(\frac{\partial U}{\partial t} \right) = \frac{\partial^2}{\partial x^2} \left(\frac{\partial^2 U}{\partial x^2} \right),$$

assuming that these derivatives exist. If we put $6k/h^2 = 1$, the expression in the brackets is then zero and $T_{i,j}$ is $O(k^2) + O(h^4)$. This is of little use in practice because $k = \frac{1}{6}h^2$ is very small for small h so the volume of arithmetic needed to advance the solution to a large time-level is substantial.

Consistency or compatibility

It is sometimes possible to approximate a parabolic or hyperbolic equation by a finite-difference scheme that is stable but which has a solution that converges to the solution of a different differential equation as the mesh lengths tend to zero. Such a difference scheme is said to be *inconsistent* or *incompatible* with the partial differential equation and an example is given in worked example 3.4.

The real importance of the concept of consistency lies in a theorem by Lax (reference 31) which states that if a linear finite-difference equation is consistent with a properly posed linear initial-value problem then stability guarantees convergence.

Consistency can be defined in either of two equivalent but slightly different ways.

The more general definition is as follows. Let $L(U) = 0$ represent the partial differential equation in the independent variables x and t, with exact solution U.

Let $F(u) = 0$ represent the approximating finite-difference equation with exact solution u.

Let v be a continuous function of x and t with a sufficient number of continuous derivatives to enable $L(v)$ to be evaluated at the point (ih, jk).

Then the truncation error $T_{i,j}(v)$ at the point (ih, jk) is defined by

$$T_{i,j}(v) = F(v_{i,j}) - L(v_{i,j}). \qquad (3.17)$$

If $T_{i,j}(v) \to 0$ as $h \to 0$, $k \to 0$, the difference equation is said to be consistent or compatible with the partial differential equation. With this definition $T_{i,j}$ gives an indication of the error resulting from the replacement of $L(v_{i,j})$ by $F(v_{i,j})$.

Most authors put $v = U$ because $L(U) = 0$. It then follows by equation (3.17) that

$$T_{i,j}(U) = F(U_{i,j})$$

and the truncation error coincides with the local truncation error. The difference equation is then consistent if the limiting value of the local truncation error is zero as $h \to 0$, $k \to 0$. This is the definition that will be adopted in this book. By the worked example 3.3 it follows that the classical explicit approximation to $\partial U/\partial t = \partial^2 U/\partial x^2$ is consistent with the differential equation.

Example 3.4

The equation

$$\frac{\partial U}{\partial t} - \frac{\partial^2 U}{\partial x^2} = 0$$

is approximated at the point (ih, jk) by the difference equation

$$\frac{u_{i,j+1} - u_{i,j-1}}{2k} - \frac{u_{i+1,j} - 2\{\theta u_{i,j+1} + (1-\theta)u_{i,j-1}\} + u_{i-1,j}}{h^2} = 0.$$

Show that the local truncation error at this point is

$$\frac{k^2}{6}\frac{\partial^3 U}{\partial t^3} - \frac{h^2}{12}\frac{\partial^4 U}{\partial x^4} + (2\theta - 1)\frac{2k}{h^2}\frac{\partial U}{\partial t} + \frac{k^2}{h^2}\frac{\partial^2 U}{\partial t^2} + O\left(\frac{k^3}{h^2}, h^4, k^4\right).$$

Discuss the consistency of this scheme with the partial differential equation when:

(i) $k = rh$ and   (ii) $k = rh^2$,

where r is a positive constant and θ a variable parameter.

Expansion of the terms $U_{i,j+1}$, $U_{i,j-1}$, $U_{i+1,j}$, and $U_{i-1,j}$ about the point (ih, jk) by Taylor's series, as in example 3.3, and substitution into

$$T_{i,j} = \frac{U_{i,j+1} - U_{i,j-1}}{2k} - \frac{U_{i+1,j} - 2\{\theta U_{i,j+1} + (1-\theta)U_{i,j-1}\} + U_{i-1,j}}{h^2}$$

leads to

$$T_{i,j} = \left(\frac{\partial U}{\partial t} - \frac{\partial^2 U}{\partial x^2}\right)_{i,j} + \left\{\frac{k^2}{6}\frac{\partial^3 U}{\partial t^3} - \frac{h^2}{12}\frac{\partial^4 U}{\partial x^4}\right.$$
$$\left. + (2\theta - 1)\frac{2k}{h^2}\frac{\partial U}{\partial t} + \frac{k^2}{h^2}\frac{\partial^2 U}{\partial t^2}\right\} + O\left(\frac{k^3}{h^2}, h^4, k^4\right).$$

Hence the result since $\partial U/\partial t - \partial^2 U/\partial x^2 = 0$.

Case (i) $k = rh$

As $h \to 0$,

$$T_{i,j} = F(U_{i,j}) \to \left\{\frac{\partial U}{\partial t} - \frac{\partial^2 U}{\partial x^2} + (2\theta - 1)\frac{2r}{h}\frac{\partial U}{\partial t} + r^2\frac{\partial^2 U}{\partial t^2}\right\}_{i,j}.$$

When $\theta \neq \frac{1}{2}$ the third term tends to infinity. When $\theta = \frac{1}{2}$ the limiting value of $T_{i,j}$ is

$$\frac{\partial U}{\partial t} - \frac{\partial^2 U}{\partial x^2} + r^2\frac{\partial^2 U}{\partial t^2}.$$

In this case the finite-difference equation is consistent with the *hyperbolic* equation

$$\frac{\partial U}{\partial t} - \frac{\partial^2 U}{\partial x^2} + r^2\frac{\partial^2 U}{\partial t^2} = 0.$$

Hence the difference equation is always inconsistent with $\partial U/\partial t - \partial^2 U/\partial x^2 = 0$ when $k = rh$.

Case (ii) $k = rh^2$

As $h \to 0$,

$$T_{i,j} \to \frac{\partial U}{\partial t} - \frac{\partial^2 U}{\partial x^2} + 2(2\theta - 1)r\frac{\partial U}{\partial t}.$$

When $\theta \neq \frac{1}{2}$ the difference scheme is consistent with the parabolic equation

$$\{1 + 2(2\theta - 1)r\}\frac{\partial U}{\partial t} - \frac{\partial^2 U}{\partial x^2} = 0.$$

It is only when $\theta = \frac{1}{2}$ that the difference scheme is consistent with the given differential equation. This is then the well-known Du Fort and Frankel three-level explicit scheme which is also stable for all $r > 0$. (See worked example 3.5.) It was devised to overcome the unconditional instability of the early Richardson explicit scheme

$$\frac{u_{i,j+1} - u_{i,j-1}}{2k} - \frac{u_{i-1,j} - 2u_{i,j} + u_{i+1,j}}{h^2} = 0,$$

but to retain the advantage of the central-difference approximation to the time-derivative which gives a local truncation error of $O(k^2) + O(h^2)$ as opposed to $O(k) + O(h^2)$ for the classical explicit approximation to $\partial U/\partial t - \partial^2 U/\partial x^2 = 0$.

A simple example demonstrating the relationship between convergence, stability, and consistency

The classical explicit approximation to the heat conduction equation provides a comparatively simple illustration of Lax's equivalence theorem which states that if a linear difference equation is consistent with a properly posed linear initial-value problem then stability is the necessary and sufficient condition for convergence.

In general, a problem is properly posed if:

(i) The solution is unique when it exists.

(ii) The solution depends continuously on the initial data.

(iii) A solution always exists for initial data that is arbitrarily close to initial data for which no solution exists. (In heat flow problems, for example, discontinuous temperature distributions can be approximated by a sum of N continuous functions whose limiting value, as N tends to infinity, equals the discontinuous distribution except at the points of discontinuity.)

Let U satisfy the equation

$$\frac{\partial U}{\partial t} = \frac{\partial^2 U}{\partial x^2}, \quad 0 < x < 1, \tag{3.18}$$

have known continuous initial values when $t = 0$, $0 \le x \le 1$, and known continuous boundary values at $x = 0$ and 1, $t > 0$.

The classical explicit approximation to (3.18) is

$$\frac{u_{i,j+1} - u_{i,j}}{k} = \frac{u_{i-1,j} - 2u_{i,j} + u_{i+1,j}}{h^2}, \quad i = 1(1)(N-1), \tag{3.19}$$

where $x = ih$, $t = jk$, and $Nh = 1$. The local truncation error of this difference scheme is defined by

$$T_{i,j} = \frac{U_{i,j+1} - U_{i,j}}{k} - \frac{U_{i-1,j} - 2U_{i,j} + U_{i+1,j}}{h^2},$$

which may be written as

$$U_{i,j+1} = kT_{i,j} + rU_{i-1,j} + (1 - 2r)U_{i,j} + rU_{i+1,j}, \tag{3.20}$$

where $r = k/h^2$. For $i = 1(1)(N-1)$, equations (3.20) can be written in matrix form as

$$\begin{bmatrix} U_{1,j+1} \\ U_{2,j+1} \\ \cdot \\ \cdot \\ \cdot \\ U_{N-1,j+1} \end{bmatrix} = k \begin{bmatrix} T_{1,j} \\ T_{2,j} \\ \cdot \\ \cdot \\ \cdot \\ T_{N-1,j} \end{bmatrix} + \begin{bmatrix} (1-2r) & r & & & \\ r & (1-2r) & r & & \\ & & \cdot & & \\ & & & \cdot & \\ & & & & \cdot \\ & & & r & (1-2r) \end{bmatrix}$$

$$\times \begin{bmatrix} U_{1,j} \\ U_{2,j} \\ \cdot \\ \cdot \\ \cdot \\ U_{N-1,j} \end{bmatrix} + \begin{bmatrix} rU_{0,j} \\ 0 \\ \cdot \\ \cdot \\ 0 \\ rU_{N,j} \end{bmatrix}$$

i.e., as

$$\mathbf{U}_{j+1} = k\mathbf{T}_j + \mathbf{A}\mathbf{U}_j + \mathbf{c}_j, \tag{3.21}$$

where $\mathbf{c}_j$ is a vector of known boundary values. Applying this recursively,

$$\begin{aligned} \mathbf{U}_{j+1} &= k\mathbf{T}_j + \mathbf{A}(k\mathbf{T}_{j-1} + \mathbf{A}\mathbf{U}_{j-1} + \mathbf{c}_{j-1}) + \mathbf{c}_j \\ &= k(\mathbf{T}_j + \mathbf{A}\mathbf{T}_{j-1}) + \mathbf{A}^2\mathbf{U}_{j-1} + (\mathbf{c}_j + \mathbf{A}\mathbf{c}_{j-1}) \\ &= \cdots \\ &= k(\mathbf{T}_j + \mathbf{A}\mathbf{T}_{j-1} + \ldots + \mathbf{A}^j\mathbf{T}_0) + \mathbf{A}^{j+1}\mathbf{u}_0 \\ &\quad + (\mathbf{c}_j + \mathbf{A}\mathbf{c}_{j-1} + \ldots + \mathbf{A}^j\mathbf{c}_0). \end{aligned} \tag{3.22}$$

As the boundary and initial values for u are the same as for U, it follows from equation (3.19), which can be written as

$$u_{i,j+1} = r u_{i-1,j} + (1-2r)u_{i,j} + r u_{i+1,j}, \; i = 1(1)(N-1),$$

that

$$\mathbf{u}_{j+1} = \mathbf{A}\mathbf{u}_j + \mathbf{c}_j.$$

This leads as before to

$$\mathbf{u}_{j+1} = \mathbf{A}^{j+1}\mathbf{u}_0 + (\mathbf{c}_j + \mathbf{A}\mathbf{c}_{j-1} + \ldots + \mathbf{A}^j \mathbf{c}_0). \tag{3.23}$$

Subtraction of (3.23) from (3.22) shows that

$$\mathbf{U}_{j+1} - \mathbf{u}_{j+1} = k(\mathbf{T}_j + \mathbf{A}\mathbf{T}_{j-1} + \ldots + \mathbf{A}^j\mathbf{T}_0) + \mathbf{A}^{j+1}(\mathbf{U}_0 - \mathbf{u}_0).$$

But $\mathbf{U}_0 = \mathbf{u}_0 =$ vector of initial values. Hence

$$\mathbf{U}_{j+1} - \mathbf{u}_{j+1} = k(\mathbf{T}_j + \mathbf{A}\mathbf{T}_{j-1} + \ldots + \mathbf{A}^j\mathbf{T}_0). \tag{3.24}$$

This equation shows that the difference between the *exact* solution of the partial differential equation and the *exact* solution of the approximating difference equation at, say, the $(i, j+1)$th mesh point depends on the local truncation errors at certain mesh points on every preceding time-level and on the difference scheme used. Whether or not the accumulative effect at the $(i, j+1)$th mesh point of these preceding errors is a catastrophic build-up or a hoped-for decay as j increases depends clearly on the matrix $\mathbf{A}$ and the nature of the $T_{i,j}$, $i = 1(1)N$, $j = 0(1)j$.

When we considered previously the stability of equations (3.23) it was shown that the error vector $\mathbf{e}$ satisfied the equation

$$\mathbf{e}_j = \mathbf{A}\mathbf{e}_{j-1} = \ldots = \mathbf{A}^{j-n}\mathbf{e}_n = \ldots = \mathbf{A}^j\mathbf{e}_0,$$

and the analysis was carried out in terms of the eigenvalues and eigenvectors of $\mathbf{A}$. In that analysis $k = \delta t$ remained fixed for all j, which is what happens in practice, and it was found that the components of $\mathbf{e}_j$ either tended to zero or remained a linear combination of the components of the initial perturbation $\mathbf{e}_0$ as j increased if $\rho(\mathbf{A}) = \max_i |\lambda_i| \leqslant 1$, where ρ is the spectral radius of the matrix $\mathbf{A}$. In order however to relate stability to convergence *at a fixed time-level* it is necessary to let $jk = t$, a fixed finite number. The condition $\rho(\mathbf{A}) \leqslant 1$ is also unnecessarily restrictive and will not cope with parabolic and hyperbolic partial differential problems whose solutions increase gradually with t. In these cases $\mathbf{u}_j$ must also increase with j if it is to converge to $\mathbf{U}_j$. (See page 106.) It is convenient therefore, at this stage, to return to the original definition of stability which defines a difference scheme as stable when there is a limit to the magnification of $\mathbf{u}_j$, or of any form of local error, as j increases indefinitely.

Mathematically, the difference scheme is stable if there exists a real positive number K such that

$$\|\mathbf{e}_j\| = \|\mathbf{A}^{j-n}\mathbf{e}_n\| \leqslant \|\mathbf{A}^{j-n}\| \|\mathbf{e}_n\| \leqslant \|\mathbf{A}\|^{j-n} \|\mathbf{e}_n\|$$
$$= C^{j-n} \|\mathbf{e}_n\| = K \|\mathbf{e}_n\|,$$

for $n = 0, 1, 2, \ldots, j$, where $jk = t$ and $j \to \infty$. (The number $C = \|\mathbf{A}\|$, called a norm of $\mathbf{A}$, is a positive number representing the 'size' of $\mathbf{A}$. For example, the infinity norm of $\mathbf{A}$ is the maximum row sum of the moduli of its elements. The infinity norm of the vector $\mathbf{e}_n$ is the maximum of the moduli of its components. Vector and associated matrix norms must be compatible, i.e., $\|\mathbf{A}\mathbf{e}_n\| \leqslant \|\mathbf{A}\| \|\mathbf{e}_n\|$. Also $\|\mathbf{AB}\| \leqslant \|\mathbf{A}\| \|\mathbf{B}\|$ and $\|\mathbf{A} + \mathbf{B}\| \leqslant \|\mathbf{A}\| + \|\mathbf{B}\|$. See reference 24.)

Assuming that the difference equations are stable in this sense it follows by equation (3.24) that

$$\|\mathbf{U}_{j+1} - \mathbf{u}_{j+1}\| \leqslant k\|\mathbf{T}_j\| + k\mathrm{K}(\|\mathbf{T}_{j-1}\| + \ldots + \|\mathbf{T}_0\|).$$

Define the norm of each vector $\mathbf{T}_s$, $s = 0(1)j$, to be the maximum of the moduli of its components. Then if M is the maximum of the moduli of all the components $T_{i,j}$, $\|\mathbf{T}_s\| \leqslant M$, $s = 0(1)j$. Hence

$$\|\mathbf{U}_{j+1} - \mathbf{u}_{j+1}\| \leqslant kM + kKjM = kM + KtM.$$

But $jk = t$ is finite, so $k \to 0$ as $j \to \infty$ and the first term on the right-hand side tends to zero irrespective of the magnitudes of M and K. The second term however tends to zero if and only if $M = \max_{i,j}|T_{i,j}|$ tends to zero as k and $h = (k/r)^{\frac{1}{2}}$ tend to zero, which, by definition, is the condition for consistency. This proves, in this particular case, that the difference scheme is convergent when it is stable and consistent.

Lax's equivalence theorem

Given a properly posed linear initial-value problem and a linear finite-difference approximation to it that satisfies the consistency condition, stability is the necessary and sufficient condition for convergence.

The proof of this theorem is beyond the scope of this book and interested readers should consult reference 31.

A comment on the condition for stability

Let us assume we are calculating the finite-difference solution within the time-interval 0 to $jk = T$, where T is finite, and are concerned with $k \to 0$, i.e., $j \to \infty$. Let the difference equation giving the solution at an intermediate time-level $t = nk$ be

$$\mathbf{u}_n = \mathbf{A}\mathbf{u}_{n-1} + \mathbf{b}_{n-1}.$$

As shown in equation (3.6), the error vector $\mathbf{e}_n$ can be expressed in terms of the $(N-1)$ eigenvalues λ_s and eigenvectors $\mathbf{v}_s$ of $\mathbf{A}$ by

$$\mathbf{e}_n = \sum_{s=1}^{N-1} c_s \lambda_s^n \mathbf{v}_s.$$

This will remain bounded if there is a constant C such that

$$|\lambda_s^n| \leq C, \quad 0 < n \leq j, \quad C \geq 1,$$

i.e.,

$$|\lambda_s| \leq C^{\frac{1}{n}} = \exp\left(\tfrac{1}{n}\log C\right)$$

$$= 1 + \tfrac{1}{n}\log C + \ldots$$
$$1 + \tfrac{k}{t}\log C + \ldots.$$

In particular this must hold for $\max_s|\lambda_s|$, i.e., for the spectral radius $\rho(\mathbf{A})$ of $\mathbf{A}$. Hence the necessary and sufficient condition for a two time-level set of difference equations $\mathbf{u}_n = \mathbf{A}\mathbf{u}_{n-1} + \mathbf{b}_{n-1}$ to be stable is that

$$\rho(\mathbf{A}) \leq 1 + O(k).$$

Case (i) If the solution of the differential equation does *not* increase exponentially with increasing t then for convergence the solution of the difference equation must behave in the same way. The condition for this is $\rho(\mathbf{A}) \leq 1$, which satisfies $\rho(\mathbf{A}) \leq 1 + O(k)$.

Case (ii) If a component of the solution of the differential equation *increases* with increasing t then $\rho(\mathbf{A}) \leq 1 + O(k)$ *must* be used if $u_{i,j}$ is to converge to $U_{i,j}$.

For example, the solution of the equation

$$\frac{\partial U}{\partial t} = \alpha \frac{\partial^2 U}{\partial x^2} + \beta U, \quad \alpha > 0, \quad \beta > 0,$$

where $U = f(x)$ when $t = 0$, $0 \leq x \leq 1$, is eventually dominated by the term $e^{\beta t} \int_0^1 f(x) \, dx$. Therefore the consistent approximation

$$\frac{1}{k}(u_{i,j+1} - u_{i,j}) = \frac{\alpha}{h^2}(u_{i-1,j} - 2u_{i,j} + u_{i+1,j}) + \beta u_{i,j}$$

will give a convergent difference replacement only if the spectral radius of its matrix of coefficients exceeds unity but is less than or equal to $1 + O(k)$. The eigenvalues of the matrix, assuming fixed boundary conditions, are

$$\lambda_s = 1 + \beta k - 4r\alpha \sin^2 \frac{s\pi}{2N}, \quad r = \frac{k}{h^2}, \quad s = 1(1)(N-1).$$

It is seen from this that the term βk allows the convergence criterion to be satisfied when the restrictions on r are the same as for $\beta = 0$, namely $r \leq 1/2\alpha$.

The stability of three or more time-level difference equations

The following theorem is useful for the matrix method of analysis of the stability of three or more time-level difference equations and is easier to use than one might at first think.

Theorem: If the matrix $\mathbf{A}$ can be written as

$$\mathbf{A} = \begin{bmatrix} \mathbf{A}_{1,1} & \mathbf{A}_{1,2} & \cdot & \cdot & \cdot & \mathbf{A}_{1,m} \\ \mathbf{A}_{2,1} & \mathbf{A}_{2,2} & \cdot & \cdot & \cdot & \mathbf{A}_{2,m} \\ \cdot & & & & & \cdot \\ \cdot & & & & & \cdot \\ \cdot & & & & & \cdot \\ \mathbf{A}_{m,1} & \mathbf{A}_{m,2} & \cdot & \cdot & \cdot & \mathbf{A}_{m,m} \end{bmatrix},$$

where each $\mathbf{A}_{i,j}$ is an $n \times n$ matrix, and all the $\mathbf{A}_{i,j}$ have a common set of n linearly independent eigenvectors, then the eigenvalues of

A are given by the eigenvalues of the matrices

$$\begin{bmatrix} \lambda_{1,1}^{(k)} & \lambda_{1,2}^{(k)} & \cdots & \lambda_{1,m}^{(k)} \\ \lambda_{2,1}^{(k)} & \lambda_{2,2}^{(k)} & \cdots & \lambda_{2,m}^{(k)} \\ \cdot & & & \cdot \\ \cdot & & & \cdot \\ \cdot & & & \cdot \\ \lambda_{m,1}^{(k)} & \lambda_{m,2}^{(k)} & \cdots & \lambda_{m,m}^{(k)} \end{bmatrix}, \quad k = 1(1)n,$$

where $\lambda_{i,j}^{(k)}$ is the kth eigenvalue of $\mathbf{A}_{i,j}$ corresponding to the kth eigenvector $\mathbf{v}_k$ common to all the $\mathbf{A}_{i,j}$'s.

Proof

Let $\mathbf{v}_k$ be an eigenvector common to all the submatrices $\mathbf{A}_{i,j}$, $i,j = 1(1)m$, and denote the corresponding eigenvalues of $\mathbf{A}_{1,1}$, $\mathbf{A}_{2,1},\ldots$ by $\lambda_{1,1}^{(k)}$, $\lambda_{2,1}^{(k)},\ldots$ respectively. For simplicity consider $i,j = 1(1)2$ and denote $\mathbf{v}_k$ by $\mathbf{v}$, $\lambda_{i,j}^{(k)}$ by $\lambda_{i,j}$. Then

$$\mathbf{A}_{1,1}\mathbf{v} = \lambda_{1,1}\mathbf{v}, \quad \mathbf{A}_{1,2}\mathbf{v} = \lambda_{1,2}\mathbf{v},$$
$$\mathbf{A}_{2,1}\mathbf{v} = \lambda_{2,1}\mathbf{v}, \quad \mathbf{A}_{2,2}\mathbf{v} = \lambda_{2,2}\mathbf{v}.$$

Multiply these equations respectively by the non-zero constants α_1, α_2, α_1, and α_2 and write them as

$$\begin{bmatrix} \mathbf{A}_{1,1} & \mathbf{A}_{1,2} \\ \mathbf{A}_{2,1} & \mathbf{A}_{2,2} \end{bmatrix} \begin{bmatrix} \alpha_1\mathbf{v} \\ \alpha_2\mathbf{v} \end{bmatrix} = \begin{bmatrix} (\lambda_{1,1}\alpha_1 + \lambda_{1,2}\alpha_2)\mathbf{v} \\ (\lambda_{2,1}\alpha_1 + \lambda_{2,2}\alpha_2)\mathbf{v} \end{bmatrix}. \tag{3.25}$$

Assume now that

$$\mathbf{A} = \begin{bmatrix} \mathbf{A}_{1,1} & \mathbf{A}_{1,2} \\ \mathbf{A}_{2,1} & \mathbf{A}_{2,2} \end{bmatrix}$$

has an eigenvalue μ corresponding to the eigenvector

$$\begin{bmatrix} \alpha_1\mathbf{v} \\ \alpha_2\mathbf{v} \end{bmatrix}$$

so that

$$\begin{bmatrix} \mathbf{A}_{1,1} & \mathbf{A}_{1,2} \\ \mathbf{A}_{2,1} & \mathbf{A}_{2,2} \end{bmatrix} \begin{bmatrix} \alpha_1\mathbf{v} \\ \alpha_2\mathbf{v} \end{bmatrix} = \mu \begin{bmatrix} \alpha_1\mathbf{v} \\ \alpha_2\mathbf{v} \end{bmatrix}. \tag{3.26}$$

By the right-hand sides of equations (3.25) and (3.26),

$$(\lambda_{1,1} - \mu)\alpha_1 + \lambda_{1,2}\alpha_2 = 0$$

and
$$\lambda_{2,1}\alpha_1 + (\lambda_{2,2} - \mu)\alpha_2 = 0.$$

These two equations will have a non-trivial solution for α_1 and α_2 if and only if

$$\det\begin{bmatrix} (\lambda_{1,1} - \mu) & \lambda_{1,2} \\ \lambda_{2,1} & (\lambda_{2,2} - \mu) \end{bmatrix} = 0,$$

i.e., if and only if μ is an eigenvalue of the matrix

$$\begin{bmatrix} \lambda_{1,1} & \lambda_{1,2} \\ \lambda_{2,1} & \lambda_{2,2} \end{bmatrix}.$$

Matrices with Common Eigenvector Systems

Proofs of the following theorems can be found in reference 36.

(i) If the $N \times N$ matrix $\mathbf{A}$ has N distinct eigenvalues λ_s it has N unique linearly independent eigenvectors $\mathbf{v}_s$, $s = 1(1)N$. As proved earlier, any polynomial $f(\mathbf{A})$ of $\mathbf{A}$ has the same set of eigenvectors $\mathbf{v}_s$ and a corresponding set of eigenvalues $f(\lambda_s)$.

(ii) All $N \times N$ Hermitian matrices, which includes real symmetric matrices, have N linearly independent eigenvectors, so the comment made in (i) applies to (ii).

(iii) If the matrices $\mathbf{A}$ and $\mathbf{B}$ commute and have linear elementary divisors then they have a common system of eigenvectors. In particular, all matrices with distinct eigenvalues, all Hermitian, and therefore all real symmetric matrices, have linear elementary divisors.

(iv) Let $\mathbf{A}$ and $\mathbf{B}$ be matrices with a common system of eigenvectors. Let λ and μ be the eigenvalues of $\mathbf{A}$ and $\mathbf{B}$ respectively corresponding to the common eigenvector $\mathbf{v}$. Then $\mathbf{v}$ is an eigenvector of $\mathbf{AB}$ and $\mathbf{A}^{-1}\mathbf{B}$ and the corresponding eigenvalues are $\lambda\mu$ and $\lambda^{-1}\mu$ respectively. These results are easily proved. By hypothesis, $\mathbf{Bv} = \mu\mathbf{v}$ and $\mathbf{Av} = \lambda\mathbf{v}$. Therefore $\mathbf{ABv} = \mu\mathbf{Av} = \mu\lambda\mathbf{v}$. Also $\mathbf{A}^{-1}\mathbf{Bv} = \mu\mathbf{A}^{-1}\mathbf{v} = (\mu/\lambda)\mathbf{v}$.

Example 3.5

Investigate the stability of the Du Fort and Frankel approximation to the equation.

$$\frac{\partial U}{\partial t} = \frac{\partial^2 U}{\partial x^2}, \quad 0 < x < 1,$$

given that U is known initially for $0 \le x \le 1$, $t = 0$ and is known on the boundaries $x = 0$ and 1 for $t > 0$. (Reference worked example 3.4.)

The approximation at the point (ih, jk) is

$$\frac{1}{2k}(u_{i,j+1} - u_{i,j-1}) = \frac{1}{h^2}\{u_{i-1,j} - (u_{i,j-1} + u_{i,j+1}) + u_{i+1,j}\},$$

which may be written as

$$(1 + 2r)u_{i,j+1} = 2r(u_{i-1,j} + u_{i+1,j}) + (1 - 2r)u_{i,j-1},$$

where $r = k/h^2$. For known boundary values and $Nh = 1$ these equations in matrix form are

$$(1+2r)\begin{bmatrix} u_{1,j+1} \\ u_{2,j+1} \\ u_{3,j+1} \\ \cdot \\ \cdot \\ u_{N-1,j+1} \end{bmatrix} = 2r\begin{bmatrix} 0 & 1 & & & \\ 1 & 0 & 1 & & \\ & 1 & 0 & 1 & \\ & & & \cdot & \\ & & & \cdot & \\ & & & 1 & 0 \end{bmatrix}\begin{bmatrix} u_{1,j} \\ u_{2,j} \\ u_{3,j} \\ \cdot \\ \cdot \\ u_{N-1,j} \end{bmatrix}$$

$$+ (1 - 2r)\begin{bmatrix} u_{1,j-1} \\ u_{2,j-1} \\ u_{3,j-1} \\ \cdot \\ \cdot \\ u_{N-1,j-1} \end{bmatrix} + 2r\begin{bmatrix} u_{0,j} \\ 0 \\ 0 \\ \cdot \\ \cdot \\ u_{N,j} \end{bmatrix}$$

giving

$$\mathbf{u}_{j+1} = \frac{2r}{1+2r}\mathbf{A}\mathbf{u}_j + \frac{1-2r}{1+2r}\mathbf{u}_{j-1} + \mathbf{c}_j, \qquad (3.27)$$

where $\mathbf{A}$ is as displayed and $\mathbf{c}_j$ is a vector of known values.

Put
$$\mathbf{v}_j = \begin{bmatrix} \mathbf{u}_j \\ \mathbf{u}_{j-1} \end{bmatrix}.$$

Then equation (3.27) and the identity $\mathbf{u}_j = \mathbf{u}_j$ can be written as

$$\begin{bmatrix} \mathbf{u}_{j+1} \\ \hdashline \mathbf{u}_j \end{bmatrix} = \begin{bmatrix} \dfrac{2r}{1+2r}\mathbf{A} & \vdots & \dfrac{(1-2r)}{(1+2r)}\mathbf{I}_{N-1} \\ \hdashline \mathbf{I}_{N-1} & \vdots & \mathbf{O} \end{bmatrix}\begin{bmatrix} \mathbf{u}_j \\ \hdashline \mathbf{u}_{j-1} \end{bmatrix} + \begin{bmatrix} \mathbf{c}_j \\ \hdashline \mathbf{O} \end{bmatrix},$$

where $\mathbf{I}_{N-1}$ is the unit matrix of order $(N-1)$,

.e., as

$$\mathbf{v}_{j+1} = \mathbf{P}\mathbf{v}_j + \mathbf{d}_j,$$

where **P** is the matrix shown and $\mathbf{d}_j$ a column vector of known constants. This technique has reduced a three-level difference equation to a two-level one. The equations will be stable when each eigenvalue of **P** has a modulus ≤ 1. The matrix **A** has $(N-1)$ different eigenvalues so it has $(N-1)$ linearly independent eigenvectors $\mathbf{v}_s$, $s = 1(1)(N-1)$. Although the matrix $\mathbf{I}_{N-1}$ has $(N-1)$ eigenvalues each equal to 1 it has $(N-1)$ linearly independent eigenvectors which may be taken as $\mathbf{v}_s$, $s = 1(1)(N-1)$, because the eigenvalue equation $\mathbf{Bx} = \lambda\mathbf{x}$ is clearly satisfied by $\mathbf{I}_{N-1}\mathbf{v}_s = 1.\mathbf{v}_s$. Hence the eigenvalues λ of P are the eigenvalues of the matrix

$$\begin{bmatrix} \dfrac{2r\lambda_k}{1+2r} & \dfrac{1-2r}{1+2r} \\ 1 & 0 \end{bmatrix},$$

where λ_k is the kth eigenvalue of **A**. For such a simple case we can work from first principles and find λ by evaluating

$$\det\begin{bmatrix} \left\{\dfrac{2r\lambda_k}{1+2r} - \lambda\right\} & \dfrac{1-2r}{1+2r} \\ 1 & -\lambda \end{bmatrix} = 0,$$

giving

$$\lambda^2 - \frac{2r\lambda_k}{1+2r}\lambda - \frac{1-2r}{1+2r} = 0.$$

By the formula on page 86,
$$\lambda_k = 2\cos(k\pi/N), \quad k = 1(1)(N-1).$$

Hence

$$\lambda = \left\{2r\cos\frac{k\pi}{N} \pm \left(1 - 4r^2\sin^2\frac{k\pi}{N}\right)^{\frac{1}{2}}\right\}\Big/(1+2r).$$

Case (*i*) $\quad 1 > 1 - 4r^2\sin^2\dfrac{k\pi}{N} \geq 0.$

Then

$$|\lambda| < \frac{2r+1}{1+2r} = 1.$$

Case (*ii*) $1 - 4r^2 \sin^2 \dfrac{k\pi}{N} < 0.$

Then

$$|\lambda|^2 = \frac{1}{(2r+1)^2} \left\{ \left(2r \cos \frac{k\pi}{N} \right)^2 + 4r^2 \sin^2 \frac{k\pi}{N} - 1 \right\}$$

$$= \frac{4r^2 - 1}{4r^2 + 4r + 1} < 1 \quad \text{since} \quad r > 0.$$

Therefore the equations are unconditionally stable for all positive *r*.

Brief introduction to the analytical solution of homogeneou finite-difference equations

Linear equations with constant coefficients

Consider the difference equation

$$u_{j+2} + a u_{j+1} + b u_j = 0, \quad j = 0, 1, 2, \cdots, \tag{3.28}$$

where *a* and *b* are real constants.

Assume that

$$u_j = Am^j$$

is a solution, where *A* and *m* are non-zero constants. Substitutio into (3.28) shows that *m* is a root of the quadratic equation

$$m^2 + am + b = 0. \tag{3.29}$$

Case (*i*) Roots real and distinct, $m = m_1$ and $m = m_2$, say.
One solution is $u_j = Am_1^j$ and another is $u_j = Bm_2^j$ where *A* and *B* are arbitrary constants. As equation (3.28) is linear in *u* its genera solution is

$$u_j = Am_1^j + Bm_2^j.$$

Case (*ii*) Repeated roots, $m = m_1$ twice, say. Clearly one solution i $u_j = Am_1^j$.

Put $u_j = m_1^j f(j)$. Substitution into (3.28) and the use of $a = -2m_1$ $b = m_1^2$ leads to

$$f(j+2) - 2f(j+1) + f(j) = 0.$$

By inspection it is seen that $f(j) = j$ satisfies this equation. Therefor a second solution of (3.28) is $u_j = Bjm_1^j$. Hence the solution o equation (3.28) in this case is

$$u_j = (A + Bj)m_1^j.$$

Case (iii) Complex roots.

Because a and b are real the roots of (3.29) will be conjugate complex numbers, $m_1 = re^{i\theta}$ and $m_2 = re^{-i\theta}$, say, where $i = \sqrt{(-1)}$.

Hence $a = -r(e^{i\theta} + e^{-i\theta}) = -2r \cos \theta$ and $b = r^2$. As in Case (i) the solution of (3.28) is

$$u_j = Ar^j e^{ij\theta} + Br^j e^{-ij\theta} = r^j \{(A + B) \cos j\theta + i(A - B) \sin j\theta\}.$$

Since A and B are arbitrary constants and $r = b^{\frac{1}{2}}$, this can be written as

$$u_j = b^{\frac{1}{2}j}(C \cos j\theta + D \sin j\theta),$$

where C and D are arbitrary constants and $\cos \theta = -a/2r = -a/2\sqrt{b}$. Methods for deriving particular integrals for non-homogeneous difference equations are given in 'Finite Difference Equations' by H. Levy and F. Lessman. (Pitman).

The eigenvalues and vectors of a common tridiagonal matrix

Let

$$\mathbf{A} = \begin{bmatrix} a & b & & & & \\ c & a & b & & & \\ & c & a & b & & \\ & & & . & & \\ & & & & . & \\ & & & & c & a \end{bmatrix}$$

be a square matrix of order N, where a, b, and c may be real or complex numbers.

Let λ represent an eigenvalue of **A** and **v** the corresponding eigenvector with components $v_1, v_2, \ldots, v_N$. Then the eigenvalue equation $\mathbf{Av} = \lambda\mathbf{v}$ gives

$$(a - \lambda)v_1 + bv_2 = 0$$
$$cv_1 + (a - \lambda)v_2 + bv_3 = 0$$
$$\cdots$$
$$cv_{j-1} + (a - \lambda)v_j + bv_{j+1} = 0$$

and

$$cv_{N-1} + (a - \lambda)v_N = 0.$$

If we define $v_0 = v_{N+1} = 0$ then these N equations can be combined into the single difference equation

$$cv_{j-1} + (a - \lambda)v_j + bv_{j+1} = 0, \quad j = 1(1)N. \tag{3.30}$$

As shown, previously, the solution of (3.30) is

$$v_j = Bm_1^j + Cm_2^j, \tag{3.31}$$

where B and C are arbitrary constants and m_1, m_2 are the roots of the equation

$$c + (a - \lambda)m + bm^2 = 0. \tag{3.32}$$

(It is proved later that the roots cannot be equal.)
By equation (3.31) it follows, since $v_0 = v_{N+1} = 0$, that,

$$0 = B + C$$

and

$$0 = Bm_1^{N+1} + Cm_2^{N+1}.$$

Hence

$$\left(\frac{m_1}{m_2}\right)^{N+1} = 1 = e^{i2s\pi}, \quad s = 1(1)N,$$

where $i = \sqrt{(-1)}$. Therefore

$$\frac{m_1}{m_2} = e^{i2s\pi/(N+1)}. \tag{3.33}$$

By equation (3.32),

$$m_1 m_2 = \frac{c}{b}, \tag{3.34}$$

and elimination of m_2 between (3.33) and (3.34) leads to

$$m_1 = \left(\frac{c}{b}\right)^{\frac{1}{2}} e^{is\pi/(N+1)}.$$

Similarly,

$$m_2 = \left(\frac{c}{b}\right)^{\frac{1}{2}} e^{-is\pi/(N+1)}.$$

Again, by equation (3.32),

$$m_1 + m_2 = (\lambda - a)/b,$$

giving that

$$\lambda = a + b\left(\frac{c}{b}\right)^{\frac{1}{2}}(e^{is\pi/(N+1)} + e^{-is\pi/(N+1)}).$$

Hence the N eigenvalues are given by

$$\lambda_s = a + 2b\left(\frac{c}{b}\right)^{\frac{1}{2}}\cos\frac{s\pi}{N+1}, \quad s = 1(1)N.$$

The jth component of the eigenvector is

$$v_j = Bm_1^j + Cm_2^j = B\left(\frac{c}{b}\right)^{\frac{1}{2}j}(e^{ijs\pi/N+1} - e^{-ijs\pi/N+1})$$

$$= 2iB\left(\frac{c}{b}\right)^{\frac{1}{2}j}\sin\frac{js\pi}{N+1},$$

so the eigenvector $\mathbf{v}_s$ corresponding to λ_s can be taken as

$$\mathbf{v}_s^T = \left\{\left(\frac{c}{b}\right)^{\frac{1}{2}}\sin\frac{s\pi}{N+1}, \frac{c}{b}\sin\frac{2s\pi}{N+1}, \left(\frac{c}{b}\right)^{\frac{3}{2}}\right.$$

$$\left. \times \sin\frac{3s\pi}{N+1}, \ldots, \left(\frac{c}{b}\right)^{\frac{N}{2}}\sin\frac{Ns\pi}{N+1}\right\}.$$

It is easily shown that the roots of equation (3.32) cannot be equal because if we assume $m_1 = m_2$ the solution of (3.32) is then

$$v_j = (B + Cj)m_1^j$$

and $v_0 = v_{N+1} = 0$ implies that $B = C = 0$, giving $\mathbf{v} = 0$, which is not possible.

An analytical solution of the classical explicit approximation to $\partial U/\partial t = \partial^2 U/\partial x^2$

Consider the equation

$$\frac{\partial U}{\partial t} = \frac{\partial^2 U}{\partial x^2}, \quad 0 < x < 1,$$

where $U = 0$ at $x = 0$ and 1, $t > 0$, and U is known when $t = 0$, $0 \le x \le 1$.

The classical explicit approximation to the differential equation is

$$u_{i,j+1} = ru_{i-1,j} + (1-2r)u_{i,j} + ru_{i+1,j}, \tag{3.35}$$

where

$$x = ih, \quad t = jk, \quad r = k/h^2, \text{ and } Nh = 1.$$

Assume that a solution of equation (3.35) is of the form

$$u_{i,j} = f_i g_j. \tag{3.36}$$

Substitution of (3.36) into (3.35) leads to

$$\frac{g_{j+1}}{g_j} = \frac{rf_{i-1} + (1 - 2r)f_i + rf_{i+1}}{f_i}. \tag{3.37}$$

Since the left-hand side of (3.37) is independent of i and the right-hand side is independent of j it follows that both sides must equal a parameter c which is independent of i and j. This gives two homogeneous difference equations for f_i and g_j, namely,

$$g_{j+1} - cg_j = 0 \tag{3.38}$$

and

$$f_{i+1} + \frac{(1 - 2r - c)}{r} f_i + f_{i-1} = 0. \tag{3.39}$$

The solution of equation (3.38) is

$$g_j = Ac^j. \tag{3.40}$$

As the solution of the partial differential equation is periodic in x it is reasonable to assume that the solution of (3.39) is periodic in i, so that

$$f_i = B \cos i\theta + D \sin i\theta, \tag{3.41}$$

where

$$\cos \theta = (2r + c - 1)/2r. \tag{3.42}$$

Then, by (3.41), the boundary condition $u_{0,j} = 0$ for all j gives that

$$f_0 = 0 = B.$$

Similarly, the condition $u_{N,j} = 0$ for all j gives that

$$f_N = 0 = D \sin N\theta,$$

showing that

$$N\theta = s\pi, \quad s \text{ an integer.}$$

Therefore

$$f_i = D \sin \frac{i s \pi}{N}. \tag{3.43}$$

By equation (3.42),

$$c = 1 - 2r(1 - \cos \theta) = 1 - 4r \sin^2 \frac{s \pi}{2N}.$$

Hence equations (3.36), (3.40), and (3.43) give that

$$u_{i,j} = E \left(1 - 4r \sin^2 \frac{s \pi}{2N} \right)^j \sin \frac{s \pi i}{N},$$

where E replaces AD and s is an integer. But equation (3.35) is linear in $u_{i,j}$ so the sum of different solutions is a solution. It follows therefore that a more general solution that will satisfy fairly general initial conditions is

$$u_{i,j} = \sum_{s=1}^{\infty} E_s \left(1 - 4r \sin^2 \frac{s \pi}{2N} \right)^j \sin \frac{s \pi i}{N}. \tag{3.44}$$

Case (i) If the initial function $u_{i,0}$ is known only at the $(N+1)$ mesh points $(i, 0)$, $i = 0(1)N$, only the first $(N+1)$ values of E_s can be found by solving the $(N+1)$ linear equations

$$u_{i,0} = \sum_{s=1}^{N+1} E_s \sin \frac{s \pi i}{N}, \quad i = 0(1)N.$$

Case (ii) If $u_{i,0} = \phi(x)$, say, is a continuous function of x in $0 < x < 1$, it follows from equation (3.44) that

$$\phi(x) = \sum_{s=1}^{\infty} E_s \sin \frac{s \pi i}{N} = \sum_{s=1}^{\infty} E_s \sin \frac{s \pi i h}{Nh}$$

$$= \sum_{s=1}^{\infty} E_s \sin s \pi x.$$

In this case E_s are the coefficients in the Fourier sine series for $\phi(x)$ and their values will be given by

$$E_s = 2 \int_0^1 \phi(x) \sin s \pi x \, dx, \quad s = 1, 2, 3, \ldots .$$

A persistent discretization error associated with derivative boundary conditions

(*This section could be omitted on a first reading.*)

The analytical solution of the equation $\partial u/\partial t = \partial^2 u/\partial x^2$ consists of two parts; a non-vanishing part called the steady-state solution, even though it is a function of x and t, and a part referred to as the transient solution which dies away as t increases. Identical definitions hold for finite-difference equations.

I. B. Parker and J. Crank, reference 29, have shown that when the solution of a parabolic equation satisfies derivative boundary conditions, the transient solution $v_{i,j}$ of the corresponding finite-difference equations may tend, for large values of t, to a small number Δ that depends only on the mesh length δx. As the differential and difference equations have the same steady-state solutions, and the transient solution of the differential equation tends to zero as t increases, it follows that the discretization error is ultimately equal to Δ for large values of t.

The analysis below shows that Δ is determined by the discontinuities in the derivatives of the Fourier representation of a function $v_{i,0}$ that is equal to the initial function for u less the steady-state solution at zero time. When the first $(p-1)$ derivatives of the graph of the Fourier series for $v_{i,0}$ are continuous, but the pth derivative discontinuous, Δ is equal to $k(\delta x)^{p+1}$. The constant k is determined by the coefficients in the Fourier series for $v_{i,0}$.

Consider the equation

$$\frac{\partial u}{\partial t} = \frac{\partial^2 u}{\partial x^2}, \qquad (3.57)$$

where u satisfies the initial condition

$$u = 0 \text{ when } t = 0, \quad 0 \leqslant x \leqslant 1,$$

and the boundary conditions

$$\frac{\partial u}{\partial x} = 0 \text{ at } x = 0, \quad t \geqslant 0; \quad \frac{\partial u}{\partial x} = 1 \text{ at } x = 1, \quad t \geqslant 0.$$

(Heat conduction in a rod that is insulated at $x = 0$ and supplied with heat at $x = 1$.)

A finite-difference approximation to equation (3.57) that includes both explicit and implicit formulae is

$$\frac{u_{i,j+1}-u_{i,j}}{\delta t}=\frac{1}{(\delta x)^2}\{\theta(u_{i-1,j+1}-2u_{i,j+1}+u_{i+1,j+1})$$

$$+(1-\theta)(u_{i-1,j}-2u_{i,j}+u_{i+1,j})\}. \quad (3.58)$$

These equations can be solved analytically by a method commonly used for solving partial differential equations, the method of separation of the variables, and leads to a Fourier-type solution of the form

$$u_{i,j}=Ae^{\alpha j}\cos(\beta i+\gamma), \quad (3.59)$$

where A, α, β, and γ are constants that may be real or complex. Substitution into (3.58) leads to

$$e^{\alpha}=\frac{1-4r(1-\theta)\sin^2(\beta/2)}{1+4r\theta\sin^2(\beta/2)}, \quad (3.60)$$

where $r=\delta t/(\delta x)^2$. As the value of e^{α} determines whether $u_{i,j}$ increases or decreases as t increases it is called the growth factor. Since equation (3.58) is linear and the sum of solutions is also a solution, another solution is

$$u_{i,j}=\sum_{n=0}^{\infty}A_n\left\{\frac{1-4r(1-\theta)\sin^2(\beta_n/2)}{1+4r\theta\sin^2(\beta_n/2)}\right\}^j\cos(\beta_n i+\gamma_n). \quad (3.61)$$

Steady state solutions of equations (3.57) and (3.58) are of the form

$$u=Bt+\tfrac{1}{2}Bx^2+Cx+D. \quad (3.62)$$

The boundary conditions give $B=1$, $C=0$. The constant D can be found from the fact that the total heat content of the bar at time $t=0$ is zero, i.e., the integral of u between $x=0$ and 1 is zero. This gives $D=-\tfrac{1}{6}$. Hence the transient solution v obtained by subtracting the steady state solution from the complete solution is given by

$$v_{i,j}=u_{i,j}+\tfrac{1}{6}-\tfrac{1}{2}x^2-t \quad (3.63)$$

This transient solution of course also satisfies equations (3.58), so

$$v_{i,j}=\sum_{n=1}^{\infty}A_n\left\{\frac{1-4r(1-\theta)\sin^2(\beta_n/2)}{1+4r\theta\sin^2(\beta_n/2)}\right\}^j\cos(\beta_n i+\gamma_n). \quad (3.64)$$

By equation (3.63) and the conditions satisfied by u it follows that v satisfies the boundary conditions

$$\frac{\partial v_{i,j}}{\partial x} = 0 \text{ at } x = 0 \text{ and } 1, \quad \text{for all } t \geq 0,$$

i.e.

$$\frac{\partial v_{i,j}}{\partial i} = 0 \text{ at } i = 0 \text{ and } \frac{1}{\delta x} \quad \text{for all } j \geq 0.$$

These give

$$\gamma_n = 0,$$

and

$$\sin(\beta_n/\delta x) = 0,$$

so that

$$\beta_n = n\pi \, \delta x.$$

Hence

$$v_{i,j} = \sum_{n=1}^{\infty} A_n \left\{ \frac{1 - 4r(1-\theta)\sin^2(\tfrac{1}{2}n\pi \, \delta x)}{1 + 4r\theta \sin^2(\tfrac{1}{2}n\pi \, \delta x)} \right\}^j \cos n\pi i \, \delta x.$$

The initial condition for $v_{i,j}$ by (3.63) is

$$v_{i,0} = \tfrac{1}{6} - \tfrac{1}{2}(i \, \delta x)^2, \quad \text{when } t = j = 0.$$

Therefore

$$v_{i,0} = \sum_{n=1}^{\infty} A_n \cos n\pi i \, \delta x = \tfrac{1}{6} - \tfrac{1}{2}(i \, \delta x)^2.$$

This shows that the A_n are the coefficients in the Fourier cosine expansion of $\tfrac{1}{6} - \tfrac{1}{2}x^2$. Hence

$$A_n = 2 \int_0^1 (\tfrac{1}{6} - \tfrac{1}{2}x^2)\cos n\pi x \, dx = -\frac{(-1)^n 2}{n^2 \pi^2},$$

and

$$v_{i,j} = -\frac{2}{\pi^2} \sum_{n=1}^{\infty} \frac{(-1)^n}{n^2} \left\{ \frac{1 - 4r(1-\theta)\sin^2(\tfrac{1}{2}n\pi \, \delta x)}{1 + 4r\theta \sin^2(\tfrac{1}{2}n\pi \, \delta x)} \right\}^j \cos n\pi i \, \delta x.$$

Now $1/\delta x$ is an integer, m say, so $\tfrac{1}{2}n\pi \, \delta x = n\pi/2m$. It follows therefore that when $n = 2mp$, $p = 0, 1, 2, \ldots$, the growth factor is unity. For other values of n it is less than unity and the corresponding terms in the series tends to zero as j increases. Hence as t increases,

$$v_{i,j} \to -\frac{2}{\pi^2} \sum_{p=1}^{\infty} \frac{1}{(2mp)^2}.$$

Since

$$\frac{1}{m} = \delta x \quad \text{and} \quad \sum_{1}^{\infty} \frac{1}{p^2} = \frac{\pi^2}{6},$$

$$v_{i,j} \to -\frac{(\delta x)^2}{12}.$$

This result is illustrated numerically in exercise 7, Chapter 2, except that the coefficient of $(\delta x)^2$ is $\frac{1}{6}$ and not $\frac{1}{12}$ because the rod is of length one-half. This particular exercise also shows, incidentally, that the major portion of the total error is the discretization error and not the stability error.

The index of δx in the limiting value of $v_{i,j}$ is the same as the index of n in the expression for A_n, the Fourier coefficients for $v_{i,0}$. Since it is known that the Fourier coefficients of $v_{i,0}$ are of order $1/n^{p+1}$ when the first $(p-1)$ derivatives of the Fourier representation of $v_{i,0}$ are continuous but the pth derivative is ordinarily discontinuous, it follows that $v_{i,j}$ is of order $(\delta x)^{p+1}$. {The Fourier cosine series for $v_{i,0}$ represents a function that is equal to $v_{i,0}$ throughout the interval $0 \leqslant x \leqslant 1$, that is even throughout the interval $-1 \leqslant x \leqslant 1$ and which is periodic with a period of 2. In the example above the graph of the cosine series for $v_{i,0} = \frac{1}{6} - \frac{1}{2}x^2$ has discontinuous *first* derivatives at $x = \pm 1$ so the discretization error is of order $(\delta x)^2$.} Further developments of this phenomenon are given in reference 23.

EXERCISES AND SOLUTIONS

1. Prove that when the explicit finite-difference scheme

$$u_{i,j+1} = ru_{i-1,j} + (1-2r)u_{i,j} + ru_{i+1,j}, \quad \text{where } r = \frac{k}{h^2},$$

is used to approximate the equation $\partial U/\partial t = \partial^2 U/\partial x^2$, and it is assumed that U possesses continuous and finite derivatives up to order three in t and order six in x, then the discretization error is the solution of the difference equation

$$e_{i,j+1} = re_{i-1,j} + (1-2r)e_{i,j} + re_{i+1,j} + k\omega(x, t),$$

where

$$\omega(x, t) = \frac{h^2}{12}\left(6r\frac{\partial^2 U}{\partial t^2} - \frac{\partial^4 U}{\partial x^4}\right)_{i,j} + \frac{k^2}{6}\frac{\partial^3 U}{\partial t^3}(x_i, t_j + \theta_j k)$$

$$- \frac{h^4}{360}\frac{\partial^6 U}{\partial x^6}(x_i + \theta_i h, t_j),$$

$-1 < \theta_i < 1$ and $0 < \theta_j < 1$.

If the maximum value of $|\omega|$ is M, deduce, for $0 < r \le \frac{1}{2}$, that $|e_{i,j}| \le tM$. Hence show that the discretization error is of order h^2, except when $r = \frac{1}{6}$ in which case it is of order h^4.

Solution

As in the text, page 79.

2. The equation

$$\frac{\partial U}{\partial t} = \alpha \frac{\partial^2 U}{\partial x^2} - \beta U, \quad 0 < x < 1,$$

where α and β are real positive constants, is approximated at the point (ih, jk) by the explicit difference scheme

$$\frac{1}{k}\Delta_t u_{i,j} = \frac{\alpha}{h^2}\delta_x^2 u_{i,j} - \beta u_{i,j}.$$

Given that U has known initial values throughout the interval $0 \le x \le 1$, $t = 0$, known boundary values at $x = 0$ and 1, $t > 0$, and that $Nh = 1$, use the matrix method of analysis to find an upper bound for $r = k/h^2$ that will be sufficient to prevent errors increasing exponentially.

Determine whether or not Gerschgorin's circle theorem is adequate for establishing sufficient conditions for stability in terms of the usual elementary condition for stability.

Solution

$$(u_{i,j+1} - u_{i,j})/k = \alpha(u_{i-1,j} - 2u_{i,j} + u_{i+1,j})/h^2 - \beta u_{i,j}.$$

Hence

$$\mathbf{u}_{j+1} = \begin{bmatrix} (1-2r\alpha - k\beta) & r\alpha & & \\ r\alpha & (1-2r\alpha - k\beta) & r\alpha & \\ & & \cdot & \cdot \\ & & r\alpha & (1-2r\alpha - k\beta) \end{bmatrix} \mathbf{u}_j + \begin{bmatrix} r\alpha u_{0,j} \\ 0 \\ 0 \\ r\alpha u_{N,j} \end{bmatrix},$$

i.e.,

$$\mathbf{u}_{j+1} = \mathbf{A}\mathbf{u}_j + \mathbf{c}_j = \mathbf{A}(\mathbf{A}\mathbf{u}_{j-1} + \mathbf{c}_{j-1}) + \mathbf{c}_j = \dots$$
$$= \mathbf{A}^{j+1}\mathbf{u}_0 + \mathbf{A}^j\mathbf{c}_0 + \mathbf{A}^{j-1}\mathbf{c}_1 + \dots + \mathbf{c}_j,$$

where the $\mathbf{c}$'s are known vectors of known boundary values. Similarly,

$$\mathbf{u}_{j+1}^* = \mathbf{A}^{j+1}\mathbf{u}_0^* + \mathbf{A}^j\mathbf{c}_0 + \mathbf{A}^{j-1}\mathbf{c}_1 + \dots + \mathbf{c}_j.$$

Hence

$$\mathbf{u}_{j+1} - \mathbf{u}_{j+1}^* = \mathbf{e}_{j+1} = \mathbf{A}^{j+1}\mathbf{e}_0 \quad \text{etc.}$$

The eigenvalues of $\mathbf{A}$ are $\lambda_s = 1 - k\beta - 4r\alpha \sin^2(s\pi/N)$, $s = 1(1)(N-1)$. Errors will not increase exponentially if $-1 \le 1 - k\beta - 4r\alpha \sin^2(s\pi/N) \le 1$, where $k = rh^2$. This gives $r \le 2/(4\alpha + \beta h^2)$. Gerschgorin's theorem applied to rows $2(1)(N-2)$ is $|\lambda - (1 - 2r\alpha - r\beta h^2)| \le 2r\alpha$, i.e., $1 - 4r\alpha - r\beta h^2 \le \lambda \le 1 - r\beta h^2$. Right-hand side shows $\lambda < 1$. For stability, $-1 \le 1 - 4r\alpha - r\beta h^2$, giving $r \le 2/(4\alpha + \beta h^2)$. Rows 1 and N give a larger value for r.

3. When the equation $\partial u/\partial t = a\partial^2 u/\partial x^2$ where a is positive is approximated by the fully implicit backward-difference scheme

$$u_{i,j+1} - u_{i,j} = ra(u_{i-1,j+1} - 2u_{i,j+1} + u_{i+1,j+1}),$$

prove

(*a*) by the matrix method that the scheme is stable for all values of r, the boundary and initial values being assumed known,

(*b*) that the truncation error is $O(k) + O(h^2)$.

Comment

This is called a backward-difference scheme because the time-difference relative to the time of the space-difference is a backward one. It is not as accurate as the Crank–Nicolson method because the truncation error for the latter is $O(k^2) + O(h^2)$.

Solution

(a) Show $\mathbf{e}_j = \{(\mathbf{I} - ra\mathbf{T}_{N-1})^{-1}\}^j\,\mathbf{e}_0$, then consider the eigenvalues of $(\mathbf{I} - ra\mathbf{T}_{N-1})^{-1}$, etc. ...

(b) $T = k\left(\dfrac{1}{2}\dfrac{\partial^2 u}{\partial t^2} - a\dfrac{\partial^3 u}{\partial x^2 \partial t}\right) - \tfrac{1}{12}ah^2\dfrac{\partial^4 u}{\partial x^4}$

$$+ k^2\left(\frac{1}{6}\frac{\partial^3 u}{\partial t^3} - \frac{1}{2}a\frac{\partial^4 u}{\partial x^2 \partial t^2}\right) + \dots$$

4. Prove by means of

(a) the matrix method,

(b) the Fourier series method,

that the finite-difference scheme

$u_{i,j+1} - u_{i,j}$
$$= r\{\theta(u_{i+1,j+1} - 2u_{i,j+1} + u_{i-1,j+1}) + (1-\theta)(u_{i+1,j} - 2u_{i,j} + u_{i-1,j})\}$$

approximating the equation $\partial u/\partial t = \partial^2 u/\partial x^2$ is stable for $0 \le \theta \le \tfrac{1}{2}$ when $r \le 1/2(1-2\theta)$, and is unconditionally stable for $\tfrac{1}{2} \le \theta \le 1$. (Assume that the boundary values are zero and the initial values are known and $0 \le \theta \le 1$.)

Solution

(a) Assuming $u_{0,j} = u_{N,j} = 0$, the equations can be written as

$$(\mathbf{I} - r\theta\mathbf{T}_{N-1})\mathbf{u}_{j+1} = \{\mathbf{I} + r(1-\theta)\mathbf{T}_{N-1}\}\mathbf{u}_j,$$

where the eigenvalues of $\mathbf{T}_{N-1}$ are $-4\sin^2(s\pi/2N)$, $s = 1, 2, \dots, N-1$. For stability

$$-1 < \frac{1 - 4r(1-\theta)\sin^2(s\pi/2N)}{1 + 4r\theta\,\sin^2(s\pi/2N)} < 1.$$

The upper inequality is automatically satisfied for $r > 0$. The lower inequality gives $2r(1-2\theta) \le 1$. Hence the result.

(b) Replacing $u_{p,q}$ by $e^{pi\beta h}\xi^q$ leads to

$$\xi = \{1 - 4r(1-\theta)\sin^2(\beta h/2)\}/\{1 + 4r\theta\,\sin^2(\beta h/2)\},$$

where $|\xi| < 1$ for stability, etc. ...

5. (a) Show that one explicit finite-difference scheme approximating the equation $\partial u/\partial t = \partial^2 u/\partial x^2$ is

$$u_{i,j+1} = ru_{i-1,j} + (1-2r)u_{i,j} + ru_{i+1,j}.$$

Given that the initial values and boundary values are known, show that Gerschgorin's first and second theorems establish stability of this scheme for $r \leq \frac{1}{2}$, but give no useful result for $r > \frac{1}{2}$.

(b) Show that Gerschgorin's first theorem is inadequate for establishing the unconditional stability of the Crank–Nicolson equations approximating the equation $\partial u / \partial t = \partial^2 u / \partial x^2$, the boundary and initial values being assumed known.

Solution

(a) Gerschgorin's theorem gives $|\lambda| \leq 2|r| + |1 - 2r|$. Consider $0 \leq r \leq \frac{1}{2}$. Then $|\lambda| \leq 2r + (1 - 2r) = 1$. Hence the finite-difference equations are stable. $r > \frac{1}{2}$ leads to $|\lambda| \leq 4r - 1 > 1$, which is inconclusive.

Brauer's theorem gives $|\lambda - 1 + 2r| \leq 2r$, since $r > 0$. Hence $-2r \leq \lambda - 1 + 2r \leq 2r$, giving $1 - 4r \leq \lambda \leq 1$. The equations will be stable when $-1 \leq 1 - 4r$, i.e., $r \leq \frac{1}{2}$. When $r > \frac{1}{2}$, $\lambda \geq 1 - 4r < -1$, which is inconclusive.

(b) As shown in the text, the necessary condition for stability is $\lambda \geq 2$, where λ is an eigenvalue of **B**. Gerschgorin's first theorem leads to $|\lambda| \leq 2 + 4r$, where $r > 0$, which is insufficient to establish $\lambda \geq 2$.

6. The function U satisfies the equation

$$\frac{\partial U}{\partial t} = \frac{\partial^2 U}{\partial x^2} + \frac{2}{x} \frac{\partial U}{\partial x}, \quad 0 < x < 1 ,$$

the initial condition $U = 1 - x^2$ when $t = 0, 0 \leq x \leq 1$, and the boundary conditions

$$\frac{\partial U}{\partial x} = 0 \text{ at } x = 0, t > 0; \quad U = 0 \text{ at } x = 1, t > 0 .$$

The partial differential equation is approximated at the point (ih, jk) in the $x - t$ plane by the explicit difference equation

$$\frac{1}{k} \Delta_t u_{i,j} = \frac{1}{h^2} \delta_x^2 u_{i,j} + \frac{1}{xh} (\Delta_x u_{i,j} + \nabla_x u_{i,j}),$$

and the limiting form of the partial differential equation at $x = 0$ is approximated in a similar manner. The derivative boundary condition is subsequently approximated by the usual central-difference formula.

Given that $Nh = 1$ and that $k = rh^2$, $r > 0$, show that the matrix A of the difference equations $u_{j+1} = Au_j$ is

$$A = \begin{bmatrix} (1-6r) & 6r & & & & \\ 0 & (1-2r) & 2r & & & \\ & r(1-\frac{1}{2}) & (1-2r) & r(1+\frac{1}{2}) & & \\ & & & \cdot & \cdot & \cdot \\ & & & r\left(1-\frac{1}{i}\right) & (1-2r) & r\left(1+\frac{1}{i}\right) \\ & & & & \cdot & \cdot & \cdot \\ & & & & & r\left(1-\frac{1}{N-1}\right) & (1-2r) \end{bmatrix}$$

Deduce that one eigenvalue of A is $(1-6r)$. Derive from this result a necessary upper bound M on r for possible stability of the equations. Deduce that for $r \leq M$ the modulus of every other eigenvalue of A cannot exceed 1. Using the largest possible value for r and taking $h = 0 \cdot 1$, calculate the solution values denoted by $u_{0,1}$, $u_{2,1}$, and $u_{9,1}$.

Solution

$$\frac{1}{k}\left(u_{i,j+1} - u_{i,j}\right) = \frac{1}{h^2}\left(u_{i-1,j} - 2u_{i,j} + u_{i+1,j}\right) + \frac{1}{ih^2}\left(u_{i+1,j} - u_{i-1,j}\right),$$

giving

$$u_{i,j+1} = r\left(1-\frac{1}{i}\right)u_{i-1,j} + (1-2r)u_{i,j} + r\left(1+\frac{1}{i}\right)u_{i+1,j}.$$

At $x = 0$, $\partial U/\partial t = 3\partial^2 U/\partial x^2$ which can be approximated by $(u_{0,j+1} - u_{0,j})/k = 3(u_{-1,j} - 2u_{0,j} + u_{1,j})$ where $u_{-1,j} = u_{1,j}$ because $\partial U/\partial x = 0$ at $x = 0$. In matrix form,

$$\mathbf{u}_{j+1} = \begin{bmatrix} (1-6r) & 6r & & & & \\ 0 & (1-2r) & 2r & & & \\ & & \cdot & & & \\ & & & \cdot & & \\ & r\left(1-\frac{1}{i}\right) & (1-2r) & r\left(1+\frac{1}{i}\right) & & \\ & & & \cdot & & \\ & & & & \cdot & \\ & & & r\left(1-\frac{1}{N-1}\right) & (1-2r) \end{bmatrix} \mathbf{u}_j, \text{ i.e.,}$$

$u_{j+1} = \mathbf{A}u_j$, where $\mathbf{u}_j^T = (u_{0,j}, u_{1,j}, \ldots, u_{N-1,j})$. Expansion of the determinant of $(\mathbf{A} - \lambda\mathbf{I})$ by the first column gives $(1 - 6r)$ as an eigenvalue. For this eigenvalue, $-1 \leq 1 - 6r \leq 1$ for stability. Hence $r \leq \frac{1}{3}$. When $0 < r \leq \frac{1}{3}$ every term in the matrix is positive and the sum of the moduli of the terms along every row except the last is 1. For the last row,

$$0 < \sum_j |u| = 1 - r\left(1 + \frac{1}{N-1}\right) < 1.$$

Taking $r = \frac{1}{3}$, $h = 0{\cdot}1$, $u_{0,j+1} = -u_{0,j} + 2u_{1,j}$ and $u_{i,j+1} = \frac{1}{3}\{(1 - (1/i))u_{i-1,j} + u_{i,j} + (1 + (1/i))u_{i+1,j}\}$, $i = 1(1)9$. Hence $u_{0,1} = 0{\cdot}98$, $u_{2,1} = 0{\cdot}94$, and $u_{9,1} = 0{\cdot}17$.

7. Use Brauer's theorem to prove that the equations of exercise 6(b), Chapter 2, are stable for $r \leq \frac{1}{2}$.

Solution

$u_{0,j+1}$ and $u_{N,j+1}$ are given in terms of $u_{1,j+1}$ and $u_{N-1,j+1}$ respectively, so any errors in their solution are not propagated forward in time. Hence we need consider only the remaining equations, namely,

$$
\begin{bmatrix} \cdot_{+1} \\ \cdot_{+1} \\ \\ \\ \\ \cdot_{1,j+1} \end{bmatrix}
=
\begin{bmatrix}
\left\{1 - 2r + \dfrac{r}{1 + h_1\,\delta x}\right\} & r & & & \\
r & (1-2r) & r & & \\
& & \cdot & & \\
& & & \cdot & \\
& & r & (1-2r) & r \\
& & & r & \left\{1 - 2r + \dfrac{r}{1 + h_2\,\delta x}\right\}
\end{bmatrix}
\begin{bmatrix} u_{i,j} \\ u_{2,j} \\ \\ \\ \\ u_{N-1,j} \end{bmatrix}
+ \mathbf{c}
$$

where $\mathbf{c}$ is a column vector of constants.
Using Brauer's theorem, the first row leads to

$$\left|\lambda - \left(1 - 2r + \frac{r}{1 + h_1\,\delta x}\right)\right| \leq r,$$

so that,

$$-r \leq \lambda - 1 + 2r - \frac{r}{1 + h_1\,\delta x} \leq r,$$

where $|\lambda| \leq 1$ for stability. Hence

$$-1 \leq 1 - r + \frac{r}{1 + h_1 \, \delta x} \leq 1, \text{ giving } 0 < r \leq 2\left(1 + \frac{1}{h_1 \, \delta x}\right),$$

and

$$-1 \leq 1 - 3r + \frac{r}{1 + h_1 \, \delta x} \leq 1, \text{ giving } 0 < r \leq \frac{2 + 2h_1 \, \delta x}{2 + 3h_1 \, \delta x}.$$

The minimum value of this is $\frac{2}{3}$ so $0 < r \leq \frac{2}{3}$.

The remaining rows show that

$$|\lambda - (1 - 2r)| \leq 2r. \text{ Hence } -2r \leq \lambda - 1 + 2r \leq 2r,$$

so that $1 - 4r \leq \lambda \leq 1$. The equations will be stable when $-1 \leq 1 - 4r$, i.e., $r \leq \frac{1}{2}$. Therefore the entire set of equations is stable for $0 < r \leq \frac{1}{2}$.

8. Prove that the solution and rounding errors of the equations in exercise 8, Chapter 2, will not increase exponentially with increasing j if $r \leq 2/(20h + 5h^2)$.

Solution

The equations in matrix form are

$$
\begin{bmatrix} u_{1,j+1} \\ u_{2,j+1} \\ u_{3,j+1} \\ u_{4,j+1} \\ u_{5,j+1} \end{bmatrix}
$$

$$
= \begin{bmatrix} (1-2rh) & rh & & & \\ 2rh & (1-4rh) & 2rh & & \\ & 3rh & (1-6rh) & 3rh & \\ & & 4rh & (1-8rh) & 4rh \\ & & & 10rh & (1-10rh-5rh^2) \end{bmatrix} \begin{bmatrix} u_{1,j} \\ u_{2,j} \\ u_{3,j} \\ u_{4,j} \\ u_{5,j} \end{bmatrix}
$$

Gerschgorin's circle theorem applied to row 5 gives that $|\lambda - (1 - 10rh - 5rh^2)| \leq 10rh$. Hence $1 - 20rh - 5rh^2 \leq \lambda \leq 1 - 5rh^2$. The right-hand inequality merely shows that $\lambda < 1$ for $r > 0$. For the solution not to increase exponentially with increasing j, $-1 \leq 1 - 20rh - 5rh^2$, i.e., $r \leq 2/(20h + 5h^2)$. The general diagonal term

$= 1 - 2irh = 1 - 4ih/(20h + 5h^2)$ for max

$$r = (20h + 5h^2 - 4ih)/(20h + 5h^2) > 0, \ i = 1(1)4.$$

Hence every element in the first four rows is positive for $r = 2/(20h + 5h^2)$ and the sum of the moduli of the elements in the first four rows $\leqslant 1$. Therefore the spectral radius of the matrix $\leqslant 1$.

9. Use the Fourier series method to prove that

(a) $\qquad u_{p,q+1} - u_{p,q} = r(u_{p-1,q} - 2u_{p,q} + u_{p+1,q})$

is stable for $0 < r \leqslant \frac{1}{2}$. (The forward-difference explicit approximation to $\partial u/\partial t = \partial^2 u/\partial x^2$.)

(b) $\qquad u_{p,q+1} - u_{p,q-1} = 2r(u_{p-1,q} - 2u_{p,q} + u_{p+1,q})$

is *unstable* for all positive values of r. (The central-difference explicit approximation to $\partial u/\partial t = \partial^2 u/\partial x^2$, often called Richardson's method.)

(c) $u_{p,q+1} - 2u_{p,q} + u_{p,q-1}$
$\qquad = \frac{1}{2}r^2\{(u_{p+1,q+1} - 2u_{p,q+1} + u_{p-1,q+1}) + (u_{p+1,q-1} - 2u_{p,q-1} + u_{p-1,q-1})\}$

is stable for all positive values of $r = \delta t/\delta x$. (An implicit approximation to the hyperbolic equation $\partial^2 u/\partial t^2 = \partial^2 u/\partial x^2$.)

Solution

(a) $\xi = 1 - 2r(1 - \cos \beta h)$ where $|\xi| \leqslant 1$. This gives $r \leqslant 1/(1 - \cos \beta h)$, The least value of the right-side is $\frac{1}{2}$.

(b) $\xi^2 + 8r\xi \sin^2(\beta h/2) - 1 = 0$, so $\xi_1 = -1/\xi_2$ and $\xi_1 + \xi_2 = -8r \sin^2(\beta h/2)$. For stability $|\xi_1| \leqslant 1$ and $|\xi_2| \leqslant 1$. When $|\xi_1| < 1$, $|\xi_2| > 1$, giving instability. When $\xi_1 = 1$, $\xi_2 = -1$, and $\xi_1 + \xi_2 = 0$ giving $r = 0$.

(c) $\xi^2 - 2A\xi + 1 = 0$ where $A = 1/\{1 + 2r^2 \sin^2(\beta h/2)\} < 1$. Hence $\xi = A \pm i(1 - A^2)^{\frac{1}{2}}$, giving $|\xi| = 1$ for all real values of r.

10. The equation

$$\frac{\partial U}{\partial t} = \frac{\partial^2 U}{\partial x^2} + \frac{1}{x}\frac{\partial U}{\partial x}, \quad 0 < x < 1,$$

is approximated at the point (ph, qk) by the difference equation

$$\frac{1}{k} \Delta_t u_{p,q} = \frac{1}{h^2} \delta_x^2 u_{p,q} + \frac{1}{2xh} (\Delta_x u_{p,q} + \nabla_x u_{p,q}).$$

Use the von Neumann method of analysis to show that the difference equations are stable for $x > 0$ when

$$\frac{k}{h^2} \leqslant \frac{2}{4 + p^{-2}}.$$

Evaluate the form of the differential equation at $x = 0$ given that $\partial U / \partial x = 0$ at $x = 0$, $t > 0$.

Given also that U is constant at $x = 1$, that $Nh = 1$, and that the derivative boundary condition at $x = 0$ is approximated by a central-difference formula, write out in matrix form the corresponding explicit difference equations approximating the partial differential equation and associated boundary conditions.

Given that $k/h^2 \leqslant \frac{1}{4}$, deduce that the errors will not increase exponentially with increasing q.

Solution

$$\frac{1}{k} (u_{p,q+1} - u_{p,q}) = \frac{1}{h^2} (u_{p+1,q} - 2u_{p,q} + u_{p-1,q}) + \frac{1}{2ph^2} (u_{p+1,q} - u_{p-1,q}).$$

The error function $E_{p,q} = e^{i\beta ph} \xi^q$, $i = \sqrt{(-1)}$, satisfies

$$\xi = (1 - 4r \sin^2 \tfrac{1}{2}\beta h) + i \frac{r}{p} \sin \beta h, \quad r = \frac{k}{h^2}, \quad p \geqslant 1.$$

$$|\xi|^2 = 1 + 16r^2 \sin^4 \tfrac{1}{2}\beta h - 4r \sin^2 \tfrac{1}{2} \beta h \left(2 - \frac{r}{p^2} \cos^2 \tfrac{1}{2}\beta h \right).$$

$$|\xi| \leqslant 1 \text{ if } 4r \sin^2 \tfrac{1}{2}\beta h \left(2 - \frac{r}{p^2} \cos^2 \tfrac{1}{2}\beta h \right) \geqslant 16r^2 \sin^4 \tfrac{1}{2}\beta h,$$

giving

$$r \leqslant \frac{2}{4 \sin^2 \tfrac{1}{2}\beta h + \dfrac{1}{p^2} \cos^2 \tfrac{1}{2}\beta h} > \frac{2}{4 + p^{-2}}.$$

At $x = 0$, $\partial U / \partial t = 2\partial^2 U / \partial x^2$ and this can be approximated by $(u_{0,q+1} - u_{0,q})/k = 2(u_{-1,q} - 2u_{0,q} + u_{1,q})/h^2$. As $\partial U / \partial x = 0$ at $x = 0$,

$u_{1,q} = u_{-1,q}$. Hence $u_{0,q+1} = (1-4r)u_{0,q} + 4ru_{1,q}$. In matrix form $\mathbf{u}_{q+1} = \mathbf{Au}_q + \mathbf{c}_q$, where $\mathbf{c}_q$ is a column of constants and

$$\mathbf{A} = \begin{bmatrix} (1-4r) & 4r \\ (1-\frac{1}{2})r & (1-2r) & (1+\frac{1}{2})r \\ & & \cdot \\ & & & \cdot \\ & \left(1-\dfrac{1}{2p}\right)r & (1-2r) & \left(1+\dfrac{1}{2p}\right)r \\ & & & & \cdot \\ & & & & & \cdot \\ & & & \left(1-\dfrac{1}{2(N-1)}\right)r & (1-2r) \end{bmatrix}.$$

When $r \leqslant \frac{1}{4}$ the sum of the moduli of the terms along each row $\leqslant 1$. Hence $\rho(\mathbf{A}) \leqslant 1$.

11. Investigate the stability of the Douglas equations, given on page 45, for known boundary values and positive values of r.

Solution

In matrix form the equations can be written as

$$\{(10+12r)\mathbf{I} + (1-6r)\mathbf{A}\}\mathbf{u}_{j+1} = \{(10-12r)\mathbf{I} + (1+6r)\mathbf{A}\}\mathbf{u}_j + \mathbf{c}_j,$$

where $\mathbf{c}_j$ is a column vector involving known boundary values and

$$\mathbf{A} = \begin{bmatrix} 0 & 1 \\ 1 & 0 & 1 \\ & 1 & 0 & 1 \\ & & & \cdot \\ & & & & \cdot \\ & & & 1 & 0 \end{bmatrix}$$

is of order $(N-1)$, assuming $Nh = 1$. Let λ be an eigenvalue of $\mathbf{A}$. The equations will be stable if

$$-1 \leqslant \frac{(10-12r)+(1+6r)\lambda}{(10+12r)+(1-6r)\lambda} \leqslant 1,$$

i.e.,

$$-(10+12r)-(1-6r)\lambda \leqslant (10-12r)+(1+6r)\lambda \leqslant (10+12r)+(1-6r)\lambda,$$

provided $(10+12r)+(1-6r)\lambda >0$. Hence $-10 \leqslant \lambda \leqslant 2$. The eigenvalues of $\mathbf{A}$ are $\lambda_s = 2 \cos s\pi/N$. This value satisfies all the necessary inequalities for $r>0$.

12. If it is assumed that every function value of $u_{i,j}$, $i = 1(1)(N-1)$, in the elements of the matrices of Richtmyer's equations (2.44) of Chapter 2, can be replaced by a 'representative' positive constant parameter c, show that the equations are stable for

$$r \leqslant \frac{1}{2c(1-4\theta)} \text{ when } 0 \leqslant \theta < \tfrac{1}{4}$$

and unconditionally stable when $\tfrac{1}{4} \leqslant \theta$. How should c be chosen?

Solution

Replacing $u_{i,j}$, $i = 1(1)(N-1)$ by c leads to $(\mathbf{I} - 2r\theta c\mathbf{T}_{N-1}) \times (\mathbf{u}_{j+1} - \mathbf{u}_j) = rc\mathbf{T}_{N-1}\mathbf{u}_j + \mathbf{b}$, where $\mathbf{T}_{N-1}$ is defined on page 84 and $\mathbf{b}$ is a column vector of known boundary values. Hence

$$\mathbf{u}_{j+1} = (\mathbf{I} - 2r\theta c\mathbf{T}_{N-1})^{-1}\{\mathbf{I} - rc(2\theta-1)\mathbf{T}_{N-1}\}\mathbf{u}_j + \mathbf{b}' = \mathbf{A}\mathbf{u}_j + \mathbf{b}'.$$

The eigenvalues of $\mathbf{T}_{N-1} = -4 \sin^2 s\pi/2N$, $s = 1(1)(N-1)$, so the eigenvalues λ_s of $\mathbf{A}$ are $\{1 - 4rc(1-2\theta)\sin^2 s\pi/2N\}/(1 + 8r\theta c \sin^2 s\pi/2N)$. For stability $-1 \leqslant \lambda_s \leqslant 1$. This gives that $rc(1-4\theta)\sin^2 s\pi/2N \leqslant \tfrac{1}{2}$, which is always satisfied by $\theta \geqslant \tfrac{1}{4}$ since $r>0$, $c>0$. For

$$0 \leqslant \theta < \tfrac{1}{4}, \quad r \leqslant \frac{1}{2c(1-4\theta)\sin^2 s\pi/2N} > \frac{1}{2c(1-4\theta)}.$$

The smallest value for r would normally be used, in which case $c = \max_{i,j} u_{i,j}$.

Comment

The conditions given in Richtmyer's book for this problem are $r \leqslant (1/4u_{i,j}(1-2\theta))$ when $0 \leqslant \theta < \tfrac{1}{2}$, and unconditional stability when $\tfrac{1}{2} \leqslant \theta$ (reference 31.) Numerical studies indicate that these are over-pessimistic and that the conditions in the exercise can sometimes be over-optimistic.

13. (a) Show that the local truncation error at the point (ih, jk) of the Crank–Nicolson approximation to $\partial U/\partial t = \partial^2 U/\partial x^2$ is $O(h^2) + O(k^2)$.

(b) The equation

$$\frac{\partial U}{\partial t} + \frac{\partial U}{\partial x} = 0$$

is approximated at the point (ph, qk) by the difference scheme

$$\frac{1}{k}(u_{p,q+1} - u_{p,q}) + \frac{1}{2h}(u_{p+1,q} - u_{p-1,q}) = 0.$$

Investigate the stability of this scheme by the von Neumann method, using $r = k/h$ as a parameter.

Solution

(a) $$T_{i,j} = \frac{1}{k}(U_{i,j+1} - U_{i,j}) - \frac{1}{2h^2}(\delta_x^2 U_{i,j+1} + \delta_x^2 U_{i,j}),$$

where, by Taylor's expansion about the point (ih, jk),

$$U_{i+1,j} = \left[U + h\frac{\partial U}{\partial x} + \frac{1}{2}h^2\frac{\partial^2 U}{\partial x^2} + \frac{1}{6}h^3\frac{\partial^3 U}{\partial x^3} + \frac{1}{24}h^4\frac{\partial^4 U}{\partial x^4} + O(h^6)\right]_{i,j}$$

etc., giving

$$\delta_x^2 U_{i,j} = \left[h^2\frac{\partial^2 U}{\partial x^2} + \frac{1}{12}h^4\frac{\partial^4 U}{\partial x^4} + O(h^6)\right]_{i,j}.$$

Again, by Taylor's expansion,

$$\delta_x^2 U_{i,j+1} = \delta_x^2 U_{i,j} + k\frac{\partial}{\partial t}\delta_x^2 U_{i,j} + O(k^2).$$

Therefore

$$T_{i,j} = \left[\frac{\partial U}{\partial t} - \frac{\partial^2 U}{\partial x^2}\right]_{i,j} + \frac{1}{2}k\frac{\partial}{\partial t}\left[\frac{\partial U}{\partial t} - \frac{\partial^2 U}{\partial x^2}\right]_{i,j}$$

$$- \frac{1}{12}h^2\left(\frac{\partial^4 U}{\partial x^4}\right)_{i,j} + O(k^2) + O(h^4) + O(kh^2),$$

where

$$\frac{\partial U}{\partial t} - \frac{\partial^2 U}{\partial x^2} = 0.$$

Hence the result.

(b) The error function $E_{p,q} = e^{i\beta ph}\xi^q$ satisfies

$$\xi = 1 - \tfrac{1}{2}r(e^{i\beta h} - e^{-i\beta h}) = 1 - ir\sin\beta h, \quad i = \sqrt{(-1)}.$$

Hence $|\xi|^2 = 1 + r^2\sin^2\beta h > 1$ for all $r > 0$.

14. The equation

$$\alpha\frac{\partial U}{\partial t} + \frac{\partial U}{\partial x} - f(x, t) = 0, \quad \alpha \text{ constant,}$$

is approximated at the point (ih, jk) in the $x-t$ plane by the difference scheme

$$\frac{\alpha}{k}\{u_{i,j+1} - \tfrac{1}{2}(u_{i+1,j} + u_{i-1,j})\} + \frac{1}{2h}(u_{i+1,j} - u_{i-1,j}) - f_{i,j} = 0.$$

Investigate the consistency of this scheme for (a) $k = rh$ and (b) $k = rh^2$, r a positive constant, where it is assumed that U is sufficiently smooth for third-order derivatives in x and second-order derivatives in t to exist.

If either is inconsistent with the differential equation obtain the equation it does approximate.

Solution

$$U_{i,j+1} = U_{i,j} + k\frac{\partial U_{i,j}}{\partial t} + \tfrac{1}{2}k^2\frac{\partial^2 U_{i,j+\theta_1}}{\partial t^2},$$

$$U_{i+1,j} = U_{i,j} + h\frac{\partial U_{i,j}}{\partial x} + \tfrac{1}{2}h^2\frac{\partial^2 U_{i,j}}{\partial x^2} + \tfrac{1}{6}h^3\frac{\partial^3 U_{i+\theta_2,j}}{\partial x^3},$$

$$U_{i-1,j} = U_{i,j} - h\frac{\partial U_{i,j}}{\partial x} + \tfrac{1}{2}h^2\frac{\partial^2 U_{i,j}}{\partial x^2} - \tfrac{1}{6}h^3\frac{\partial^3 U_{i-\theta_3,j}}{\partial x^3},$$

where $0 < \theta_1, \theta_2, \theta_3 < 1$. Hence

$$T_{i,j} = F(U_{i,j}) = \left(\alpha\frac{\partial U}{\partial t} + \frac{\partial U}{\partial x} - f\right)_{i,j} + \tfrac{1}{2}\alpha k\frac{\partial^2 U_{i,j+\theta_1}}{\partial t^2}$$

$$- \frac{\alpha h^2}{2k}\frac{\partial^2 U_{i,j}}{\partial x^2} - \frac{\alpha h^3}{12k}\left(\frac{\partial^3 U_{i+\theta_2,j}}{\partial x^3} - \frac{\partial^3 U_{i-\theta_3,j}}{\partial x^3}\right) + O(h^2).$$

(a) $k = rh$. Scheme is consistent.

(b) $k = rh^2$. The difference equation approximates

$$\alpha \frac{\partial U}{\partial t} + \frac{\partial U}{\partial x} - \frac{\alpha}{2r} \frac{\partial^2 U}{\partial x^2} - f = 0$$

as $h \to 0$.

15. The equation

$$\frac{\partial U}{\partial t} - \frac{\partial^2 U}{\partial x^2} = 0$$

is approximated at the point (ih, jk) by the difference equation

$$\theta \left(\frac{u_{i,j+1} - u_{i,j-1}}{2k} \right) + (1 - \theta) \left(\frac{u_{i,j} - u_{i,j-1}}{k} \right) - \frac{1}{h^2} \delta_x^2 u_{i,j} = 0.$$

Show that the truncation error at this point is given by

$$-\tfrac{1}{2}k(1 - \theta) \frac{\partial^2 U}{\partial t^2} - \tfrac{1}{12}h^2 \frac{\partial^4 U}{\partial x^4} + O(k^2, h^4).$$

Hence find the value of θ that will reduce this error to one of order k^2 and h^4.

Solution

$$T_{i,j} = \theta(U_{i,j+1} - U_{i,j-1})/2k + (1 - \theta)(U_{i,j} - U_{i,j-1})/k$$
$$- (U_{i-1,j} - 2U_{i,j} + U_{i+1,j})/h^2.$$

Expand each term by Taylor's series about (ih, jk) to get

$$T_{i,j} = -\tfrac{1}{2}k(1 - \theta) \frac{\partial^2 U_{i,j}}{\partial t^2} - \tfrac{1}{12}h^2 \frac{\partial^4 U}{\partial x^4} + O(k^2) + O(h^4),$$

where

$$\frac{\partial U}{\partial t} = \frac{\partial^2 U}{\partial x^4}$$

so that

$$\frac{\partial^2 U}{\partial t^2} = \frac{\partial^4 U}{\partial x^4}.$$

Hence

$$T_{i,j} = \{-\tfrac{1}{2}k(1 - \theta) - \tfrac{1}{12}h^2\} \frac{\partial^4 U_{i,j}}{\partial x^4} + O(k^2) + O(h^4).$$

Therefore

$$\tfrac{1}{2}k(1-\theta) = -\tfrac{1}{12}h^2, \text{ i.e., } \theta = 1 + \frac{h^2}{6k}$$

gives the required result.

16. The Crank–Nicolson method approximates the equation

$$\left(\frac{\partial u}{\partial t}\right)_{i,j-\frac{1}{2}} = \left(\frac{\partial^2 u}{\partial x^2}\right)_{i,j-\frac{1}{2}} \quad \text{by} \quad \left(\frac{\partial u}{\partial t}\right)_{i,j-\frac{1}{2}} = \frac{1}{2}\left\{\left(\frac{\partial^2 u}{\partial x^2}\right)_{i,j} + \left(\frac{\partial^2 u}{\partial x^2}\right)_{i,j-1}\right\}.$$

Assuming the following central-difference formulae for the derivatives f' and f'' for a mesh length h, namely,

$$hf'_{\frac{1}{2}} = (\delta - \tfrac{1}{24}\delta^3 + \tfrac{3}{640}\delta^5 - \ldots)f_{\frac{1}{2}},$$

and

$$h^2 f''_0 = (\delta^2 - \tfrac{1}{12}\delta^4 + \tfrac{1}{90}\delta^6 - \ldots)f_0,$$

show that the Crank–Nicolson equation leads to

$$(u_{i,j} - u_{i,j-1})/k$$
$$= (u_{i+1,j} - 2u_{i,j} + u_{i-1,j} + u_{i+1,j-1} - 2u_{i,j-1} + u_{i-1,j-1})/2h^2 + C$$

where the correction term C is given by

$$C = \frac{1}{k}\left(\tfrac{1}{24}\delta^3 u_{i,j-\frac{1}{2}} - \tfrac{3}{640}\delta^5 u_{i,j-\frac{1}{2}} + \ldots\right)_t$$

$$+ \frac{1}{2h^2}\left\{\left(-\tfrac{1}{12}\delta^4 u_{i,j} + \tfrac{1}{90}\delta^6 u_{i,j} - \ldots\right)_x + \left(-\tfrac{1}{12}\delta^4 u_{i,j-1} + \tfrac{1}{90}\delta^6 u_{i,j-1} - \ldots\right)_x\right\}.$$

Comment

The correction term can be used to improve the accuracy of the finite-difference solution obtained initially by neglecting it. The function values of this first approximation are differenced in the t direction to give $\delta^3 u_{i,j-\frac{1}{2}}$ and in the x direction to give

$$\delta^4 u_{i,j}, \ \delta^4 u_{i,j-1}, \ \delta^6 u_{i,j} \text{ and } \delta^6 u_{i,j-1}.$$

The correction term for each equation is then calculated from these differences, which of course are numbers, and the corrected finite-difference equations re-solved for the $u_{i,j}$.

17. The equation

$$\frac{\partial U}{\partial t} = \frac{\partial^2 U}{\partial x^2}, \quad 0 < x < 1,$$

is approximated at the point (ih, jk) by the difference equation

$$\frac{3}{2}\left(\frac{u_{i,j+1} - u_{i,j}}{k}\right) - \frac{1}{2}\left(\frac{u_{i,j} - u_{i,j-1}}{k}\right) = \frac{1}{h^2}\delta_x^2 u_{i,j+1},$$

where $x = ih$, $t = jk$, and $Nh = 1$. Investigate the stability of this system of equations, for known boundary values and positive values of $r = k/h^2$, by the matrix method.

Solution

Equations are $-2ru_{i-1,j+1} + (3+4r)u_{i,j+1} - 2ru_{i+1,j+1} = 4u_{i,j} - u_{i,j-1}$, $i = 1(1)(N-1)$. For known boundary values they can be written as

$$\mathbf{u}_{j+1} = 4\mathbf{A}^{-1}\mathbf{u}_j - \mathbf{A}^{-1}\mathbf{u}_{j-1} + \mathbf{A}^{-1}\mathbf{c}_{j+1},$$

where $\mathbf{c}_{j+1}$ is a vector of known boundary values and

$$\mathbf{A} = \begin{bmatrix} (3+4r) & -2r & & \\ -2r & (3+4r) & -2r & \\ & & \cdot & \\ & & & \cdot \\ & & -2r & (3+4r) \end{bmatrix}$$

is of order $(N-1)$. This equation and $\mathbf{u}_j = \mathbf{u}_j$ can be written as

$$\begin{bmatrix} \mathbf{u}_{j+1} \\ \hline \mathbf{u}_j \end{bmatrix} = \begin{bmatrix} 4\mathbf{A}^{-1} & -\mathbf{A}^{-1} \\ \hline \mathbf{I} & \mathbf{O} \end{bmatrix} \begin{bmatrix} \mathbf{u}_j \\ \hline \mathbf{u}_{j-1} \end{bmatrix} + \begin{bmatrix} \mathbf{A}^{-1}\mathbf{c} \\ \hline \mathbf{O} \end{bmatrix},$$

i.e., $\mathbf{v}_{j+1} = \mathbf{P}\mathbf{v}_j + \mathbf{c}'$. $\mathbf{A}$ has distinct eigenvalues and all the submatrices of $\mathbf{P}$ commute with each other so the eigenvalues λ of $\mathbf{P}$ are the eigenvalues of

$$\begin{bmatrix} \dfrac{4}{\mu_k} & -\dfrac{1}{\mu_k} \\ 1 & 0 \end{bmatrix}$$

where μ_k is the kth eigenvalue of $\mathbf{A}$. As $\det(\mathbf{P} - \lambda\mathbf{I}) = 0 = \lambda^2 - (4/\mu_k)\lambda + (1/\mu_k)$, $\lambda = \{2 \pm \sqrt{(4 - \mu_k)}\}/\mu_k$ where

$$\mu_k = 3 + 8r\sin^2(k\pi/2N), \quad k = 1(1)(N-1)$$

by page 86. Hence

$$\lambda = \left\{ 2 \pm \left(1 - 8r \sin^2 \frac{k\pi}{2N} \right)^{\frac{1}{2}} \right\} \Big/ \left(3 + 8r \sin^2 \frac{k\pi}{2N} \right).$$

When the roots are real, $|\lambda| < (2+1)/(3+\delta)$, $\delta > 0$. Hence $|\lambda| < 1$. When the roots are complex, $|\lambda| = \mu_k^{-\frac{1}{2}} < 1$. Therefore the equations are unconditionally stable.

18 (a) Prove that a real tridiagonal matrix with either all its off-diagonal elements positive or all its off-diagonal elements negative is similar to a real symmetric tridiagonal matrix with non-zero off-diagonal elements. Deduce that the eigenvalues of such a matrix are real.

(b) Show that the components of the eigenvector $\mathbf{v}$, corresponding to the eigenvalue λ of the matrix $\mathbf{A}$ of order N defined by

$$\mathbf{A} = \begin{bmatrix} -2 & 2 & & & & \\ 1 & -2 & 1 & & & \\ & 1 & -2 & 1 & & \\ & & & \cdot & & \\ & & & 1 & -2 & 1 \\ & & & & 1 & -2 \end{bmatrix}$$

are given by the solution of the difference equations

$$v_{j-1} - (2+\lambda)v_j + v_{j+1} = 0, \quad j = 2(1)N,$$

satisfying the conditions

$$-(2+\lambda)v_1 + 2v_2 = 0 = v_{N+1}.$$

Hence prove that the jth component of $\mathbf{v}$ can be expressed as

$$v_j = B \cos j\theta + C \sin j\theta,$$

where B and C are constants. Deduce that the rth eigenvalue of $\mathbf{A}$ is $-4 \sin^2((2r-1)\pi/4N)$.

Solution

(a) Let

$$\mathbf{A} = \begin{bmatrix} a_1 & c_2 & & & & \\ b_2 & a_2 & c_3 & & & \\ & b_3 & a_3 & c_4 & & \\ & & & \cdot & & \\ & & & & \cdot & \\ & & & & b_n & a_n \end{bmatrix}$$

Let **D** be a real diagonal matrix with elements $d_1, d_2, \ldots, d_n$, i.e., $\mathbf{D} = \mathrm{diag}(d_1, d_2, \ldots, d_n)$. Calculate $\mathbf{DAD}^{-1}$. This matrix will be symmetric if

$$\frac{d_1^2}{d_2^2} = \frac{b_2}{c_2}, \quad \frac{d_2^2}{d_3^2} = \frac{b_3}{c_3}, \quad \ldots, \quad \frac{d_{n-1}^2}{d_n^2} = \frac{b_n}{c_n}.$$

Each right-hand side is positive. Assign a real value to d_1, then $d_2, d_3, \ldots, d_n$ are determined. Matrix **A** and the real symmetric $\mathbf{DAD}^{-1}$ are similar so they have the same eigenvalues. Hence the eigenvalues of **A** are real because the eigenvalues of a real symmetric matrix are real.

(b) Expand $\mathbf{Av} = \lambda \mathbf{v}$. The solution of $v_{j-1} - (2 + \lambda)v_j + v_{j+1} = 0$ is $v_j = Dm_1^j + Em_2^j$ where m_1, m_2 are the roots of $m^2 - (2 + \lambda)m + 1 = 0$. Therefore $m_1 m_2 = 1$, $m_1 + m_2 = 2 + \lambda$, λ real by part (a). Put $m_1 = re^{i\theta}$. Then $r = 1$, giving $\lambda = 2(-1 + \cos \theta)$, $v_j = B \cos j\theta + C \sin j\theta$. As $v_{N+1} = 0$, $B/C = -\tan(N+1)\theta$. Substitute for v_1, v_2, and λ in terms of θ into $-(2 + \lambda)v_1 + 2v_2 = 0$ to get $C \sin \theta \cos N\theta = 0$. Hence $\theta = (2r - 1)\pi/2N$, r an integer.

19. Verify that $U = e^{-\pi^2 t} \sin \pi x$ is a solution of the equation

$$\frac{\partial U}{\partial t} = \frac{\partial^2 U}{\partial x^2}, \quad 0 < x < 1,$$

which satisfies the boundary conditions $U = 0$ at $x = 0$ and 1, $t > 0$, and the initial condition $U = \sin \pi x$ when $t = 0$, $0 \leqslant x \leqslant 1$. The differential equation is approximated at the point (ih, jk) by the explicit equation

$$\frac{1}{k} \Delta_t u_{i,j} = \frac{1}{h^2} \delta_x^2 u_{i,j}.$$

Show that the analytical solution of the difference equation satisfying the same boundary and initial values is

$$u_{i,j} = \left(1 - 4r \sin^2 \frac{\pi h}{2}\right)^j \sin \pi x_i,$$

where $r = k/h^2$.

Given that $0 < r \leqslant \frac{1}{2}$, deduce that $u_{i,j}$ converges to $U_{i,j}$ as h tends to zero, for finite values of t.

Solution

As in the text. Because of the initial function, only the first term of the series solution is needed.

$$|u_{i,j} - U_{i,j}| = \left|\left(1 - 4r \sin^2 \frac{\pi h}{2}\right)^j - e^{-\pi^2 k j}\right| |\sin \pi x| = |a^j - b^j| \, |\sin \pi x|,$$

where

$$|a| = \left|1 - 4r \sin^2 \frac{\pi h}{2}\right| \leqslant 1$$

since $r \leqslant \frac{1}{2}$ and $|b| = |e^{-\pi^2 k}| < 1$. Hence

$$|u_{i,j} - U_{i,j}| = |a - b| \, |a^{j-1} + a^{j-2}b + \ldots + b^{j-1}| \, |\sin \pi x| < j|a - b|,$$

since there are j terms in the second series, $= j|1 - 2r + 2r \cos \pi h - e^{-\pi^2 k}|$. Replace $\cos \pi h$ and $e^{-\pi^2 k}$ by their Maclaurin expansions to get that

$$|u_{i,j} - U_{i,j}| < t\pi^4 h^2 A,$$

where A is bounded because it cannot exceed $|a - b|$. Hence the result.

20. Use equations (3.62) and (3.63) to prove that the discretization error $v_{i,j}$ of the transient component of the finite-difference equation (3.58) tends to zero as t increases when u satisfies the following boundary conditions.

(a) $\partial u/\partial x = 0$ at $x = 0$, and $u = 0$ at $x = 1$, $t \geqslant 0$.
(b) $u = 0$ at $x = 0$ and 1, $t \geqslant 0$.

Solution

(a) Boundary conditions give $B = C = D = 0$, i.e., $u_{i,j} = v_{i,j}$; $\partial v_{i,j}/\partial x = 0$ at $i = 0$ for all j gives $\gamma = 0$; $v_{i,j} = 0$ at $i = 1/\delta x$ for all j gives $\beta_n = (2n+1)\frac{1}{2}\pi$, $(n = 0, 1, 2, ...)$, so the growth factor is less than unity and $v_{i,j} \rightarrow 0$ as $j \rightarrow \infty$.

(b) Conditions give $u_{i,j} = v_{i,j}$; $v_{i,j} = 0$ at $i = 0$ gives $\gamma = -\frac{1}{2}\pi$; $v_{i,j} = 0$ at $i = 1/\delta x = m$ gives $\beta_n = n\pi/m$. Therefore $\sin \frac{1}{2}\beta_n = 0$ for $n = pm$, $p = 0, 1, 2, \ldots$, and $v_{i,j} \rightarrow \sum A_p \sin pi\pi = 0$, as $j \rightarrow \infty$.

4 Hyperbolic equations and characteristics

First-order quasi-linear equations

CONSIDER the equation

$$a\frac{\partial U}{\partial x} + b\frac{\partial U}{\partial y} = c,$$

where a, b and c are, in general, functions of x, y and U but not of $\partial U/\partial x$ and $\partial U/\partial y$, i.e., the first-order derivatives occur only to the first degree although the equation need not be linear in U. It is customary to put $\partial U/\partial x = p$ and $\partial U/\partial y = q$ and to write the equation as

$$ap + bq = c. \tag{4.1}$$

The solution of equation (4.1) can be obtained by investigating the possibility of finding a direction at each point of the $x - y$ plane along which the integration of equation (4.1) transforms to the integration of an ordinary differential equation. In other words, *in this direction*, the expression to be integrated will be independent of partial derivatives in other directions, such as p and q.

Assume we have solved the problem and know the solution values of equation (4.1) at every point on a curve C in the $x - y$ plane, Fig. 4.1, where C does *not* coincide with the curve Γ on which initial values of U are specified. Let $D(x, y)$ and $E(x + \delta x,\ y + \delta y)$ be adjacent points on C. Then $U(x, y)$ at D, together with the associated partial derivatives

$$p = \lim_{E \to D}\frac{\delta U}{\delta x} \text{ and } q = \lim_{E \to D}\frac{\delta U}{\delta y}$$

derivable from the values of U on C, satisfy equation (4.1). Along the direction of the tangent at D to the curve C we also have that

$$dU = \frac{\partial U}{\partial x}\,dx + \frac{\partial U}{\partial y}\,dy = p\,dx + q\,dy, \tag{4.2}$$

where dy/dx is the slope of C at D. Eliminate p between (4.1) and

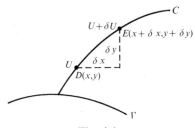

Fig. 4.1

(4.2) to give

$$dU = \frac{(c - bq)}{a} dx + q\, dy$$

which can be written as

$$q(a\, dy - b\, dx) + (c\, dx - a\, dU) = 0. \tag{4.3}$$

This equation is independent of p because the coefficients a, b and c are functions of x, y and U only. It can also be made independent of q by choosing the curve C so that its slope dy/dx satisfies the equation

$$a\, dy - b\, dx = 0. \tag{4.4}$$

By equations (4.4) and (4.3) it then follows that

$$c\, dx - a\, dU = 0. \tag{4.5}$$

Equation (4.4) is a differential equation for the curve C and (4.5) is a differential equation for the solution values of U along C. The curve C is called a characteristic curve or characteristic. These equations are easy to remember because they can be written as

$$\frac{dx}{a} = \frac{dy}{b} = \frac{dU}{c}.$$

This also shows that U may be found from either the equation $dU = (c/a)\, dx$ or the equation $dU = (c/b)\, dy$.

An alternative derivation of these results is given in exercise 3.

Example 4.1

Consider the equation

$$y \frac{\partial U}{\partial x} + \frac{\partial U}{\partial y} = 2 \tag{4.6}$$

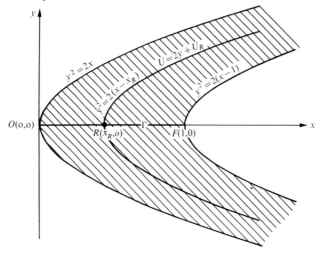

Fig. 4.2

where U is known along the initial segment Γ defined by $y = 0$, $0 \leqslant x \leqslant 1$.

The differential equation of the family of characteristic curves is

$$\frac{dx}{y} = \frac{dy}{1}.$$

Hence the equation of this family is $x = \frac{1}{2}y^2 + A$, where the parameter A is a constant for each characteristic. For the characteristic through $R(x_R, 0)$, $A = x_R$, so the equation of this particular characteristic is $y^2 = 2(x - x_R)$.

The solution along a characteristic curve is given by

$$\frac{dy}{1} = \frac{dU}{2},$$

which integrates to $U = 2y + B$, where B is constant along a particular characteristic. If $U = U_R$ at $R(x_R, 0)$ then $B = U_R$ and the solution along the characteristic $y^2 = 2(x - x_R)$ is $U = 2y + U_R$.

Since initial values for U are known only on the line segment OF, Fig. 4.2, where $0 \leqslant x_R \leqslant 1$, it follows that the solution is defined only in the region bounded by, and including, the terminal characteristics $y^2 = 2x$ and $y^2 = 2(x - 1)$. In this region the solution is clearly unique. Outside this region the solution is undefined.

If the initial curve Γ concides with a characteristic, say, for example, the characteristic $y^2 = 2x$ through $O(0, 0)$, then the solution along this characteristic is $U = 2y + U_0$, where U_0 is the specified initial value of U at O. In other words, the initial values for U on the initial curve $y^2 = 2x$ cannot now be arbitrarily prescribed as is obviously possible when Γ does not coincide with a characteristic curve. It is also easily shown in this case that the solution is not unique at points off $y^2 = 2x$. Consider, for example,

$$U = 2y + U_0 + A(y^2 - 2x),$$

where A is an arbitrary constant. This is clearly the solution along $y^2 = 2x$, whatever the value of A. It is also a solution of equation (4.6) when $y^2 \neq 2x$ as can easily be verified by direct differentiation. Since A is an arbitrary constant there is an infinite number of different solutions at points off the characteristic $y^2 = 2x$.

A method for numerical integration along a characteristic

Let U be specified on the initial curve Γ which must not be a characteristic curve.

Let $R(x_R, y_R)$ be a point on Γ and $P(x_P, y_P)$ be a point on the characteristic curve C through R such that $x_P - x_R$ is small, Fig. 4.3. The differential equation for the characteristic is

$$a\,dy = b\,dx,$$

which gives either dy or dx when the other quantities are known.

The differential equation for the solution along a characteristic is either

$$a\,dU = c\,dx \quad \text{or} \quad b\,dU = c\,dy,$$

which gives dU for known dx or dy and known a, b and c.

Denote a first approximation to U by $u^{(1)}$, a second approximation by $u^{(2)}$, etc.

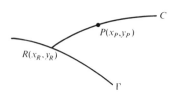

Fig. 4.3

First approximations

Assume that x_P is known. Then by the equations opposite,

$$a_R\{y_P^{(1)} - y_R\} = b_R(x_P - x_R)$$

gives a first approximation $y_P^{(1)}$ to y_P and

$$a_R\{u_P^{(1)} - u_R\} = c_R(x_P - x_R) \text{ gives } u_P^{(1)}.$$

Second and subsequent approximations

Replace the coefficients a, b and c by known mean values over the arc RP. Then

$$\tfrac{1}{2}(a_R + a_P^{(1)})(y_P^{(2)} - y_R) = \tfrac{1}{2}(b_R + b_P^{(1)})(x_P - x_R) \text{ gives } y_P^{(2)},$$

and

$$\tfrac{1}{2}(a_R + a_P^{(1)})(u_P^{(2)} - u_R) = \tfrac{1}{2}(c_R + c_P^{(1)})(x_P - x_R) \text{ gives } u_P^{(2)}.$$

This second procedure can be repeated iteratively until successive iterates agree to a specified number of decimal places.

Example 4.2

The function U satisfies the equation

$$\sqrt{x}\frac{\partial U}{\partial x} + U\frac{\partial U}{\partial y} = -U^2$$

and the condition $U = 1$ on $y = 0$, $0 < x < \infty$.

Show that the Cartesian equation of the characteristic through the point $R(x_R, 0)$, $x_R > 0$, is $y = \log(2\sqrt{x} + 1 - 2\sqrt{x_R})$. Use a finite-difference method to calculate a first approximation to the solution and to the value of y at the point $P(1\cdot1, y)$, $y > 0$, on the characteristic through the point $R(1, 0)$.

Calculate a second approximation to these values by an iterative method. Compare the results with those given by the analytical formulae for y and U. We have that

$$\frac{dx}{\sqrt{x}} = \frac{dy}{U} = \frac{dU}{-U^2}.$$

Hence $y = -\log AU$. As $U = 1$ at $(x_R, 0)$, $A = 1$, so

$$y = \log\frac{1}{U}. \tag{4.7}$$

Similarly, $2\sqrt{x} = 1/U + B$. As $U = 1$ at $(x_R, 0)$, $B = 2\sqrt{x_R} - 1$. Therefore

$$\frac{1}{U} = 2\sqrt{x} + 1 - 2\sqrt{x_R}. \tag{4.8}$$

Elimination of the parameter U between equations (4.7) and (4.8) shows that the Cartesian equation of the characteristic through $(x_R, 0)$ is

$$y = \log(2\sqrt{x} + 1 - 2\sqrt{x_R}). \tag{4.9}$$

The solution along the characteristic is given either by

$$U = e^{-y} \text{ or } U = \frac{1}{(2\sqrt{x} + 1 - 2\sqrt{x_R})}. \tag{4.10}$$

First approximations at $(1.1, y)$. *(Fig. 4.3)*

We have that $\sqrt{x}\, dy = U\, dx$ and $\sqrt{x}\, dU = -U^2\, dx$. Hence

$$\sqrt{x_R}(y_P^{(1)} - 0) = U_R\, dx \text{ giving } y_P^{(1)} = \frac{1}{\sqrt{1}}(0 \cdot 1) = 0 \cdot 1,$$

and

$$\sqrt{x_R}(u_P^{(1)} - 1) = -U_R^2\, dx \text{ giving } u_P^{(1)} = 1 - 0 \cdot 1 = 0 \cdot 9.$$

Second approximations

Using average values for the coefficients,

$$\tfrac{1}{2}(u_R + u_P^{(1)})\, dx = \tfrac{1}{2}(\sqrt{x_R} + \sqrt{x_P})(y_P^{(2)} - y_R),$$

giving

$$(1 + 0 \cdot 9)(0 \cdot 1) = (1 + 1 \cdot 0488)(y_P^{(2)} - 0),$$

from which $y_P^{(2)} = 0 \cdot 0927$. Also

$$\tfrac{1}{2}(\sqrt{x_R} + \sqrt{x_P})(u_P^{(2)} - u_R) = -\tfrac{1}{2}\{u_R^2 + (u_P^{(1)})^2\}\, dx,$$

giving

$$(1 + 1 \cdot 0488)(u_P^{(2)} - 1) = -(1 + 0 \cdot 81)(0 \cdot 1),$$

from which $U_P^{(2)} = 0 \cdot 9117$.

Note: The differential equations could have been written as

$$dy = \frac{U}{\sqrt{x}}\, dx \quad \text{and} \quad dU = -\frac{U^2}{\sqrt{x}}\, dx$$

and the second approximations obtained from

$$y_P^{(2)} = \frac{1}{2}\left\{\left(\frac{u}{\sqrt{x}}\right)_R + \left(\frac{u}{\sqrt{x}}\right)_P\right\} dx$$

and

$$u_P^{(2)} - u_R = \frac{1}{2}\left\{-\left(\frac{u^2}{\sqrt{x}}\right)_R - \left(\frac{u^2}{\sqrt{x}}\right)_P\right\} dx.$$

These approximations yield $y_P^{(2)} = 0{\cdot}0929$ and $u_P^{(2)} = 0{\cdot}9114$.

Analytical values

By equation (4.9),

$$y_P = \log_e \{2(1{\cdot}0488) + 1 - 2\} = 0{\cdot}0934.$$

By equation (4.10),

$$U_P = \frac{1}{1{\cdot}0976} = 0{\cdot}9111.$$

Finite-difference methods on a rectangular mesh for first-order equations

Lax-Wendroff explicit method

In the theory of fluid flow the equations of motion, of continuity, and of energy can be combined into one conservation equation of the form

$$\frac{\partial \mathbf{U}}{\partial t} + \frac{\partial \mathbf{F}(\mathbf{U})}{\partial x} = 0, \tag{4.11}$$

where $\mathbf{U}$ and $\mathbf{F}$ are each column vectors with three components. The Lax-Wendroff method, as illustrated below for a single dependent variable, can be used to approximate equation (4.11) by an explicit difference equation of second-order accuracy.

Consider

$$\frac{\partial U}{\partial t} + a\,\frac{\partial U}{\partial x} = 0,$$

a a positive constant. By Taylor's expansion,

$$U_{i,j+1} = U(x_i, t_j + k) = U_{i,j} + k\left(\frac{\partial U}{\partial t}\right)_{i,j} + \tfrac{1}{2}k^2\left(\frac{\partial^2 U}{\partial t^2}\right)_{i,j} + \dots,$$

where $x_i = ih$ and $t_j = jk$, $i = 0, \pm1, \pm2, \dots$, $j = 0, 1, 2, \dots$.

The differential equation can now be used to eliminate the t-derivatives because it gives that

$$\frac{\partial}{\partial t} \equiv -a\frac{\partial}{\partial x},$$

so

$$U_{i,j+1} = U_{i,j} - ka\left(\frac{\partial U}{\partial x}\right)_{i,j} + \tfrac{1}{2}k^2a^2\left(\frac{\partial^2 U}{\partial x^2}\right)_{i,j} + \dots.$$

Finally, the replacement of the x-derivatives by central-difference approximations gives, to terms in k^2, the explicit difference equation

$$\begin{aligned}
u_{i,j+1} &= u_{i,j} - \frac{ka}{2h}(u_{i+1,j} - u_{i-1,j}) + \frac{k^2a^2}{2h^2}(u_{i-1,j} - 2u_{i,j} + u_{i+1,j}) \\
&= \tfrac{1}{2}ap(1+ap)u_{i-1,j} + (1 - a^2p^2)u_{i,j} - \tfrac{1}{2}ap(1-ap)u_{i+1,j},
\end{aligned} \tag{4.12}$$

where $p = k/h$. This may be used for both initial-value and initial-value-boundary-value problems. It is probably used more often than any other difference method to obtain numerical solutions to differential equations in fluid flow problems when the dependent variables change rapidly with time. In such problems k must be kept small so the advantages of implicit methods are lost, namely stable equations and accurate results for fairly large values of k. As shown in exercise 4, equation (4.12) is stable for $0 < ap \leq 1$ and its local truncation error is

$$\tfrac{1}{6}k^2\frac{\partial^3 U}{\partial t^3} + \tfrac{1}{6}ah^2\frac{\partial^3 U}{\partial x^3} + \dots.$$

Example 4.3

The function U satisfies the equation

$$\frac{\partial U}{\partial t} + \frac{\partial U}{\partial x} = 0, \quad 0 < x < \infty, \quad t > 0,$$

the boundary condition

$$U(0, t) = 2t, \quad t > 0,$$

and the initial conditions

$$U(x, 0) = x(x - 2), \quad 0 \leq x \leq 2,$$
$$U(x, 0) = 2(x - 2), \quad 2 \leq x.$$

Calculate (i) the analytical solution and (ii) a numerical solution using the explicit Lax-Wendroff equation.

(i) $dt = dx = dU/0$. Hence U is constant along the straight line characteristics $t = x + \text{constant}$. Therefore the equation of the characteristic from $R(x_R, 0)$ is $t = x - x_R$ and if $U(x, 0)$ is $\phi(x)$, say, then the solution along this characteristic is $U(x, t) = \phi(x_R) = \phi(x - t)$. Similarly, if $U(0, t) = \psi(t)$, the solution along the characteristic $t - t_S = x$ from $S(0, t_s)$ is $U(x, t) = \psi(t - x)$. Hence the conditions of the problem give the solution shown in Fig. 4.4.

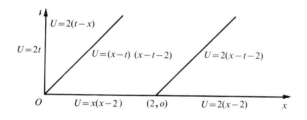

Fig. 4.4

(ii) Equation (4.12) for $h = \frac{1}{4}$ and $k = \frac{1}{8}$ is

$$u_{i,j+1} = \tfrac{3}{8} u_{i-1,j} + \tfrac{3}{4} u_{i,j} - \tfrac{1}{8} u_{i+1,j}.$$

Table 4.1 displays the analytical solution of the differential equation below each mesh point and the corresponding Lax–Wendroff solution to $3D$ above each mesh point for $x = 0(1)7$ and $t = 0(\frac{1}{2})3$. The boundary and initial conditions make $U(x, t)$ continuous at $(0, 0)$ and $(2, 0)$, $\partial U/\partial x$ continuous at $(2, 0)$, and ensures that $\partial U/\partial x$ and $\partial U/\partial t$ satisfy the differential equation at $(0, 0)$.

TABLE 4.1

t	$x=0$	$x=1$	$x=2$	$x=3$	$x=4$	$x=5$	$x=6$	$x=7$
		6·000	3·996	2·008	−0·060			
(0, 4)	8·0	6·0	4·0	2·0	0	−1·0	0	2·0
		5·001	3·005	0·946	−0·763	−0·695		
(0, 3·5)	7·0	5·0	3·0	1·0	−0·75	−0·75	1·0	3·0
		3·999	2·010	−0·050	−1·011	0·054	2·000	4·000
(0, 3)	6·0	4·0	2·0	0	−1·0	0	2·0	4·0
		3·000	0·965	−0·754	−0·713	1·004	3·000	5·000
(0, 2·5)	5·0	3·0	1·0	−0·75	−0·75	1·0	3·0	5·0
		2·007	−0·039	−1·009	0·041	2·000	4·000	6·000
(0, 2)	4·0	2·0	0	−1·0	0	2·0	4·0	6·0
		0·985	−0·750	−0·732	1·001	3·000	5·000	7·000
(0, 1·5)	3·0	1·0	−0·75	−0·75	1·0	3·0	5·0	7·0
		−0·022	−1·002	0·024	2·000	4·000	6·000	8·000
(0, 1)	2·0	0	−1·0	0	2·0	4·0	6·0	8·0
		−0·750	−0·750	1·000	3·000	5·000	7·000	9·000
(0, 0·5)	1·0	−0·75	−0·75	1·0	3·0	5·0	7·0	9·0
		−1·0	0	2·0	4·0	6·0	8·0	10·0
(0, 0)		(1, 0)	(2, 0)	(3, 0)	(4, 0)	(5, 0)	(6, 0)	(7, 0)

The Lax–Wendroff method for a set of simultaneous equations

The Lax–Wendroff approximation is easily extended to the set of simultaneous equations

$$\frac{\partial \mathbf{U}}{\partial t} + \mathbf{A}\frac{\partial \mathbf{U}}{\partial x} = 0, \tag{4.13}$$

where

$$\mathbf{U} = \begin{bmatrix} U_1 \\ U_2 \\ U_3 \end{bmatrix}, \quad \frac{\partial \mathbf{U}}{\partial t} = \begin{bmatrix} \dfrac{\partial U_1}{\partial t} \\ \dfrac{\partial U_2}{\partial t} \\ \dfrac{\partial U_3}{\partial t} \end{bmatrix}, \quad \frac{\partial \mathbf{U}}{\partial x} = \begin{bmatrix} \dfrac{\partial U_1}{\partial x} \\ \dfrac{\partial U_2}{\partial x} \\ \dfrac{\partial U_3}{\partial x} \end{bmatrix},$$

and $\mathbf{A}$ is a 3×3 real matrix, because equation (4.12) may be written as

$$u_{i,j+1} = \{1 - \tfrac{1}{2}pa(\Delta_x + \nabla_x) + \tfrac{1}{2}p^2 a^2(\Delta_x - \nabla_x)\} u_{i,j},$$

where

$$\Delta_x u_{i,j} = u_{i+1,j} - u_{i,j} \text{ and } \nabla_x u_{i,j} = u_{i,j} - u_{i-1,j}.$$

The corresponding approximation to equation (4.13) is

$$\mathbf{u}_{i,j+1} = \{\mathbf{I} - \tfrac{1}{2}p\mathbf{A}(\Delta_x + \nabla_x) + \tfrac{1}{2}p^2\mathbf{A}^2(\Delta_x - \nabla_x)\}\mathbf{u}_{i,j}.$$

(See exercise 6.)

The formal development of an approximation to (4.11) is as follows. By Taylor's series,

$$\mathbf{U}_{i,j+1} = \mathbf{U}_{i,j} + k\frac{\partial \mathbf{U}}{\partial t} + \tfrac{1}{2}k^2 \frac{\partial}{\partial t}\left(\frac{\partial \mathbf{U}}{\partial t}\right)_{i,j} + \dots.$$

In virtue of equation (4.11) it follows that

$$\mathbf{U}_{i,j+1} = \mathbf{U}_{i,j} - k\left(\frac{\partial \mathbf{F}}{\partial x}\right)_{i,j} - \tfrac{1}{2}k^2 \frac{\partial}{\partial t}\left(\frac{\partial \mathbf{F}}{\partial x}\right)_{i,j} + \dots. \tag{4.14}$$

But

$$\frac{\partial}{\partial t}\frac{\partial \mathbf{F}}{\partial x} = \frac{\partial}{\partial x}\frac{\partial \mathbf{F}}{\partial t} = \frac{\partial}{\partial x}\left\{\frac{\partial \mathbf{F}}{\partial \mathbf{U}}\frac{\partial \mathbf{U}}{\partial t}\right\} = -\frac{\partial}{\partial x}\left\{\frac{\partial \mathbf{F}}{\partial \mathbf{U}}\frac{\partial \mathbf{F}}{\partial x}\right\},$$

where

$$\frac{\partial \mathbf{F}}{\partial \mathbf{U}} = A(\mathbf{U}),$$

the Jacobian matrix of $\mathbf{F}$ with respect to $\mathbf{U}$, is defined by

$$A_{m,n} = \frac{\partial F_m}{\partial U_n}.$$

For example, if

$$\mathbf{U} = \begin{bmatrix} U_1 \\ U_2 \end{bmatrix} \text{ and } \mathbf{F}(\mathbf{U}) = \begin{bmatrix} F_1(U) \\ F_2(U) \end{bmatrix} \text{ then } A(\mathbf{U}) = \begin{bmatrix} \partial F_1/\partial U_1 & \partial F_1/\partial U_2 \\ \partial F_2/\partial U_1 & \partial F_2/\partial U_2 \end{bmatrix}.$$

Therefore equation (4.14) can be written as

$$\mathbf{U}_{i,j+1} = \mathbf{U}_{i,j} - k\left(\frac{\partial \mathbf{F}}{\partial x}\right)_{i,j} + \tfrac{1}{2}k^2 \frac{\partial}{\partial x}\left\{A(\mathbf{U})\frac{\partial \mathbf{F}}{\partial x}\right\}_{i,j} + \dots,$$

and using the central-difference approximation for the last term as shown in Chapter 2, exercise 13(a), we obtain the Lax–Wendroff approximation

$$\mathbf{u}_{i,j+1} = \mathbf{u}_{i,j} - \tfrac{1}{2}p(\mathbf{F}_{i+1,j} - \mathbf{F}_{i-1,j})$$
$$+ \tfrac{1}{2}p^2\{A_{i+\frac{1}{2},j}(\mathbf{F}_{i+1,j} - \mathbf{F}_{i,j}) - A_{i-\frac{1}{2},j}(\mathbf{F}_{i,j} - \mathbf{F}_{i-1,j})\}.$$

To avoid mid-point evaluations it is usual to approximate $A_{i+\frac{1}{2},j}$ by $\tfrac{1}{2}(A_{i,j} + A_{i+1,j})$ and $A_{i-\frac{1}{2},j}$ by $\tfrac{1}{2}(A_{i-1,j} + A_{i,j})$.

Further details concerning stability, well posedness, and the solution of equation (4.11) are given in references 31 and 23.

The Courant-Friedrichs-Lewy (C.F.L) condition for first-order equations

Assume that a first-order hyperbolic differential equation has been approximated by a difference equation of the form

$$u_{i,j+1} = au_{i-1,j} + bu_{i,j} + cu_{i+1,j}.$$

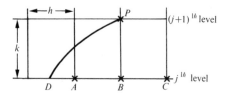

Fig. 4.5

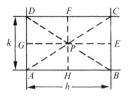

Fig. 4.6

Then u_P, Fig. 4.5, depends on the values of u at the mesh points A, B and C. Assume now that the characteristic curve through P of the hyperbolic equation meets the line AC at D and consider AC as an initial line segment. If the initial values along AC are altered then the solution value at P of the finite-difference equation will change, but these alterations will not affect the solution value at P of the differential equation which depends on the initial value at D. In this case u_P cannot converge to U_P as $h \to 0$, $k \to 0$. For convergence D must lie between A and C. (The C.F.L. condition.) Considering the Lax–Wendroff equation (4.12), the slope dt/dx of the characteristic of the corresponding differential equation is given by $dt/1 = dx/a$. For convergence of the difference equation, slope of $PD \geqslant$ slope of PA, i.e., $1/a \geqslant k/h$, giving $ap \leqslant 1$, which coincides with the condition for stability, namely, $0 < ap \leqslant 1$, since $a > 0$, $p > 0$.

Wendroff's implicit approximation

An implicit approximation of second-order accuracy to the equation

$$a \frac{\partial U}{\partial x} + b \frac{\partial U}{\partial y} = c$$

at the point P, Fig. 4.6, is given by approximating $(\partial U/\partial x)_P$ and $(\partial U/\partial y)_P$ by

$$\frac{1}{2}\left\{ \left(\frac{\partial U}{\partial x}\right)_H + \left(\frac{\partial U}{\partial x}\right)_F \right\} \text{ and } \frac{1}{2}\left\{ \left(\frac{\partial U}{\partial y}\right)_G + \left(\frac{\partial U}{\partial y}\right)_E \right\}$$

respectively, then approximating these derivatives by central-difference formulae to give the Wendroff equation

$$\frac{a}{2}\left\{ \frac{u_B - u_A}{h} + \frac{u_C - u_D}{h} \right\} + \frac{b}{2}\left\{ \frac{u_D - u_A}{k} + \frac{u_C - u_B}{k} \right\} = c,$$

which can be written as

$$(b - ap)u_D + (b + ap)u_C = (b + ap)u_A + (b - ap)u_B + 2kc.$$

This is unconditionally stable. (See exercise 5.) It cannot be used for pure initial-value problems, i.e., conditions on $t = 0$ only, because it would give an infinite number of simultaneous equations. If, however, initial values are known on Ox, $x \geqslant 0$, and boundary values on Ot, $t \geqslant 0$, the equation can be used explicitly by writing it as

$$u_C = u_A + \frac{b - ap}{b + ap}(u_B - u_D) + \frac{2kc}{b + ap}, \tag{4.15}$$

and will give the solution in the quarter plane $x > 0$, $t > 0$. (See page 159 for a numerical example.) It will also give approximation values in the strip $0 < x \leqslant 1$, $t > 0$, when initial values are known on Ox, $0 \leqslant x \leqslant 1$, and boundary values are known on Ot, $t > 0$.

Propagation of discontinuities in first-order equations

Consider the equation

$$\frac{\partial U}{\partial x} + \frac{\partial U}{\partial y} = 1, \quad y \geqslant 0, \quad -\infty < x < \infty,$$

where U is known at points $P(x_P, 0)$ on the x axis. The characteristic direction is given by $dx = dy$ and along the characteristics, $dU = dy$. Hence the characteristic through P is $y = x - x_P$ and the solution along it is $U = U_P + y$.

Discontinuous initial values

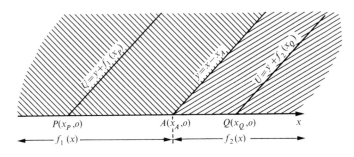

Fig. 4.7

Let

$$U(x, 0) = f_1(x), \quad -\infty < x < x_A,$$

and

$$U(x, 0) = f_2(x), \quad x_A < x < \infty.$$

To the left of the characteristic $y = x - x_A$ the solution $U_{(L)}$ is $U_{(L)} = f_1(x_P) + y$ along $y = x - x_P$. To the right of the straight line $y = x - x_A$ the solution $U_{(R)}$ is $U_{(R)} = f_2(x_Q) + y$ along $y = x - x_Q$.

Hence, for the same value of y in both solutions,

$$U_{(L)} - U_{(R)} = f_1(x_P) - f_2(x_Q).$$

Clearly, as x_P and x_Q both tend to x_A, $U_{(L)} - U_{(R)}$ is discontinuous along $y = x - x_A$ when

$$\lim_{x_P \to x_A} f_1(x_P) \neq \lim_{x_Q \to x_A} f_2(x_Q).$$

This shows that when the initial values are discontinuous at a particular point A, say, then the solution is discontinuous along the characteristic curve C from A. Moreover, the effect of this initial discontinuity does not diminish as we move away from A along C. With parabolic and elliptic differential equations the effect of an initial discontinuity is quite different as it tends to be localized and to diminish fairly rapidly with distance from the point of discontinuity.

Discontinuous initial derivatives

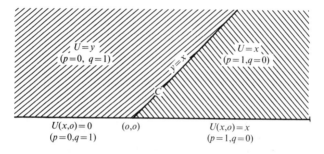

Fig. 4.8

Let

$$U(x, 0) = 0, \quad -\infty \leqslant x \leqslant 0,$$

and

$$U(x, 0) = x, \quad 0 < x < \infty.$$

This makes the initial derivative $p(x, 0) = \partial U(x, 0)/\partial x$ discontinuous at $(0, 0)$. The initial derivative $q(x, 0) = [\partial U(x, y)/\partial y]_{(x,0)}$ is also discontinuous at $(0, 0)$ because $p + q = 1$ from the differential equation. $U(x, 0)$ itself is clearly continuous at $(0, 0)$. As before, $U - U_P = y = x - x_P$ along the characteristic $y = x - x_P$ from $P(x_P, 0)$. Therefore the solution $U_{(L)}$ to the left of $y = x$ is

$$U_{(L)} = y, \quad -\infty < x \leqslant 0, \quad y \geqslant 0,$$

and the solution $U_{(R)}$ to the right of $y = x$ is

$$U_{(R)} = x, \quad 0 < x < \infty, \quad y \geqslant 0$$

It is seen from this that $U_{(L)} = U_{(R)}$ along $y = x$ but that $p_{(L)} = \partial U_{(L)}/\partial x = 0$ and $p_{(R)} = \partial U_{(R)}/\partial x = 1$, i.e., the solution is continuous along the characteristic C from the point of discontinuity but the initial discontinuities in the partial derivatives are propagated undiminished across the solution domain along this characteristic. In many cases discontinuous partial derivatives arise from sudden changes in the direction of the boundary-initial curve as illustrated in the next section.

Discontinuities and finite difference methods

As shown above, a discontinuity in the initial data of a first-order equation is propagated across the solution domain along the characteristic from the point of discontinuity. In such a case one would expect the 'method of characteristics' on page 146 to give a more accurate numerical solution than finite-difference methods because the solution corresponding to a particular initial value is developed along a characteristic that does not intersect the characteristic from a point of discontinuity. But the programming of the method of characteristics, especially for problems involving a set of simultaneous first-order equations, is much more difficult than the programming of difference methods. For this reason a great deal of research has been devoted to the formulation of finite-difference schemes that simulate the propagation of discontinuities along characteristics in the sense that rapid changes are confined to narrow regions. (See references 18 and 31.)

The "blurring" of discontinuities that occurs with difference methods is illustrated in the following example. Consider the equation

$$\frac{\partial U}{\partial x} + \frac{\partial U}{\partial y} = 1, \quad 0 < x < \infty, \quad y > 0, \tag{4.16}$$

where $U(0, y) = 0, \quad 0 < y < \infty,$
$\qquad U(x, 0) = 0, \quad 0 \leqslant x \leqslant 3,$
$\qquad U(x, 0) = x - 3, \quad 3 \leqslant x < \infty.$

In general, if $U = U_R$ at the initial point $R(x_R, y_R)$ then the characteristic through R and the solution along it are given by $U - U_R =$

$x - x_R = y - y_R$. Therefore:

(i) $U = x$ along the characteristic from $(0, y_R)$, $y_R > 0$.

(ii) $U = y$ along the characteristic from $(x_R, 0)$ $0 \leq x_R \leq 3$, and

(iii) $U = x - 3$ along the characteristic from $(x_R, 0)$, $x_R \geq 3$.

The numerical values for this analytical solution are shown in Table 4.2 for $x = 0(1)8$ and $y = 0(1)5$. Although the boundary values and initial values are continuous at $(0, 0)$ and $(3, 0)$ it is seen that $p = \partial U/\partial x$ and $q = \partial U/\partial y$ are discontinuous at these points. The discontinuities at $(0, 0)$ arise from the sudden change in the direction of the boundary curve yOx at O. (N.B. As $p + q = 1$ from the differential equation, $\delta p = -\delta q$.) The table also illustrates numerically how the discontinuities in p and q in the solution domain occur on the characteristics from $(0, 0)$ and $(3, 0)$, i.e., the discontinuities are propagated cleanly along these characteristics.

TABLE 4.2

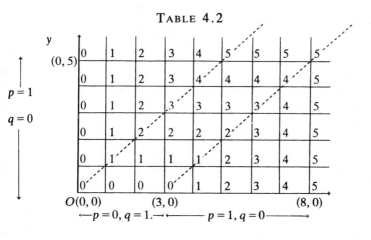

Let us now approximate equation (4.16) by Wendroff's equation (4.15). By Fig. 4.6,

$$u_C = u_A + \frac{h - k}{h + k}(u_B - u_D) + \frac{2hk}{h + k}.$$

Case (i) Take $h = k = 1$. Then

$$u_C = u_A + 1, \text{ i.e., } u_{i+1,j+1} = u_{i,j} + 1,$$

and the application of this equation to the boundary and initial values of Table 4.2 shows that the numerical solution coincides with the differential equation solution. This happens because $u_{i+1,j+1} = u_{i,j} + 1$ propagates the finite-difference solution forward along the characteristics from the initial points and boundary points and because the local truncation error at the point $(i+\frac{1}{2}, j+\frac{1}{2})$ is zero in this case.

Case (ii) Take $h = 1$, $k = \frac{1}{2}$. Then

$$u_C = u_A + \tfrac{1}{3}(u_B - u_D) + \tfrac{2}{3}.$$

The numerical solution to $2D$ is displayed in Table 4.3. Every second row corresponds to a row of Table 4.2. The values of $q_{i,j} = (u_{i,j+1} - u_{i,j})/k$ are shown in Table 4.4. The analytical values for $q_{i,j}$ for the three different solution domains are indicated at the top of Table 4.4. The values for u clearly improve in accuracy as one moves away from the characteristics through $(0, 0)$ and $(3, 0)$ and

TABLE 4.3

y									
(0, 5)	0	1·00	2·00	2·99	3·95	4·77	5·23	5·22	5·13
	0	1·00	2·00	2·98	3·90	4·57	4·76	4·61	4·80
(0, 4)	0	1·00	1·00	2·96	3·79	4·23	4·17	4·09	4·74
	0	1·00	1·99	2·91	3·58	3·73	3·54	3·82	4·87
(0, 3)	0	1·00	1·98	2·81	3·23	3·09	3·05	3·82	4·97
	0	1·00	1·94	2·60	2·68	2·46	2·87	3·91	5·01
(0, 2)	0	0·99	1·84	2·22	2·00	2·06	2·89	3·93	5·14
	0	0·96	1·63	1·60	1·41	1·96	2·85	4·15	4·90
(0, 1)	0	0·89	1·19	0·89	1·16	1·79	3·12	3·94	5·02
	0	0·67	0·44	0·52	0·83	2·06	2·98	4·01	5·00
	0	0	0	0	1	2	3	4	5 → x

(0, 0)   (3, 0)   (8, 0)

TABLE 4.4

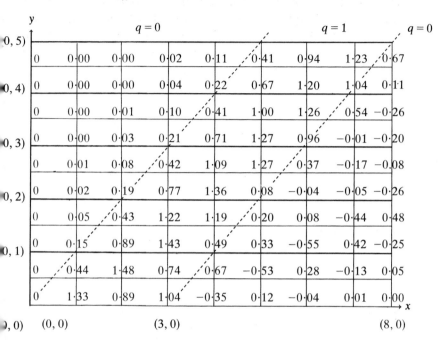

Table 4.4 shows that the discontinuity in q along these characteristics is 'diffused' over an area each side of the characteristics.

Second-order quasi-linear hyperbolic equations

Consider the second-order quasi-linear partial differential equation

$$a \frac{\partial^2 u}{\partial x^2} + b \frac{\partial^2 u}{\partial x\,\partial y} + c \frac{\partial^2 u}{\partial y^2} + e = 0, \qquad (4.17)$$

where a, b, c and e may be functions of x, y, u, $\partial u/\partial x$ and $\partial u/\partial y$ but not of $\partial^2 u/\partial x^2$, $\partial^2 u/\partial x\,\partial y$ and $\partial^2 u/\partial y^2$, i.e., the second-order derivatives occur only to the first degree.

It will be shown that at every point of the $x - y$ plane there are two directions in which the integration of the partial differential equation reduces to the integration of an equation involving total differentials only; in other words, in these directions the equation to be integrated is not complicated by the presence of partial derivatives in other directions. Furthermore it will be seen that this leads to a natural classification of partial differential equations.

Denote the first and second derivatives by

$$\frac{\partial u}{\partial x} = p; \quad \frac{\partial u}{\partial y} = q; \quad \frac{\partial^2 u}{\partial x^2} = r; \quad \frac{\partial^2 u}{\partial x \, \partial y} = s \text{ and } \frac{\partial^2 u}{\partial y^2} = t.$$

Let C be a curve in the $x - y$ plane on which the values of u, p and q are such that they, and the second-order derivatives r, s and t derivable from them, satisfy equation (4.17). (C is *not* a curve on which *initial* values of u, p and q are given. The reason for this will be apparent later.) Therefore the differentials of p and q in directions tangential to C satisfy the equations

$$dp = \frac{\partial p}{\partial x} \, dx + \frac{\partial p}{\partial y} \, dy = r \, dx + s \, dy, \qquad (4.18)$$

and

$$dq = \frac{\partial q}{\partial x} \, dx + \frac{\partial q}{\partial y} \, dy = s \, dx + t \, dy, \qquad (4.19)$$

where

$$ar + bs + ct + e = 0, \qquad (4.20)$$

and dy/dx is the slope of the tangent to C. Eliminate r and t from equation (4.20) by means of equations (4.18) and (4.19), giving

$$\frac{a}{dx} (dp - s \, dy) + bs + \frac{c}{dy} (dq - s \, dx) + e = 0,$$

i.e.,

$$s \left\{ a \left(\frac{dy}{dx} \right)^2 - b \left(\frac{dy}{dx} \right) + c \right\} - \left\{ a \frac{dp}{dx} \frac{dy}{dx} + c \frac{dq}{dx} + e \frac{dy}{dx} \right\} = 0 \quad (4.21)$$

Now choose the curve C so that the slope of the tangent at every point on it is a root of the equation

$$a\left(\frac{dy}{dx}\right)^2 - b\left(\frac{dy}{dx}\right) + c = 0 \qquad (4.22)$$

so that s also is eliminated.

By equation (4.21) it follows that in this direction

$$a\frac{dp}{dx}\frac{dy}{dx} + c\frac{dq}{dx} + e\frac{dy}{dx} = 0. \qquad (4.23)$$

This shows that at every point $P(x, y)$ of the solution domain there are two directions, given by the roots of equation (4.22), along which there is a relationship, given by equation (4.23), between the total differentials dp and dq. As will be seen later this relationship can be used to solve the original differential equation by a series of step-by-step integrations.

The directions given by the roots of equation (4.22) are called the *characteristic directions* and the partial differential equation is said to be *hyperbolic*, *parabolic* or *elliptic* according to whether these roots are real and distinct, equal, or complex, respectively, i.e., according to whether $b^2 - 4ac \gtrless 0$. The best known examples in these classes are the hyperbolic 'wave'-equation $\partial^2 u/\partial t^2 = \partial^2 u/\partial x^2$, the parabolic 'heat-conduction' or 'diffusion' equation $\partial u/\partial t = \partial^2 u/\partial x^2$, and the elliptic 'Laplace' equation $\partial^2 u/\partial x^2 + \partial^2 u/\partial y^2 = 0$.

Assume equation (4.17) is hyperbolic and that the roots of equation (4.22) are $dy/dx = f$ and $dy/dx = g$. Then the curve through the point P whose slope at every point is f is said to be an f characteristic. Clearly there are two different characteristic curves through every point of the solution domain of a second-order hyperbolic equation.

It should be noted that the classification of a partial differential equation, and consequently its method of solution, may depend on the region in which the solution is to be found. For example, the characteristic directions of the equation

$$y\frac{\partial^2 u}{\partial x^2} + x\frac{\partial^2 u}{\partial x\,\partial y} + y\frac{\partial^2 u}{\partial y^2} = F(x, y, u, p, q),$$

are given by the roots m_1, m_2 of the quadratic

$$ym^2 - xm + y = 0, \quad m = dy/dx,$$

which are real, equal or complex according as $x^2 \gtreqqless 4y^2$. Thus, the equation is hyperbolic when $|x| > 2\,|y|$, parabolic along $|x| = 2\,|y|$ and elliptic for $|x| < 2\,|y|$.

Solution of hyperbolic equations by the method of characteristics

Summarizing the previous work, the slopes of the characteristic directions associated with the equation

$$a\frac{\partial^2 u}{\partial x^2} + b\frac{\partial^2 u}{\partial x\,\partial y} + c\frac{\partial^2 u}{\partial y^2} + e = 0 \qquad (4.24)$$

are given by the roots of the quadratic equation

$$a\left(\frac{dy}{dx}\right)^2 - b\left(\frac{dy}{dx}\right) + c = 0, \qquad (4.25)$$

and along these characteristic directions the differentials dp and dq are related by the equation

$$a\frac{dy}{dx}\frac{dp}{dx} + c\frac{dq}{dx} + e\frac{dy}{dx} = 0,$$

which can be written as

$$a\frac{dy}{dx}\,dp + c\,dq + e\,dy = 0. \qquad (4.26)$$

Assuming that equation (4.24) is hyperbolic the roots of equation (4.25) will be real and distinct. Let them be

$$\frac{dy}{dx} = f \text{ and } \frac{dy}{dx} = g. \qquad (4.27)$$

Let Γ be a *non-characteristic curve* along which initial values for u, p and q are known. Let P and Q be points on Γ that are close together and let the f characteristic through P intersect the g characteristic through Q at the point $R(x_R, y_R)$, Fig. 4.9.

As a first approximation we may regard the arcs PR and QR as straight lines of slopes f_P and g_Q respectively. Then equations (4.27) can be approximated by

$$y_R - y_P = f_P(x_R - x_P) \qquad (4.28)$$

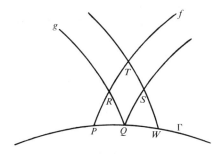

Fig. 4.9

and

$$y_R - y_Q = g_Q(x_R - x_Q), \tag{4.29}$$

giving two equations for the two unknowns x_R, y_R.

By equation (4.26) the differential relationships along the characteristics are

$$af\,dp + c\,dq + e\,dy = 0 \tag{4.30}$$

and

$$ag\,dp + c\,dq + e\,dy = 0. \tag{4.31}$$

The first one can be approximated along PR by the equation

$$a_P f_P(p_R - p_P) + c_P(q_R - q_P) + e_P(y_R - y_P) = 0, \tag{4.32}$$

and the second along QR by the equation

$$a_Q g_Q(p_R - p_Q) + c_Q(q_R - q_Q) + e_Q(y_R - y_Q) = 0. \tag{4.33}$$

These are two equations for the two unknowns p_R, q_R, as soon as x_R, y_R have been calculated from (4.28) and (4.29). The value of u at R can then be obtained from

$$du = \frac{\partial u}{\partial x}\,dx + \frac{\partial u}{\partial y}\,dy = p\,dx + q\,dy,$$

by replacing the values of p and q along PR by their average values and approximating the last equation by

$$u_R - u_P = \tfrac{1}{2}(p_P + p_R)(x_R - x_P) + \tfrac{1}{2}(q_P + q_R)(y_R - y_P). \tag{4.34}$$

This first approximation for u_R can now be improved by replacing the pivotal values of the various coefficients by average values. Equations (4.28) and (4.29) for improved values of x_R and y_R then become

$$y_R - y_P = \tfrac{1}{2}(f_P + f_R)(x_R - x_P) \qquad (4.35)$$

and

$$y_R - y_Q = \tfrac{1}{2}(g_Q + g_R)(x_R - x_Q), \qquad (4.36)$$

and equations (4.32), (4.33) for improved values of p_R, q_R become

$$\tfrac{1}{2}(a_P + a_R)\tfrac{1}{2}(f_P + f_R)(p_R - p_P) + \tfrac{1}{2}(c_P + c_R)(q_R - q_P)$$
$$+ \tfrac{1}{2}(e_P + e_R)(y_R - y_P) = 0 \quad (4.37)$$

and

$$\tfrac{1}{2}(a_Q + a_R)\tfrac{1}{2}(g_Q + g_R)(p_R - p_Q) + \tfrac{1}{2}(c_Q + c_R)(q_R - q_Q)$$
$$+ \tfrac{1}{2}(e_Q + e_R)(y_R - y_Q) = 0. \quad (4.38)$$

An improved value for u_R can then be found from equation (4.34). Repetition of this last cycle of operations will eventually yield u_R to the accuracy warranted by these finite-difference approximations. Provided Q is close to P the number of iterations will usually be small.

In this way we can calculate solution values at the grid points R and S, Fig. 4.9, and thence proceed to the grid point T, and so on.

Example 4.4

Use the method of characteristics to derive a solution of the quasi-linear equation

$$\frac{\partial^2 u}{\partial x^2} - u^2 \frac{\partial^2 u}{\partial y^2} = 0,$$

at the first characteristic grid point between $x = 0\cdot2$ *and* $0\cdot3$, $y > 0$, where u satisfies the conditions

$$u = 0\cdot2 + 5x^2 \text{ and } \frac{\partial u}{\partial y} = 3x,$$

along the initial line $y = 0$, for $0 \leqslant x \leqslant 1$.

Since u is given as a continuous function of x along Ox the initial value of $p = \partial u / \partial x$ is $10x$. The slopes of the characteristics are the roots of the equation $m^2 - u^2 = 0$. Hence

$$f = u = -g.$$

In this example the characteristics depend on the solution so the network of characteristics can be built-up only as the solution unfolds.

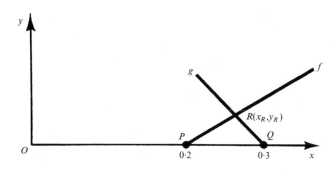

Fig. 4.10

Initially,

$$u = 0 \cdot 2 + 5x^2 = f = -g,$$
$$p = 10x \text{ and } q = 3x.$$

Also $a = 1$, $b = e = 0$, $c = -u^2$, therefore, Fig. 4.10,

$$f_P = 0 \cdot 4, \quad g_Q = -0 \cdot 65, \quad p_P = 2 \cdot 0, \quad p_Q = 3 \cdot 0, \quad u_P = 0 \cdot 4,$$
$$u_Q = 0 \cdot 65, \quad q_P = 0 \cdot 6, \quad q_Q = 0 \cdot 9, \quad c_P = -0 \cdot 16, \quad c_Q = -0 \cdot 4225.$$

By equations (4.28) and (4.29),

$$y_R = 0 \cdot 4(x_R - 0 \cdot 2)$$

and

$$y_R = -0 \cdot 65(x_R - 0 \cdot 3),$$

giving, as a first approximation,

$$x_R = 0 \cdot 26190, \quad y_R = 0 \cdot 024762,$$

to five significant figures.

The differential relationships along the characteristics are, by equations (4.32) and (4.33),

$$0\cdot4(p_R - 2\cdot0) - 0\cdot16(q_R - 0\cdot6) = 0,$$

and

$$-0\cdot65(p_R - 3\cdot0) - 0\cdot4225(q_R - 0\cdot9) = 0.$$

Their solution is

$$p_R = 2\cdot45524; \quad q_R = 1\cdot73810.$$

By equation (4.34),

$$u_R = 0\cdot4 + \tfrac{1}{2}(2\cdot0 + 2\cdot45524)(0\cdot0619) + \tfrac{1}{2}(1\cdot73810 + 0\cdot6)(0\cdot024762)$$
$$= 0\cdot56684.$$

For the second approximation,

$$f_R = -g_R = u_R = 0\cdot56684; \quad c_R = -u_R^2 = -0\cdot32131.$$

By equations (4.35) and (4.36), more accurate values for x_R, y_R are given by

$$y_R = \tfrac{1}{2}(0\cdot4 + 0\cdot56684)(x_R - 0\cdot2)$$

and

$$y_R = -\tfrac{1}{2}(0\cdot65 + 0\cdot56684)(x_R - 0\cdot3),$$

from which,

$$x_R = 0\cdot25572, \quad y_R = 0\cdot026938.$$

By equations (4.37) and (4.38),

$$\tfrac{1}{2}(0\cdot4 + 0\cdot56684)(p_R - 2\cdot0) - \tfrac{1}{2}(0\cdot16 + 0\cdot32131)(q_R - 0\cdot6) = 0,$$
$$-\tfrac{1}{2}(0\cdot65 + 0\cdot56684)(p_R - 3\cdot0) - \tfrac{1}{2}(0\cdot4225 + 0\cdot32131)(q_R - 0\cdot9) = 0.$$

These equations give the improved values

$$p_R = 2\cdot53117; \quad q_R = 1\cdot66700.$$

Hence the second approximation to u_R, by equation (4.34), is

$$u_R = 0\cdot4 + \tfrac{1}{2}\{(2 + 2\cdot53117)(0\cdot05572) + (0\cdot6 + 1\cdot6670)(0\cdot026938)\}$$
$$= 0\cdot55677.$$

It is left to the reader to show that the next iteration gives

$$x_R = 0.25578, \quad y_R = 0.02668,$$
$$p_R = 2.52876, \quad q_R = 1.67637,$$

and

$$u_R = 0.55667.$$

Since, to four decimal places,

$$u_R^{(1)} = 0.5668, \quad u_R^{(2)} = 0.5568 \text{ and } u_R^{(3)} = 0.5567,$$

it is obvious that the solution of the finite-difference equations for u_R is 0.5567, to this degree of accuracy. A fourth iteration does, in fact, give $u_R = 0.55666$ to five decimal places.

An alternative derivation of the characteristic equations

The previous method found unique directions, at each point of the x–y plane, in which relationships between the total differentials of p and q were independent of partial derivatives in other directions. Consequently we could integrate in these directions to find p and q and hence calculate u.

The same final result can be obtained by seeking to extend the solution of the differential equation from points on a given curve C in the x–y plane to points off C.

Assume the values of u, p and q are known at all points on C. Then the extension of the solution from the point $P(x, y)$ on C to an adjacent point with co-ordinates $(x + \delta x, y + \delta y)$ can be made by means of Taylor's expansion provided we can find unique values for r, s and t and all third and higher order partial derivatives of u at P, because Taylor's expansion states that

$$u(x + \delta x, y + \delta y) = u(x, y) + p\,\delta x + q\,\delta y$$
$$+ \tfrac{1}{2}\{r(\delta x)^2 + 2s\,\delta x\,\delta y + t(\delta y)^2\} + \text{higher order derivative terms.}$$

Since all third and higher order derivatives at P can be found in terms of u, p, q, r, s and t at P by successive differentiations of the differential equation

$$ar + bs + ct + e = 0,$$

it follows that we need only find r, s and t. Hence the problem reduces to finding the conditions under which the known values of

u, p and q are sufficient for the determination of unique values of r, s and t along C that satisfy the differential equation.

Before proceeding further it must be mentioned that the following analysis can be made in terms of the differentials of p and q in directions tangential to C, as previously, but as differentials in three dimensions have not been explicitly defined in this book the following approach has been adopted because it replaces differentials by derivatives. The previous theory can easily be re-written in terms of derivatives if preferred.

Since p and q are known at all points on C their arc rates of change dp/dl, dq/dl, where δl represents the length of an element of the curve C, can be considered as known quantities that are related to the values of r, s and t along C by the equations

$$\frac{dp}{dl} = \frac{\partial p}{\partial x}\frac{dx}{dl} + \frac{\partial p}{\partial y}\frac{dy}{dl} = r\frac{dx}{dl} + s\frac{dy}{dl}, \tag{4.39}$$

$$\frac{dq}{dl} = \frac{\partial q}{\partial x}\frac{dx}{dl} + \frac{\partial q}{\partial y}\frac{dy}{dl} = s\frac{dx}{dl} + t\frac{dy}{dl}. \tag{4.40}$$

The derivatives dx/dl and dy/dl are, of course, the direction cosines of the tangent to C with slope dy/dx. As these values of r, s and t must satisfy the differential equation

$$ar + bs + ct + e = 0, \tag{4.41}$$

it is seen that equations (4.39), (4.40) and (4.41) are three equations for the three unknowns r, s and t. Rewriting them as

$$ar + bs + ct + e = 0,$$

$$r\frac{dx}{dl} + s\frac{dy}{dl} + 0 - \frac{dp}{dl} = 0,$$

$$0 + s\frac{dx}{dl} + t\frac{dy}{dl} - \frac{dq}{dl} = 0,$$

their solution can be exhibited in the form

$$\frac{r}{\Delta_1} = \frac{-s}{\Delta_2} = \frac{t}{\Delta_3} = \frac{-1}{\Delta_4} \tag{4.42}$$

where

$$\Delta_1 = \begin{vmatrix} b & c & e \\ \dfrac{dy}{dl} & 0 & -\dfrac{dp}{dl} \\ \dfrac{dx}{dl} & \dfrac{dy}{dl} & -\dfrac{dq}{dl} \end{vmatrix}, \quad \Delta_2 = \begin{vmatrix} a & c & e \\ \dfrac{dx}{dl} & 0 & -\dfrac{dp}{dl} \\ 0 & \dfrac{dy}{dl} & -\dfrac{dq}{dl} \end{vmatrix},$$

$$\Delta_3 = \begin{vmatrix} a & b & e \\ \dfrac{dx}{dl} & \dfrac{dy}{dl} & -\dfrac{dp}{dl} \\ 0 & \dfrac{dx}{dl} & -\dfrac{dq}{dl} \end{vmatrix} \quad \text{and } \Delta_4 = \begin{vmatrix} a & b & c \\ \dfrac{dx}{dl} & \dfrac{dy}{dl} & 0 \\ 0 & \dfrac{dx}{dl} & \dfrac{dy}{dl} \end{vmatrix}.$$

When $\Delta_4 \neq 0$ the values of r, s and t are finite and unique and the solution can be extended from points on C to points off C by means of Taylor's expansion, derivatives higher than the second being obtained by differentiation of the differential equation.

At first sight this would appear to be the answer to our initial problem but odd as it may seem it is the exceptional case given by $\Delta_4 = 0$ that is of most use to us in the solution of hyperbolic equations.

When $\Delta_4 = 0$ the values of r, s and t will usually be infinite, in which case the known values of u, p and q on C will not satisfy the differential equation. If, however, Δ_1, Δ_2 and Δ_3 are also zero then r, s and t can be finite and satisfy equation (4.41). (It can be proved that when $\Delta_4 = 0$ and any one of the other three determinants is also zero then all the determinants are zero.) The expansion of $\Delta_4 = 0$ gives

$$a\left(\frac{dy}{dl}\right)^2 - b\frac{dy}{dl}\frac{dx}{dl} + c\left(\frac{dx}{dl}\right)^2 = 0$$

and is equivalent to the equation

$$a\left(\frac{dy}{dx}\right)^2 - b\left(\frac{dy}{dx}\right) + c = 0. \tag{4.43}$$

Similarly, $\Delta_2 = 0$ leads to the equation

$$a\frac{dy}{dl}\frac{dp}{dl} + c\frac{dx}{dl}\frac{dq}{dl} + e\frac{dx}{dl}\frac{dy}{dl} = 0,$$

which can be written in differential form as

$$a\frac{dy}{dx}\,dp + c\,dq + e\,dy = 0. \tag{4.44}$$

As previously explained, equations (4.43) and (4.44) show that at every point of the solution domain there are two directions, the characteristic directions, along which there is a relationship between the differentials of p and q that is independent of derivatives in other directions. When the slope at every point on C is equal to the slope of one of the characteristic directions, i.e., C is a characteristic curve, the values of u, p and q along C will be a solution provided they satisfy equation (4.44). In practice, as demonstrated in worked example 4.4, we make use of the differential relationship to propagate the solution forward along the characteristic curves C from known initial values along a non-characteristic curve Γ.

Additional comments on characteristics

A characteristic as an initial curve

When the curve on which initial values are given is itself a characteristic the equation can have no solution unless the initial conditions satisfy the necessary differential relationship for this characteristic. If they do, the solution will be unique along the initial curve but nowhere else, as is illustrated in the example below. It is also impossible in this case to use the method of characteristics to extend the solution from points on the initial curve to points off it, because the locus of all points such as R in Fig. 4.10 is the initial curve itself.

Consider the equation

$$\frac{\partial^2 u}{\partial x^2} - \frac{\partial^2 u}{\partial x\,\partial y} - 6\frac{\partial^2 u}{\partial y^2} = 0.$$

The characteristic directions are given by

$$\left(\frac{dy}{dx}\right)^2 + \left(\frac{dy}{dx}\right) - 6 = 0 = \left(\frac{dy}{dx} + 3\right)\left(\frac{dy}{dx} - 2\right),$$

so the characteristics are the straight lines $y + 3x = \text{constant}$, and $y - 2x = \text{constant}$. Let the initial curve be the characteristic $y - 2x = 0$. The differential relationship along this line by equation (4.26) is

$2\,dp - 6\,dq = 0$, i.e., $p - 3q = $ constant, and is obviously satisfied by the initial conditions $u = 2$, $p = -2$, $q = 1$. It is easily verified that one solution satisfying these conditions is

$$u = 2 + (y - 2x) + A(y - 2x)^2,$$

where A is an arbitrary constant. This is unique along $y - 2x = 0$ but nowhere else in the x–y plane.

Propagation of discontinuities

It can be proved that the solutions of elliptic and parabolic equations are analytic even when the boundary or initial conditions are discontinuous. Hyperbolic equations however are different in that discontinuities in initial conditions are propagated as discontinuities into the solution domain along the characteristics.

Let Γ, Fig. 4.11, be a non-characteristic curve along which initial values for u, p and q are known. Let P and Q be two distinct points on Γ and let the f characteristic through P meet the g characteristic through Q at R. Then the solution at R can be calculated in terms of the initial conditions at P and Q. Assuming no two characteristics of the same family intersect it follows that the solution at every point such as S inside the curvilinear triangle PQR is determined by the initial conditions between the points P and Q. Similarly the

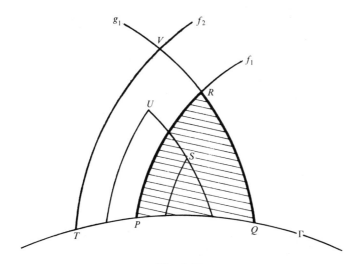

Fig. 4.11

solution at each point U inside the curvilinear strip $PRVT$ is determined by the initial conditions at a point along the arc TP (propagated along an f characteristic) and the initial conditions at a point along the arc PQ (propagated along a g characteristic). When the initial conditions along TP are analytically different from the initial conditions along PQ then the solution inside the strip $PRVT$ will be analytically different from the solution inside the curvilinear triangle PQR. As T tends to P the strip tends to the characteristic PR proving that the discontinuity in the initial conditions at P is propagated along a characteristic.

The argument above shows that a characteristic can separate two different solutions. *An extremely important feature of second-order equations is that these two solutions together with their first-order derivatives can be continuous across the dividing characteristic, but their second- and higher-order derivatives discontinuous across the same characteristic.* This is easily seen by recalling that the characteristic directions and the differential relationships in these directions were originally defined from equation (4.21), namely,

$$s\left\{a\left(\frac{dy}{dx}\right)^2 - b\left(\frac{dy}{dx}\right) + c\right\} - \left\{a\frac{dp}{dx}\frac{dy}{dx} + c\frac{dq}{dx} + e\frac{dy}{dx}\right\} = 0,$$

by making the expressions inside both pairs of braces zero. Then $s = \partial^2 u/\partial x\,\partial y$ is indeterminate and can be given an arbitrary value along a characteristic. Once, however, a value has been assigned to s the values of r and t are not indeterminate but can be calculated uniquely from any pair of equations (4.18), (4.19) and (4.20). In general any one of r, s and t can be chosen arbitrarily and the other two calculated uniquely. (Algebraically this implies that equations (4.18), (4.19) and (4.20) are not linearly independent but that any one can be written as a linear combination of the other two.) Now assume that continuous values of u, p and q are prescribed along a characteristic C. Also prescribe a continuous value for s, say, along C. Continuous values for r and t can then be calculated from the equations above, and all the higher order derivatives found by differentiating the differential equation. Taylor's expansion then gives a solution at points off C. Assume this solution is confined to one side of the characteristic. Repetition of this argument with the same values of u, p and q but a different value for s leads to the possible existence of a second solution on the other side of the characteristic. It will differ from the first solution but be related to it through the common continuous values of u, p and q along C. The indeterminacy associated with one of the second-order derivatives when it is not explicitly specified also explains why the solution of a hyperbolic equation is not unique when the initial values for u, p and q are given on a characteristic. As an illustration consider the wave equation

$$\frac{\partial^2 u}{\partial x^2} - \frac{\partial^2 u}{\partial y^2} = 0,$$

for which the characteristic directions and differential relationships are given by $dy/dx = dp/dq = \pm 1$. Hence $p - q$ is constant along $y - x = $ constant, and $p + q$ is constant along $y + x = $ constant. A possible solution along the characteristic $y - x = 0$ is therefore $p = 1$, $q = -1$ and $u = 3$. It is easily verified

$$u_1 = 2 + \sin(x - y) + \cos(x - y)$$

and

$$u_2 = 3 + (x - y)^2 + (x - y)$$

both satisfy the differential equation and give the stipulated solution along $y = x$, but that $\partial^2 u_1/\partial x^2 = -1$ and $\partial^2 u_2/\partial x^2 = 2$ on this characteristic.

In terms of the physics of steady supersonic flow of compressible fluids, for which the associated equations are hyperbolic, the characteristics are the Mach lines, along which small disturbances in velocity, pressure and density are propagated, or which separate different flow patterns possessing a common continuous velocity across the geometrical boundary of separation. (Discontinuous changes of state can also be propagated along shock lines. These differ from Mach lines both in position and the associated physics. Whereas the flow across a Mach line satisfies the equations of motion despite possible discontinuities in the normal derivatives of velocity, pressure, density and entropy, the flow across a shock line does not. Along a shock line the differential equation is replaced by relationships between finite jumps in velocity, pressure, density and entropy, the Rankine–Hugoniot equations, and these serve as boundary conditions for the differential equations used to calculate the continuous flow on each side of the shock. See reference 22.)

Rectangular nets and finite-difference methods for second order hyperbolic equations

In general the method of characteristics provides the most accurate process for solving hyperbolic equations. It is probably the most convenient method as well when the initial data are discontinuous because the propagation of the discontinuities into the solution domain along the characteristics is difficult to deal with on any grid other than a grid of characteristics. Problems involving no discontinuities however, can be solved satisfactorily by convergent and stable finite-difference methods using rectangular grids, and the organization of the computations for evaluation on a digital computer is certainly easier than for the method of characteristics.

Explicit methods and the Courant–Friedrichs–Lewy (C.F.L) condition

Consider the wave-equation

$$\frac{\partial^2 U}{\partial x^2} = \frac{\partial^2 U}{\partial t^2}, \quad t > 0, \tag{4.4}$$

where initially $U(x, 0) = f(x)$ and $\partial U(x, 0)/\partial t = g(x)$.

By equation (4.22) the slopes dt/dx of the characteristic curves are given by $(dt/dx)^2 = 1$, so the characteristics through $P(x_P, t_P)$ Fig. 4.12, are the straight lines $t - t_P = \pm(x - x_P)$ meeting the x-axis

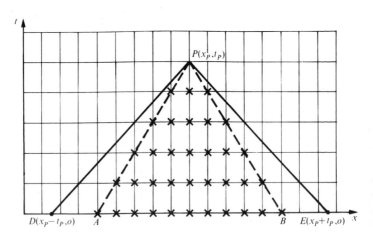

Fig. 4.12

at $D(x_P - t_P, 0)$ and $E(x_P + t_P, 0)$. The solution to equation (4.45) satisfying the given initial conditions is, (See exercise 16(b).),

$$U(x, t) = \frac{1}{2}\left\{ f(x+t) + f(x-t) + \int_{x-t}^{x+t} g(\zeta)\, d\zeta \right\}.$$

Hence the solution at $P(x_P, t_P)$ is

$$U(x_P, t_P) = \frac{1}{2}\left\{ f(x_P + t_P) + f(x_P - t_P) + \int_{x_P - t_P}^{x_P + t_P} g(\zeta)\, d\zeta \right\},$$

which shows that it depends upon the values of $f(x)$ at D and E and upon the value of $g(x)$ at every point of the closed interval DE, i.e., it depends on the initial data along the *interval of dependence DE*. The area PDE is called the *domain of dependence* of the point P. A central-difference approximation to equation (4.45) at the mesh points $(x_i, t_j) = (ih, jk)$ of a rectangular mesh covering the solution domain is

$$\frac{u_{i+1,j} - 2u_{i,j} + u_{i-1,j}}{h^2} = \frac{u_{i,j+1} - 2u_{i,j} + u_{i,j-1}}{k^2},$$

i.e.,

$$u_{i,j+1} = r^2 u_{i-1,j} + 2(1 - r^2)u_{i,j} + r^2 u_{i+1,j} - u_{i,j-1}, \qquad (4.46)$$

where $r = k/h$. This is an explicit formula giving approximation values at mesh points along $t = 2k, 3k, \ldots$, as soon as the mesh values along $t = k$ have been determined. Putting $j = 0$ in equation (4.46) yields

$$u_{i,1} = r^2 u_{i-1,0} + 2(1 - r^2)u_{i,0} + r^2 u_{i+1,0} - u_{i,-1}$$
$$= r^2 f_{i-1} + 2(1 - r^2)f_i + r^2 f_{i+1} - u_{i,-1}$$

and a central difference approximation to the initial derivative condition gives that

$$\frac{1}{2k}(u_{i,1} - u_{i,-1}) = g_{i,0}.$$

Eliminating $u_{i,-1}$ between these two equations shows that mesh values along $t = k$ can be calculated from the equation

$$u_{i,1} = \frac{1}{2}\{ r^2 f_{i-1} + 2(1 - r^2)f_i + r^2 f_{i+1} + 2k g_{i,0} \}. \qquad (4.47)$$

Equations (4.46) and (4.47) show that u_P at the mesh point P depends on the values of $u_{i,j}$ at the mesh points marked with crosses in Fig. 4.12. This set of mesh points is called the numerical domain of dependence of the point P, and the lines PA, PB are often termed the numerical characteristics. Assume now that the initial conditions along DA and BE are changed. These changes will alter the analytical solution of the partial differential equation at P but *not* the numerical solution given by equations (4.46) and (4.47). In this case u_P cannot possibly converge to U_P for all such arbitrary changes, as $h \to 0$ and $k \to 0$, with r remaining constant. When, however, the numerical characteristics PA, PB lie outside the domain of dependence PDE, Courant, Friedrichs and Lewy, reference 7, have shown that the effect of the initial data along DA and BE upon the solution at P of the finite-difference equations tends to zero as h and k both tend to zero, r remaining constant, P remaining fixed. This C.F.L condition for convergence is usually expressed by saying that the numerical domain of dependence of the difference equation must include the domain of dependence of the differential equation. Using this condition it is clear that the difference equation (4.46) is convergent for $0 < r \le 1$. (A proof for $r = 1$ is given in exercise 22(b) and proofs for $r \le 1$ are given in references 11 and 20. A numerical example is given in exercise 17 and stability is considered in worked example 3.2.)

Implicit difference methods

Implicit methods cannot be used without simplifying assumptions to solve pure initial value problems because they give an infinite number of simultaneous equations. They can, however, be used effectively for initial-boundary value problems when, for example initial conditions on $0 \le x \le 1$ for $t = 0$ and boundary conditions on $x = 0$ and 1, $t > 0$, are given. Convergence is usually dealt with via Lax's equivalence theorem.

One satisfactory scheme approximating the wave equation (4.45) at the point (ih, jk) is

$$\frac{1}{k^2} \delta_t^2 u_{i,j} = \frac{1}{h^2} \left(\tfrac{1}{4} \delta_x^2 u_{i,j+1} + \tfrac{1}{2} \delta_x^2 u_{i,j} + \tfrac{1}{4} \delta_x^2 u_{i,j-1} \right), \qquad (4.48)$$

where $\delta_t^2 u_{i,j} = u_{i,j+1} - 2u_{i,j} + u_{i,j-1}$ etc. This gives a tridiagonal system of equations that can be solved by the algorithm in Chapter 2. As shown in exercise 21 it is unconditionally stable for all $r = k/h > 0$

s truncation error is

$$h^2\left\{-\tfrac{1}{12}(4r^2+1)\frac{\partial^4 U}{\partial x^4}-\tfrac{1}{720}h^2(13r^4+15r^2+1)\frac{\partial^6 U}{\partial x^6}+\ldots\right\}_{i,j},$$

hich tends to zero for finite r as $h\to 0$. Hence the scheme is onvergent. A more general scheme is given by Mitchell in reference 23. He also uses the following method for deriving implicit approximations to equation (4.45) that are accurate to fourth-order differences. Expand $U_{i,j+1}$ and $U_{i,j-1}$ about the point (i,j) by Taylor's series to obtain

$$U_{i,j+1}+U_{i,j-1}=2U_{i,j}+k^2\left(\frac{\partial^2 U}{\partial t^2}\right)_{i,j}+\tfrac{1}{12}k^4\left(\frac{\partial^4 U}{\partial t^4}\right)_{i,j}+\ldots.$$

U is a solution of

$$\frac{\partial^2 U}{\partial t^2}=\frac{\partial^2 U}{\partial x^2}$$

en

$$\frac{\partial^2}{\partial t^2}\equiv\frac{\partial^2}{\partial x^2}$$

ad therefore

$$\left(\frac{\partial^4 U}{\partial t^4}\right)_{i,j}=\left(\frac{\partial^4 U}{\partial x^4}\right)_{i,j},\quad\left(\frac{\partial^6 U}{\partial t^6}\right)_{i,j}=\left(\frac{\partial^6 U}{\partial x^6}\right)_{i,j},\text{ etc.}$$

ence

$$U_{i,j+1}-2U_{i,j}+U_{i,j-1}=k^2\left(\frac{\partial^2 U}{\partial x^2}\right)_{i,j}+\tfrac{1}{12}k^4\left(\frac{\partial^4 U}{\partial x^4}\right)_{i,j}+\ldots\quad(4.49)$$

s

$$\frac{\partial^2 U}{\partial x^2}=\frac{1}{h^2}\,(\delta_x^2-\tfrac{1}{12}\delta_x^4+\tfrac{1}{90}\delta_x^6+\ldots)U,$$

follows that to fourth-order differences,

$$\frac{\partial^2 U}{\partial x^2}=\frac{1}{h^2}\,(\delta_x^2-\tfrac{1}{12}\delta_x^4)U$$

ad

$$\frac{\partial^4 U}{\partial x^2}=\frac{\partial^2}{\partial x^2}\left(\frac{\partial^2 U}{\partial x^2}\right)=\frac{1}{h^4}\,\delta_x^4 U.$$

Substitution of these approximations into (4.49) shows that to this order of accuracy

$$U_{i,j+1} - 2U_{i,j} + U_{i,j-1} = r^2\{1 + \tfrac{1}{12}(r^2 - 1)\delta_x^2\}\delta_x^2 U_{i,j}, \qquad (4.50)$$

where $r = k/h$. If we now operate on both sides of this equation with $\{1 + \tfrac{1}{12}(r^2 - 1)\delta_x^2\}^{-\frac{1}{2}}$ and expand each operator up to terms in δ_x^4 by the binomial expansion, we obtain the following implicit difference approximation to the wave equation,

$$u_{i,j+1} - 2u_{i,j} + u_{i,j-1} = \tfrac{1}{24}(r^2 - 1)(\delta_x^2 u_{i,j+1} + \delta_x^2 u_{i,j-1})$$
$$+ \tfrac{1}{12}(11r^2 + 1)\delta_x^2 u_{i,j} + \tfrac{1}{192}(r^2 - 1)(9r^2 - 1)\delta_x^4 u_{i,j}$$
$$- \tfrac{1}{384}(r^2 - 1)^2(\delta_x^4 u_{i,j+1} + \delta_x^4 u_{i,j-1})$$

Similarly, if both sides of equation (4.50) are operated on by $\{1 + \tfrac{1}{12}(r^2 - 1)\delta_x^2\}^{-1}$ the corresponding difference equation is

$$u_{i,j+1} - 2u_{i,j} + u_{i,j-1} = \tfrac{1}{12}(r^2 - 1)(\delta_x^2 u_{i,j+1} + \delta_x^2 u_{i,j-1})$$
$$+ \tfrac{1}{6}(5r^2 + 1)\delta_x^2 u_{i,j} + \tfrac{1}{144}(r^2 - 1)^2\{2\delta_x^4 u_{i,j} - (\delta_x^4 u_{i,j+1} + \delta_x^4 u_{i,j-1})\}$$

These high-order difference approximations would be difficult to implement in practice because of the difficulties associated with the boundary conditions.

Simultaneous first-order equations and their stability

All partial differential equations of second and higher order in the independent variable t, say, can be reduced to a system of simultaneous equations of first-order in t which may then be approximated by stable and convergent difference schemes, such as the Lax-Wendroff scheme. A range of problems for which this is convenient method of solution is given in reference 31. For example, the displacement Y of a point a distance x from one end of thin beam vibrating transversely satisfies the equation

$$\frac{\partial^2 Y}{\partial t^2} = -a^2 \frac{\partial^4 Y}{\partial x^4}, \quad a \text{ constant.}$$

Let $(\partial Y/\partial t) = v$ and $a(\partial^2 Y/\partial x^2) = w$. Then the original equation can be replaced by the pair of equations

$$\frac{\partial v}{\partial t} = -a^2 \frac{\partial^2 w}{\partial x^2} \quad \text{and} \quad \frac{\partial w}{\partial t} = a \frac{\partial^2 v}{\partial x^2},$$

each of first-order in t.

The wave equation can be dealt with in a similar manner by putting $(\partial U/\partial x) = p$ and $(\partial U/\partial t) = q$. Then equation (4.45) and the relationship $\partial^2 U/\partial x\, \partial t = \partial^2 U/\partial t\, \partial x$ lead to

$$\frac{\partial q}{\partial t} = \frac{\partial p}{\partial x} \quad \text{and} \quad \frac{\partial p}{\partial t} = \frac{\partial q}{\partial x}. \tag{4.51}$$

One explicit difference approximation to (4.51) which is stable for $k/h \leqslant 1$ *is*

$$\frac{1}{k}\{q_{i,j+1} - \tfrac{1}{2}(q_{i+1,j} + q_{i-1,j})\} = \frac{1}{2h}(p_{i+1,j} - p_{i-1,j})$$

and

$$\frac{1}{k}\{p_{i,j+1} - \tfrac{1}{2}(p_{i+1,j} + p_{i-1,j})\} = \frac{1}{2h}(q_{i+1,j} - q_{i-1,j}).$$

A numerical example, including the calculation of $u_{i,j}$, is given in exercise 20 and the stability of the equations is considered in exercise 19(a). Another approximation to equations (4.51) is

$$\frac{1}{k}(q_{i,j+1} - q_{i,j}) = \frac{1}{h}(p_{i+\frac{1}{2},j} - p_{i-\frac{1}{2},j})$$

and

$$\frac{1}{k}(p_{i-\frac{1}{2},j+1} - p_{i-\frac{1}{2},j}) = \frac{1}{h}(q_{i,j+1} - q_{i-1,j+1})$$

and is shown to be stable for $k/h \leqslant 1$ in exercise 19(b). This scheme, sometimes called the Courant, Friedrichs and Lewy scheme, is equivalent to the explicit difference equation (4.46) if $p_{i-\frac{1}{2},j}$ is replaced by $(u_{i,j} - u_{i-1,j})/h$ and $q_{i,j}$ by $(u_{i,j} - u_{i,j-1})/k$.

An implicit scheme that is unconditionally stable and equivalent to (4.48) is

$$\frac{1}{k}(q_{i,j+1} - q_{i,j}) = \frac{1}{2h}(p_{i+\frac{1}{2},j} - p_{i-\frac{1}{2},j} + p_{i+\frac{1}{2},j+1} - p_{i-\frac{1}{2},j+1})$$

and

$$\frac{1}{k}(p_{i-\frac{1}{2},j+1} - p_{i-\frac{1}{2},j}) = \frac{1}{2h}(q_{i,j+1} - q_{i-1,j+1} + q_{i,j} - q_{i-1,j}).$$

To illustrate the von Neumann method for investigating stability and to demonstrate that apparently reasonable schemes can be useless, consider the following explicit scheme which approximates the

x-derivatives of equation (4.51) by central-differences and the t-derivatives by forward differences, namely,

$$\frac{1}{k}(q_{i,j+1} - q_{i,j}) = \frac{1}{2h}(p_{i+1,j} - p_{i-1,j})$$

and

$$\frac{1}{k}(p_{i,j+1} - p_{i,j}) = \frac{1}{2h}(q_{i+1,j} - q_{i-1,j}). \qquad (4.52)$$

Let the initial perturbations in p and q along $t = 0$ be $Ae^{\beta x \sqrt{(-1)}}$ and $Be^{\beta x \sqrt{(-1)}}$ respectively, where A and B are different constants and $x = ih$. Then we can assume that the perturbations in the calculated values of p and q will be $Ae^{\beta x \sqrt{(-1)}} \xi^j$ and $Be^{\beta x \sqrt{(-1)}} \xi^j$ respectively. Substitution of these perturbations into equations (4.52) leads to

$$2B(\xi - 1) = rA(e^{\beta h \sqrt{(-1)}} - e^{-\beta h \sqrt{(-1)}})$$

and

$$2A(\xi - 1) = rB(e^{\beta h \sqrt{(-1)}} - e^{-\beta h \sqrt{(-1)}}),$$

where $r = k/h$. Elimination of A/B gives that

$$(\xi - 1)^2 = -r^2 \sin^2 \beta h.$$

Hence

$$\xi = 1 \pm (\sqrt{-1})r \sin \beta h$$

and

$$|\xi| = (1 + r^2 \sin^2 \beta h)^{\frac{1}{2}} = 1 + \tfrac{1}{2}r^2 \sin^2 \beta h + \ldots = 1 + O(r^2).$$

On page 106 it was shown that the perturbations at the finite time-level $t = jk$ would be unbounded as the mesh lengths tend to zero if $|\xi| > 1 + O(k)$, which is so in this case.

Simulation of shock waves

Physically, shock waves in supersonic flow are very narrow regions separating different continuous flow patterns and across which there are very rapid changes in pressure, velocity, density, temperature and entropy.

Mathematically, within the framework of classical hydrodynamics dealing with the isentropic flow of non-viscous, non heat-conducting, compressible fluid, this state of affairs can be represented by a reduction of the narrow region to a surface, and a linking of the two solutions giving the different continuous flow patterns each side of the shock surface by different, but related, boundary conditions. The equations connecting the different boundary conditions are known as the Rankine–Hugoniot equations. They give the relationships that must exist between finite jumps in pressure, velocity, density and internal energy, and are derived from the principles of conservation of mass, momentum and energy (reference 22). As these boundary conditions have to be applied over a moving shock front whose future position is not known it is no surprise to learn that the solution of shock-problems by this approach is difficult.

The generation of an actual shock wave in a gas appears to be associated with viscosity and heat-conduction, so a direct approach to shock wave calculations would appear to lie in the solution of the equations of motion taking these properties into account. Unfortunately, however, they are much too complicated to integrate, either analytically or numerically. An examination of the equations for gas flow shows that all the derivatives representing dissipative effects, i.e., viscosity and heat-conduction, are of higher order than the other terms and all their coefficients are small. This feature is very significant when it is borne in mind that differential equations of this character frequently have solutions that vary rapidly but continuously in narrow regions. Von Neumann and Richtmyer (1950), references 12, 26 and 31, suggested that shock waves could be simulated mathematically by the addition of a small term representing viscosity to the equations of non-viscous flow. This term would be of higher order than the others but simpler than the viscosity terms in the equations for viscous flow. They applied the idea successfully to a number of problems and since then it has been developed in several ways. The first-order simultaneous equations of this 'pseudo-viscosity' method are generally solved numerically by explicit formulae similar to those indicated in the text.

Lax, reference 18, developed an extremely ingenious method in 1954 which avoids the explicit introduction of terms representing viscosity and heat-conduction, but which introduces their effects entirely through the manner in which derivatives are replaced by finite-differences in special forms of the equations for non-viscous flow. The method virtually solves the exact equations for viscous heat-conducting flow but with incorrect viscosity and heat-conduction coefficients. His scheme, devoid of all detail, replaces

$$\frac{\partial f(x,t)}{\partial x} \quad \text{by} \quad \frac{f_{i+1,j} - f_{i-1,j}}{2\delta x}$$

and

$$\frac{\partial f(x,t)}{\partial t} \quad \text{by} \quad \frac{1}{\delta t}\{f_{i,j+1} - \tfrac{1}{2}(f_{i+1,j} + f_{i-1,j})\}.$$

The last approximation introduces a heat-conduction or diffusion term into every equation where it occurs because it can be written identically as

$$\frac{1}{\delta t}(f_{i,j+1} - f_{i,j}) - \frac{1}{2\delta t}(f_{i+1,j} - 2f_{i,j} + f_{i-1,j}),$$

which is a finite-difference representation of

$$\frac{\partial f}{\partial t} - K\frac{\partial^2 f}{\partial x^2} \quad \text{where} \quad K = \frac{(\delta x)^2}{2\delta t}.$$

EXERCISES AND SOLUTIONS

1. The function U satisfies the equation

$$x^2 U \frac{\partial U}{\partial x} + e^{-y}\frac{\partial U}{\partial y} = -U^2$$

and the condition $U = 1$ on $y = 0$, $0 < x < \infty$.

Calculate the Cartesian equation of the characteristic through the point $R(x_R, 0)$, $x_R > 0$, and the solution along this characteristic.

Use a finite-difference method to calculate first approximations to the solution and to the value of y at the point $P(1.1, y)$, $y > 0$, on the characteristic through the point $R(1, 0)$.

Calculate second approximations to these values to $4D$ by an iterative method. Compare your final results with those given by the analytical solution.

Solution

$dx/x^2 U = e^y\, dy = -U^{-2}\, dU$. Hence $1/x = A + \log U$. As $U = 1$ at $(x_R, 0)$, $A = 1/x_R$ so $\log U = x^{-1} - x_R^{-1}$. Similarly $e^y = B + U^{-1}$. As $U = 1$ at $(x_R, 0)$, $B = 0$ so $U = e^{-y}$. Eliminating U, $y = x_R^{-1} - x^{-1}$.

First approximations. $dy = dx/x^2 U e^y$, therefore $y_P^{(1)} - 0 = (0.1)/1 = 0.1$. $dU = -U dx/x^2$, therefore $u_P^{(1)} - 1 = -(0.1)$ giving $u_P^{(1)} = 0.9$.

Second approximations

$$y_P^{(2)} = \frac{dx}{2}\left\{\left(\frac{1}{x^2 u e^y}\right)_R + \left(\frac{1}{x^2 u e^y}\right)_P\right\} = 0.05(1 + 0.831) = 0.0915.$$

N.B. If we use $x^2 U dy = e^{-y} dx$ and approximate by

$$\tfrac{1}{2}\{(x^2 u)_R + (x^2 u)_P\}(y_P^{(2)} - 0) = \tfrac{1}{2}\{e_R^{-y} + e_P^{-y}\}(0 \cdot 1),$$

then $y_P^{(2)} = 0 \cdot 0912$.

$$u_P^{(2)} - 1 = -\frac{1}{2}\left\{\left(\frac{u}{x^2}\right)_R + \left(\frac{u}{x^2}\right)_P\right\} dx = -\tfrac{1}{2}(0 \cdot 1 + 0 \cdot 0744)(0 \cdot 1)$$

giving $u_P^{(2)} = 0.9128$.

Analytical values: $y_P = x_R^{-1} - x_P^{-1} = 0 \cdot 0909$ and $u_P = e^{-y_P} = \exp(-0 \cdot 0909)$ giving $u_P = 0 \cdot 9131$.

2. The function U satisfies the equation

$$\frac{\partial U}{\partial x} + \frac{x}{\sqrt{U}}\frac{\partial U}{\partial y} = 2x$$

and the conditions $U = 0$ on $x = 0$, $y \geqslant 0$ and $U = 0$ on $y = 0$, $x > 0$.

Calculate the analytical solutions at the points $(2, 5)$ and $(5, 4)$. Sketch the characteristics through these two points. If the initial condition along $y = 0$ is replaced by $U = x$, calculate approximations to the solution and to the value of y at the point $P(4 \cdot 05, y)$ on the characteristic through the point $R(4, 0)$. Compare with the analytical values.

Solution

$dx = \sqrt{U} \, dy/x = dU/2x$. Hence $U = x^2 + A$, $y = B + \sqrt{U}$. As $U = 0$ at $R(x_R, 0)$, $A = -x_R^2$ and $B = 0$. Eliminating U between these equations gives that the solution along the characteristic $x^2 - y^2 = x_R^2$ from $(x_R, 0)$ is $U = y^2$. Similarly, as $U = 0$ at $S(0, y_s)$, $A = 0$ and $B = y_s$ and elimination of U shows that the solution along the characteristic $y - y_s = x$ from $S(0, y_s)$ is $U = x^2$. Therefore $U(5, 4) = 4^2 = 16$ and $U(2, 5) = 2^2 = 4$.

Approximations: $dU = 2x \, dx$. Therefore $u_P^{(1)} - 4 = 2 \cdot \tfrac{1}{2}(x_R + x_P) \, dx = (8 \cdot 05)(0 \cdot 05) = 0 \cdot 4025$. $\sqrt{U} \, dy = x \, dx$ may be approximated by $\tfrac{1}{2}(\sqrt{u_R} + \sqrt{u_P})(y_P^{(1)} - 0) = \tfrac{1}{2}(x_P + x_R) \, dx$ giving $(2 + \sqrt{4 \cdot 4025}) \, y_P^{(1)} = (4 + 4 \cdot 05)(0 \cdot 05)$, from which $y_P^{(1)} = 0 \cdot 0982$.

Analytical values: As $U_R = x_R$ at $(x_R, 0)$, $A = x_R - x_R^2$, $B = -\sqrt{x_R}$. Hence $y = \sqrt{U} - \sqrt{x_R}$ and $U = x^2 + x_R - x_R^2$. Therefore $U_P = 4 \cdot 4025$ and $y_P = 0 \cdot 0982$.

3. The function U satisfies the quasi-linear equation

$$a\frac{\partial U}{\partial x} + b\frac{\partial U}{\partial y} = c.$$

Prove that the characteristic direction is given by the equation $dx/a = dy/b$, and that U satisfies the relationship $dx/a = dU/c$ along a characteristic curve, by finding the conditions that U and its derivatives must satisfy if U is to assume prescribed values on a given curve C in the $x-y$ plane and also satisfy the differential equation on this curve. (The alternative approach.)

(i) Calculate analytically the solution of the equation

$$\frac{\partial U}{\partial x} + 3x^2\frac{\partial U}{\partial y} = x + y$$

at the point $(3, 19)$ given that $U(x, 0) = x^2$.

(ii) Calculate to $4D$, by an iterative method, the solution of the equation

$$(x - y)\frac{\partial U}{\partial x} + U\frac{\partial U}{\partial y} = x + y$$

at the point $P(1 \cdot 1, y)$ on the characteristic through $R(1, 0)$, given that $U(x, 0) = 1$.

Solution

The arc rate of change of U along C is known, and must be related to the values of p and q satisfying the differential equation by

$$\frac{dU}{ds} = \frac{\partial U}{\partial x}\frac{dx}{ds} + \frac{\partial U}{\partial y}\frac{dy}{ds} = p\frac{dx}{ds} + q\frac{dy}{ds},$$

where $ap + bq = c$. Solve for p and q in the form $p/\Delta_1 = q/\Delta_2 = 1/\Delta_3$. When $\Delta_3 = 0$ there is no solution unless $\Delta_1 = \Delta_2 = 0$, etc.

(i) The characteristic through $(3, 19)$ is $y = x^3 - 8$. This meets $y = 0$ at $(2, 0)$. As $dU = (x + y)\,dx$, $U(3, 19) - U(2, 0) = \int_2^3 (x + x^3 - 8)\,dx$ giving $U(3, 19) = 14\frac{3}{4}$.

(ii) Using $(x - y)\,dy = U\,dx$ and $(x - y)\,dU = (x + y)\,dx$ iteratively as in the text, gives that $y_P^{(1)} = 0 \cdot 1$, $u_P^{(1)} = 1 \cdot 1$, $y_P^{(2)} = 0 \cdot 105$, $u_P^{(2)} =$

1·11, $y_P^{(3)} = 0·105764$, $u_P^{(3)} = 1·110526$, $y_P^{(4)} = 0·105831$, $u_P^{(4)} = 1·110607$, $y_P^{(5)} = 0·105839$, $u_P^{(5)} = 1·110614$. Clearly $u_P = 1·1106$ to 4D.

4 (a). Use the von Neumann method to prove that the Lax-Wendroff difference equations (4.12) are stable for $0 < ap \le 1$.

(b). Show that the principal part of the local truncation error of the Lax-Wendroff equation (4.12) is

$$\left[\tfrac{1}{6}k^2 \frac{\partial^3 U}{\partial t^3} + \tfrac{1}{6}ah^2 \frac{\partial^3 U}{\partial x^3}\right]_{i,j}.$$

(c). Prove that the solution of the differential equation

$$\frac{\partial U}{\partial t} + a \frac{\partial U}{\partial x} = 0,$$

a constant, is the solution of the approximating Lax-Wendroff equation (4.12) when $k/h = 1/a$. Comment on this result in relation to the characteristics of the differential equation.

Solution

(a) The substitution of $E_{i,j} = e^{\sqrt{(-1)}\beta x}\xi^j$ into (4.12) leads to $\xi = (1 - 2a^2p^2 \sin^2 \tfrac{1}{2}\beta h) - 2(\sqrt{-1})ap \sin \tfrac{1}{2}\beta h \cos \tfrac{1}{2}\beta h$. Hence $|\xi|^2 = 1 - 4a^2p^2(1 - a^2p^2) \sin^4 \beta h$. Errors will not increase exponentially with j if $|\xi|^2 \le 1$, i.e., $0 \le 4a^2p^2(1 - a^2p^2)$, giving $0 < ap \le 1$.

(b) $T_{i,j} = \{U_{i,j+1} - \tfrac{1}{2}ap(1 + ap)U_{i-1,j} - (1 - a^2p^2)U_{i,j} + \tfrac{1}{2}ap(1 - ap)U_{i+1,j}\}/k.$
Expand each term by Taylor's series about the point (i, j), etc.

(c) When $k/h = 1/a$, i.e., $ap = 1$,

$$T_{i,j} = (U_{i,j+1} - U_{i-1,j})/k = k\left(\frac{\partial U}{\partial t} + a \frac{\partial U}{\partial x}\right)_{i,j} + \tfrac{1}{2}k^2\left(\frac{\partial^2 U}{\partial t^2} - a^2 \frac{\partial^2 U}{\partial x^2}\right)_{i,j}$$

$$+ \tfrac{1}{6}k^3\left(\frac{\partial^3 U}{\partial t^3} + a^3 \frac{\partial^3 U}{\partial x^3}\right)_{i,j} + \dots.$$

Each bracketed term is zero because $(\partial/\partial t) \equiv -a(\partial/\partial x)$ by the differential equation. Hence the result. By the differential equation, $dt/1 = dx/a = dU/0$. Hence U is constant along the straight-line characteristics of slope $1/a$. The difference equation gives $u_{i,j+1} = u_{i-1,j}$ along the line of slope $1/a$.

5. The equation

$$a\frac{\partial U}{\partial x} + b\frac{\partial U}{\partial t} = c$$

is approximated at the point $(i+\frac{1}{2}, j+\frac{1}{2})$ by the Wendroff implicit scheme

$$(b+ap)u_{i+1,j+1} + (b-ap)u_{i,j+1} - (b-ap)u_{i+1,j} - (b+ap)u_{i,j} - 2kc = 0,$$

where $p = k/h$.

Prove that : (a) The scheme is unconditionally stable, and

(b) the principal part of the local truncation error *at the point*

$(i+\frac{1}{2}, j+\frac{1}{2})$ is $\frac{1}{12}h^2\left(3b\dfrac{\partial^3 U}{\partial x^2 \partial t} + a\dfrac{\partial^3 U}{\partial x^3}\right)_{i+\frac{1}{2},j+\frac{1}{2}}$

$$+ \frac{1}{12}k^2\left(b\frac{\partial^3 U}{\partial t^3} + 3a\frac{\partial^3 U}{\partial x \partial t^2}\right)_{i+\frac{1}{2},j+\frac{1}{2}}.$$

Solution

(a) The error function $E_{i,j} = e^{\sqrt{(-1)}\beta x}\xi^j$ is a solution of the equation

$$(b+ap)E_{i+1,j+1} + (b-ap)E_{i,j+1} - (b-ap)E_{i+1,j} - (b+ap)E_{i,j} = 0$$

if

$$\xi = (b\cos\beta h - (\sqrt{-1})ap\sin\beta h)/(b\cos\beta h + (\sqrt{-1})ap\sin\beta h).$$

Hence $|\xi| = 1$ for all real values of a, b and p.

(b) $T_{i,j} = \{(b+ap)U_{i+1,j+1} + (b-ap)U_{i,j+1} - (b-ap)U_{i+1,j}$

$$- (b+ap)U_{i,j} - 2kc\}/k.$$

Expand each term about the point $(i+\frac{1}{2}, j+\frac{1}{2})$ by Taylor's series.

6. Show that the following second-order equations can each be written as a system of first-order equations in terms of the transformations indicated. Express each system in matrix form.

(a) The wave equation

$$\frac{\partial^2 U}{\partial t^2} - \frac{\partial^2 U}{\partial x^2} = 0,$$

putting

$$\frac{\partial U}{\partial t} = U_1 \quad \text{and} \quad \frac{\partial U}{\partial x} = U_2.$$

(b) The heat conduction equation

$$\frac{\partial U}{\partial t} - \frac{\partial^2 U}{\partial x^2} = 0,$$

putting

$$U = U_1 \quad \text{and} \quad \frac{\partial U}{\partial x} = U_2.$$

Solution

(a)

$$\frac{\partial^2 U}{\partial t^2} - \frac{\partial^2 U}{\partial x^2} = 0 = \frac{\partial U_1}{\partial t} - \frac{\partial U_2}{\partial x} = 0$$

A second equation is given by the identity

$$\frac{\partial^2 U}{\partial x \, \partial t} - \frac{\partial^2 U}{\partial t \, \partial x} = 0 = \frac{\partial U_1}{\partial x} - \frac{\partial U_2}{\partial t}.$$

These two equations for U_1 and U_2 can be written as

$$\begin{bmatrix} 1 & 0 \\ 0 & 1 \end{bmatrix} \frac{\partial}{\partial t} \begin{bmatrix} U_1 \\ U_2 \end{bmatrix} + \begin{bmatrix} 0 & -1 \\ -1 & 0 \end{bmatrix} \frac{\partial}{\partial x} \begin{bmatrix} U_1 \\ U_2 \end{bmatrix} = 0,$$

i.e., as

$$\mathbf{A} \frac{\partial \mathbf{U}}{\partial t} + \mathbf{B} \frac{\partial \mathbf{U}}{\partial x} = \mathbf{O}, \quad \text{where} \quad \mathbf{U} = \begin{bmatrix} U_1 \\ U_2 \end{bmatrix}$$

(b)

$$\frac{\partial U}{\partial t} - \frac{\partial^2 U}{\partial x^2} = 0 = \frac{\partial U_1}{\partial t} - \frac{\partial U_2}{\partial x}.$$

A second equation is given by the identity

$$-\frac{\partial U}{\partial x} + \frac{\partial U}{\partial x} = 0 = -\frac{\partial U_1}{\partial x} + U_2$$

These two first-order equations for U_1 and U_2 can be written as

$$\begin{bmatrix} 1 & 0 \\ 0 & 0 \end{bmatrix} \frac{\partial}{\partial t} \begin{bmatrix} U_1 \\ U_2 \end{bmatrix} + \begin{bmatrix} 0 & -1 \\ -1 & 0 \end{bmatrix} \frac{\partial}{\partial x} \begin{bmatrix} U_1 \\ U_2 \end{bmatrix} + \begin{bmatrix} 0 & 0 \\ 0 & 1 \end{bmatrix} \begin{bmatrix} U_1 \\ U_2 \end{bmatrix} = 0,$$

i.e., as

$$\mathbf{A} \frac{\partial \mathbf{U}}{\partial t} + \mathbf{B} \frac{\partial \mathbf{U}}{\partial x} + \mathbf{C} \mathbf{U} = \mathbf{O}.$$

7. Write a computer program to obtain a numerical solution to the following problem by the Lax-Wendroff explicit method, taking $h = \frac{1}{2}$, $k = \frac{1}{4}$, $x = 0(\frac{1}{2})8$, and $t = 0(\frac{1}{4})4$.

Compare with the analytical solution at mesh points along $t = 3$.

$$\frac{\partial U}{\partial t} + \frac{1}{2}\frac{\partial U}{\partial x} = 0,$$

where $U(0, t) = -\frac{1}{2}t$, $t > 0$; $U(x, 0) = x$, $0 \leqslant x \leqslant 2$; $U(x, 0) = 4 - x$, $2 < x \leqslant 4$; $U(x, 0) = 0$, $4 < x$.

Solution

$$u_{i,j+1} = \frac{1}{32}(5u_{i-1,j} + 30u_{i,j} - 3u_{i+1,j})$$

The analytical solution between $x = 0$ and $t = 2x$ is $U = -\frac{1}{2}(t - 2x)$, between $t = 2x$ and $t = 2(x - 2)$ it is $U = x - \frac{1}{2}t$, between $t = 2(x - 2)$ and $t = 2(x - 4)$ it is $U = 4 - x + \frac{1}{2}t$, and to the right of $t = 2(x - 4)$ it is $U = 0$. Along $t = 3$ the two solutions for $x = 1(1)8$ to $4D$ are as shown.

Lax–Wendroff	−0·4562	0·4602	1·6848	1·4564
P.D.E.	−0·5	0·5	1·5	1·5
Lax–Wendroff	0·3951	0·0314	0·0008	0
P.D.E.	0·5	0	0	0

8. The function U satisfies the equation

$$\frac{\partial^2 U}{\partial x^2} - 4x^2\frac{\partial^2 U}{\partial y^2} = 0$$

and the initial conditions

$$U = x^2, \frac{\partial U}{\partial y} = 0, \quad \text{on} \quad y = 0, -\infty < x < \infty.$$

Show from first principles, with the usual notation, that $dp - 2x\,dq = 0$ along the characteristic of slope $2x$, and that $dp + 2x\,dq = 0$ along the characteristic of slope $(-2x)$.

The characteristic with positive slope through the point $A(0·3, 0)$ intersects the characteristic with negative slope through the point $B(0·4, 0)$ at R. Calculate to $3D$ an approximation to the solution at R.

Solution

First part is bookwork. The characteristic with slope $2x$ through A is $y = x^2 - 0.09$. The other characteristic through B is $y = 0.16 - x^2$. Hence $x_R^2 = 1/8$, so $x_R = 0.3536$, $y_R = 0.035$. The equations, $p_R - 0.6 - (x_A + x_R)q_R = 0 = p_R - 0.8 + (x_B + x_R)q_R$ give $p_R = 0.6929$, $q_R = 0.1421$. Then

$$dU \simeq u_R - u_A = \tfrac{1}{2}(p_A + P_R)(x_R - x_A) + \tfrac{1}{2}(q_A + q_R)(y_R - y_A)$$

gives $u_R = 0.127$.

9. The equation

$$\frac{\partial^2 u}{\partial x^2} + (1 - 2x)\frac{\partial^2 u}{\partial x\,\partial y} + (x^2 - x - 2)\frac{\partial^2 u}{\partial y^2} = 0,$$

satisfies the initial conditions $u = x$, $\partial u/\partial y = 1$, on $y = 0, 0 \leqslant x \leqslant 1$.

Use the method of characteristics to calculate the solution at the points R, S and T of Fig. 4.9, $y > 0$, the co-ordinates of the initial points P, Q and W being $(0.4, 0)$, $(0.5, 0)$, (0.6) respectively.

Solution

Equations (4.30) and (4.31) give $p = q = 1$. The points of intersection of the characteristics can be calculated either iteratively using equations (4.35) and (4.36), or directly by integrating $f = 2 - x$, $g = -1 - x$. The complete iterative solution for two y-steps forward between $x = 0$ and 1 is as follows. Hence the solution values at the mesh points R, S and T specified above are 0.5258, 0.6290 and 0.6575.

First step

x	y	p	q	u
0·035	0·071	1	1	0·1065
0·139	0·074	1	1	0·2123
0·242	0·075	1	1	0·3174
0·345	0·077	1	1	0·4219
0·448	0·077	1	1	0·5258
0·552	0·077	1	1	0·6290
0·655	0·077	1	1	0·7316
0·758	0·075	1	1	0·8335
0·861	0·074	1	1	0·9348
0·965	0·071	1	1	1·0355

Second step

x	y	p	q	u
0·074	0·147	1	1	0·2212
0·181	0·152	1	1	0·3322
0·287	0·155	1	1	0·4420
0·394	0·157	1	1	0·5504
0·500	0·158	1	1	0·6575
0·606	0·157	1	1	0·7633
0·713	0·155	1	1	0·8678
0·819	0·152	1	1	0·9710
0·926	0·147	1	1	1·0728

10. Calculate the solution of the problem in the worked example 4.4 at the first characteristic grid point between $x = 0.3$ and 0.4, $y > 0$.

Solution

The complete solution for two y-steps forward between $x = 0$ and 1 is as follows. The solution value at the point specified is 0.8852.

First step

x	y	p	q	u
0·053	0·011	0·5066	2·3653	0·2267
0·156	0·016	1·4970	1·9537	0·3379
0·256	0·027	2·5287	1·6765	0·5567
0·355	0·042	3·5824	1·6414	0·8852
0·454	0·063	4·6526	1·7495	1·3288
0·554	0.090	5·7386	1·9342	1·8933
0·653	0·122	6·8406	2·1609	2·5843
0·753	0·160	7·9594	2·4119	3·4073
0·853	0·204	9·0957	2·6778	4·3682
0·952	0·254	10·2501	2·9532	5·4727

Second step

x	y	p	q	u
0·117	0·029	0·9011	3·7464	0·3266
0·222	0·045	1·8975	2·8486	0·5190
0·321	0·072	2·9954	2·3266	0·8263
0·418	0·110	4·1394	2·1461	1·2574
0·516	0·160	5·3172	2·1608	1·8252
0·614	0·223	6·5283	2·2827	2·5430
0·712	0·299	7·7750	2·4659	3·4244
0·811	0·389	9·0602	2·6858	4·4833
0·910	0·493	10·3868	2·9288	5·7343

11. The function U is a solution of the equation

$$\frac{\partial^2 U}{\partial x^2} + \left(\frac{\partial U}{\partial x} - U\right)\frac{\partial^2 U}{\partial x\,\partial y} - U\frac{\partial U}{\partial x}\frac{\partial^2 U}{\partial y^2} + x = 0$$

and satisfies the initial conditions $U = 1 + x^2$,

$$\frac{\partial U}{\partial y} = 1, \quad \text{on} \quad y = 0, \ -\infty < x < \infty.$$

Show from first principles, for this particular example, that $p\,dp - U p\,dq + x\,dy = 0$ along the characteristic of slope p, and that $-U\,dp - U p\,dq + x\,dy = 0$ along the characteristic of slope $(-U)$, where

$$p = \frac{\partial U}{\partial x} \quad \text{and} \quad q = \frac{\partial U}{\partial y}.$$

The characteristic with positive slope through the point $A(0.5, 0)$ intersects the characteristic with negative slope through the point $B(0.6, 0)$ at the point $R(x_R, y_R)$. Calculate first approximation values for the co-ordinates of R.

Write down the equations giving first approximation values for p and q at R, but do not solve them. Explain how to calculate a first approximation value for U at R.

Given that the first approximation values for U, p and q at R are $u_R^{(1)} = 1.3711$, $p_R^{(1)} = 1.1009$ and $q_R^{(1)} = 1.1038$, calculate second approximation values for the co-ordinates of R.

Solution

First part is bookwork. First approximation equations to (x_R, y_R) are $y_R^{(1)} = x_R^{(1)} - 0.5 = -1.36(x_R^{(1)} - 0.6)$. Hence $x_R^{(1)} = 0.5576$. $y_R^{(1)} = 0.0576$. First approximations to (p_R, q_R) are given by

$$p_A(p_R^{(1)} - p_A) - p_A u_A(q_R - q_A) + x_A(y_R^{(1)} - y_A) = 0,$$
$$-u_B(p_R^{(1)} - p_B) - p_B u_B(q_R - q_B) + x_B(y_R^{(1)} - y_B) = 0,$$

i.e.,

$$p_R^{(1)} - 1.25 q_R^{(1)} + 0.2788 = 0 - 1.36 p_R^{(1)} - 1.632 q_R^{(1)} + 3.2986,$$

from which

$p_R^{(1)} = 1.1009$ and $q_R^{(1)} = 1.1038$. Usual approximation to $dU = p\,dx + q\,dy$ gives that $u_R^{(1)} = 1.3711$.

$$y_R^{(2)} = \tfrac{1}{2}(2.1009)(x_R^{(2)} - 0.5) = -\tfrac{1}{2}(1.36 + 1.3711)(x_R^{(2)} - 0.6).$$

Therefore $x_R^{(2)} = 0.5565$ and $y_R^{(2)} = 0.0594$.

12. The transverse displacement U of a point at a distance X from one end of a vibrating string of length L at time T satisfies the equation $\partial^2 U/\partial T^2 = c^2 \partial^2 U/\partial X^2$. Show this can be reduced to the non-dimensional form

$$\frac{\partial^2 u}{\partial x^2} = \frac{\partial^2 u}{\partial t^2}, \quad 0 \le x \le 1,$$

by putting $x = X/L$, $u = U/L$ and $t = cT/L$.

A solution of the latter equation satisfies the boundary conditions

$$u = 0 \text{ at } x = 0 \text{ and } 1, \ t \ge 0,$$

and the initial conditions

$$u = \tfrac{1}{2}x(1-x) \text{ and } \frac{\partial u}{\partial t} = 0, \text{ for } 0 \le x \le 1 \text{ when } t = 0.$$

Use the method of characteristics to calculate the non-dimensional velocities and displacements at time $t = 0.3$ of the points on the string defined by $x = 0(0.1)0.5$.

Use the method of separation of the variables to show that the analytical solution to this problem is

$$u = \frac{2}{\pi^3} \sum_1^\infty \frac{1}{n^3} (1 - \cos n\pi)\cos n\pi t \sin n\pi x.$$

Compare the two solutions at $x = 0.5$, $t = 0.3$.

Solution

The characteristics are $t \pm x = $ constant, on which $p \pm q = $ constant, where $p = \partial u/\partial x$, $q = \partial u/\partial t = $ velocity. Construct the characteristics through C and B as shown in Fig. 4.13. From the initial conditions, $p_A = 0.3$, $q_A = 0$, $p_D = 0.1$, $q_D = 0$.

Along DC, $p_D + q_D = 0.1 = p_C + q_C$.
Along AB, $p_A + q_A = 0.3 = p_B + q_B = p_B$.
Along BC, $p_B - q_B = p_B = p_C - q_C$.

Hence $p_C = 0.2$ and $q_C = -0.1 = $ speed of C. The displacement u can be found by step-wise integration of p with respect to x along $t = 0.3$, i.e., from $du = (\partial u/\partial x)dx$, which can be approximated by

$$u_C - u_F = \tfrac{1}{2}(p_F + p_C)(x_C - x_F) = \tfrac{1}{2}(p_F + 0.2)(0.1).$$

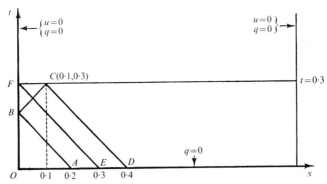

Fig. 4.13

Along EF, $p_E + q_E = 0 \cdot 2 = p_F + q_F = p_F$.
As $u_F = 0$, $u_C = 0 \cdot 02$. Similarly for the other points.

	$x =$	0·1	0·2	0·3	0·4	0·5
Numerical solution	q	−0·1	−0·2	−0·3	−0·3	−0·3
	u	0·02	0·04	0·06	0·075	0·08
Analytical solution	q	−0·1	−0·2	−0·3	−0·3	−0·3
	u	0·02	0·04	0·06	0·075	0·08

13. The equations of motion for steady two-dimensional isentropic flow of compressible fluid can be written as

$$u \frac{\partial u}{\partial x} + v \frac{\partial u}{\partial y} = -\frac{a^2}{\rho} \frac{\partial \rho}{\partial x},$$

$$u \frac{\partial v}{\partial x} + v \frac{\partial v}{\partial y} = -\frac{a^2}{\rho} \frac{\partial \rho}{\partial y},$$

and the continuity equation as

$$u \frac{\partial \rho}{\partial x} + v \frac{\partial \rho}{\partial y} + \rho \left(\frac{\partial u}{\partial x} + \frac{\partial v}{\partial y} \right) = 0,$$

where u, v are the Cartesian components of velocity, ρ the density and a the local speed of sound. Eliminate ρ between these equations and show that for irrotational flow, defined by $u = -\partial \phi / \partial x$, $v = -\partial \phi / \partial y$, the eliminant reduces to

$$(a^2 - u^2) \frac{\partial^2 \phi}{\partial x^2} - 2uv \frac{\partial^2 \phi}{\partial x \, \partial y} + (a^2 - v^2) \frac{\partial^2 \phi}{\partial y^2} = 0,$$

where ϕ is the velocity potential.

Deduce that the fluid must be moving supersonically if it is possible for two different flows to exist side by side with continuous values for u and v across the dividing curve C but with a discontinuity in one of the space-derivatives of u or v. (C defines a characteristic.)

Prove that the differentials of u and v along C are related by the equation $dv/du = -1/m$, where m is the slope of C.

Solution

Substitute for $\partial \rho / \partial x$, $\partial \rho / \partial y$ from the first two equations into the third, etc. Then

$$0 = (a^2 - u^2)\phi_{xx} - 2uv\phi_{xy} + (a^2 - v^2)\phi_{yy},$$

$$du = u_x \, dx + u_y \, dy = -\phi_{xx} \, dx - \phi_{yx} \, dy,$$

$$dv = v_x \, dx + v_y \, dy = -\phi_{yx} \, dx - \phi_{yy} \, dy,$$

where the suffixes denote differentiations. Solve in the form

$$\phi_{xx}/\Delta_1 = -\phi_{xy}/\Delta_2 = \phi_{yy}/\Delta_3 = -1/\Delta_4.$$

When $\Delta_1 = \Delta_2 = \Delta_3 = \Delta_4 = 0$, any one of the second-order ϕ derivatives can be discontinuous. $\Delta_4 = 0$ gives a quadratic for the slope of C, namely

$$(a^2 - u^2)\left(\frac{dy}{dx}\right)^2 + 2uv\frac{dy}{dx} + (a^2 - v^2) = 0.$$

These directions are real and different when $u^2 v^2 > (a^2 - u^2) \times (a^2 - v^2)$, giving $u^2 + v^2 > a^2$.

$\Delta_2 = 0$ gives $-\dfrac{dy}{dx} = \dfrac{(a^2 - v^2)}{(a^2 - u^2)}\dfrac{dv}{du}$. Substitution into the quadratic

gives

$$(a^2 - v^2)\left(\frac{dv}{du}\right)^2 - 2uv\frac{dv}{du} + (a^2 - u^2) = 0.$$

Hence the result because the roots of this quadratic are the negative reciprocals of the roots of the quadratic for dy/dx.

14. Prove that the characteristic directions associated with the simultaneous first-order equations

$$a_1 \frac{\partial u}{\partial x} + b_1 \frac{\partial u}{\partial y} + c_1 \frac{\partial v}{\partial x} + d_1 \frac{\partial v}{\partial y} + e_1 = 0,$$

$$a_2 \frac{\partial u}{\partial x} + b_2 \frac{\partial u}{\partial y} + c_2 \frac{\partial v}{\partial x} + d_2 \frac{\partial v}{\partial y} + e_2 = 0,$$

are given by

$$0 = \begin{vmatrix} a_1 & b_1 & c_1 & d_1 \\ a_2 & b_2 & c_2 & d_2 \\ dx & dy & 0 & 0 \\ 0 & 0 & dx & dy \end{vmatrix}$$

$$= (a_1 c_2 - a_2 c_1) \left(\frac{dy}{dx} \right)^2 - (a_1 d_2 - a_2 d_1 + b_1 c_2 - b_2 c_1) \frac{dy}{dx}$$
$$+ (b_1 d_2 - b_2 d_1).$$

Show also that the differential relationship between u and v in these directions is

$$0 = \begin{vmatrix} a_1 & b_1 & c_1 & e_1 \\ a_2 & b_2 & c_2 & e_2 \\ dx & dy & 0 & -du \\ 0 & 0 & dx & -dv \end{vmatrix}$$

$$= du(a_1 b_2 - a_2 b_1) + dv\{(a_1 c_2 - a_2 c_1) \frac{dy}{dx} - (b_1 c_2 - b_2 c_1)\} +$$
$$+ (a_1 e_2 - a_2 e_1)dy - (b_1 e_2 - b_2 e_1)dx.$$

Solution

Solve the given equations and

$$du = \frac{\partial u}{\partial x} dx + \frac{\partial u}{\partial y} dy, \quad dv = \frac{\partial v}{\partial x} dx + \frac{\partial v}{\partial y} dy,$$

for the four partial derivatives, then apply the argument given in the 'alternative approach to characteristics' section.

15. The equation of motion for unsteady one-dimensional isentropic flow of compressible fluid is

$$\frac{\partial u}{\partial t} + u \frac{\partial u}{\partial x} + \frac{a^2}{\rho} \frac{\partial \rho}{\partial x} = 0,$$

and the continuity equation is

$$\rho \frac{\partial u}{\partial x} + \frac{\partial \rho}{\partial t} + u \frac{\partial \rho}{\partial x} = 0.$$

Prove that the characteristic directions are given by $dt/dx = 1/(u \pm a)$, and that the differential relationship between u and ρ in the characteristic directions is $du/d\rho = \pm a/\rho$.

Solution

As in exercise 14.

16. (a) Prove that the differential relationship between $p = \partial u/\partial x$ and $q = \partial u/\partial y$ along the coincident characteristics of the parabolic equation $\partial^2 u/\partial x^2 = \partial u/\partial y$ is the parabolic equation itself.

Solution

Eliminate r and s, say, between equations (4.18), (4.19) and (4.20) for this problem, etc., or use the 'alternative approach to characteristics' argument.

(b) Show that the change of independent variables defined by $\xi = x + y$, $\eta = x - y$ transforms the equation $\partial^2 u/\partial x^2 = \partial^2 u/\partial y^2$ to $\partial^2 u/\partial \xi \partial \eta = 0$. Hence deduce that the solution of the wave-equation satisfying the initial conditions, $u = f(x)$ and $\partial u/\partial y = g(x)$ on $y = 0$, is

$$u = \tfrac{1}{2} \left\{ f(x+y) + f(x-y) + \int\limits_{x-y}^{x+y} g(t) \, dt \right\}.$$

Also deduce that the solution at (x_0, y_0) depends only on the data on the segment of the initial line between the characteristics $x + y = $ constant, $x - y = $ constant, that pass through (x_0, y_0).

Solution

Integration of $\partial^2 u/\partial\xi\partial\eta = 0$ gives $u = \phi(x+y) + \psi(x-y)$. Hence $\partial u/\partial y = \phi'(x+y) - \psi'(x-y)$. The initial conditions along $y = 0$ gives $f(x) = \phi(x) + \psi(x)$; $g(x) = \phi'(x) - \psi'(x)$. Integrate the last equation to get

$$\phi(x) - \psi(x) + \text{constant} = \int_0^x g(x)\,dx.$$

Solve for $\phi(x)$ and $\psi(x)$ in terms of f and g then replace x in $\phi(x)$ by $x+y$ and x in $\psi(x)$ by $x-y$, etc.

17. The function u satisfies the equation

$$\frac{\partial^2 u}{\partial x^2} = \frac{\partial^2 u}{\partial t^2},$$

the boundary conditions

$$u = 0 \text{ at } x = 0 \text{ and } 1, \quad t \geqslant 0,$$

and the initial conditions

$$u = \tfrac{1}{8}\sin \pi x, \quad \frac{\partial u}{\partial t} = 0, \text{ when } t = 0, \quad 0 \leqslant x \leqslant 1.$$

Use the explicit finite-difference formula of equation (4.46), and a central-difference approximation for the derivative condition, to calculate a solution for $x = 0(0\cdot1)1$ and $t = 0(0\cdot1)0\cdot5$.

Derive the analytical solution

$$u = \tfrac{1}{8}\sin \pi x \cos \pi t$$

and compare with the numerical solution at several points.

Solution $(r = 1)$

$$u_{i,j+1} = u_{i-1,j} + u_{i+1,j} - u_{i,j-1}, \quad (j \geqslant 1).$$

$$\frac{\partial u}{\partial t} = \frac{u_{i,1} - u_{i,-1}}{2\delta t} = 0, \text{ and } j = 0 \text{ in the previous equation, gives}$$

$$u_{i,1} = \tfrac{1}{2}(u_{i-1,0} + u_{i+1,0}).$$

The problem is symmetric with respect to $x = \tfrac{1}{2}$. The following values of u were obtained by working to $5D$ and rounding to $4D$.

The analytical solution equals the finite-difference solution to this degree of accuracy.

$x = 0$	0·1	0·2	0·3	0·4	0·5
$t = 0\cdot1$ 0	0·0367	0·0699	0·0962	0·1131	0·1189
0·2 0	0·0312	0·0594	0·0818	0·0962	0·1011
0·3 0	0·0227	0·0432	0·0594	0·0699	0·0735
0·4 0	0·0119	0·0227	0·0312	0·0368	0·0386
0·5 0	0	0	0	0	0

18. The equation $(\partial^2 U/\partial t^2) = (\partial^2 U/\partial x^2)$, $0 < x < 1$, $t > 0$, is approximated at the point (ih, jk) by the difference equation

$$\frac{1}{k^2}\,\delta_t^2 u_{i,j} = \frac{1}{h^2}\,(\tfrac{1}{2}\delta_x^2 u_{i,j+1} + \tfrac{1}{2}\delta_x^2 u_{i,j-1}).$$

Given that U has known initial values throughout $0 \leqslant x \leqslant 1$, $t = 0$, known boundary values at $x = 0$ and 1, $t > 0$, and that $Nh = 1$, use the matrix method of analysis to prove that the equations are unconditionally stable in the sense that errors do not increase exponentially with increasing t.

Solution

The equations in matrix form are $\mathbf{A}\mathbf{u}_{j+1} = 2\mathbf{u}_j - \mathbf{A}\mathbf{u}_{j-1} - \mathbf{b}_j$, where $\mathbf{b}_j$ is a column vector of known values,

$$\mathbf{A} = \begin{bmatrix} (1+r) & -\tfrac{1}{2}r & & & \\ -\tfrac{1}{2}r & (1+r) & -\tfrac{1}{2}r & & \\ & & \cdot & & \\ & & & \cdot & \\ & & & -\tfrac{1}{2}r & (1+r) \end{bmatrix}, \quad \text{an } (N-1)\times(N-1) \text{ matrix,}$$

and $r = k^2/h^2$. Hence a perturbation $\mathbf{e}_0$ of the initial values satisfies $\mathbf{e}_{j+1} = 2\mathbf{A}^{-1}\mathbf{e}_j - \mathbf{e}_{j-1}$, which, with $\dot{\mathbf{e}}_j = \mathbf{e}_j$ can be expressed as

$$\begin{bmatrix} \mathbf{e}_{j+1} \\ \mathbf{e}_j \end{bmatrix} = \begin{bmatrix} 2\mathbf{A}^{-1} & -\mathbf{I} \\ \mathbf{I} & \mathbf{O} \end{bmatrix} \begin{bmatrix} \mathbf{e}_j \\ \mathbf{e}_{j-1} \end{bmatrix},$$

i.e., as $\mathbf{v}_{j+1} = \mathbf{P}\mathbf{v}_j$. The eigenvalues λ of P are given by

$$\det \begin{vmatrix} (2\lambda_k^{-1} - \lambda) & -1 \\ 1 & -\lambda \end{vmatrix} = 0,$$

where

$$\lambda_k = (1+r) - 2(\tfrac{1}{2}r)\cos\frac{k\pi}{N}, \quad k = 1(1)(N-1).$$

This leads to $\lambda = \{1 \pm \sqrt{-1}\sqrt{(\lambda_k^2-1)}\}/\lambda_k$. As $\lambda_k = 1 + 2r \times \sin^2(k\pi/2N) > 1$, $|\lambda| = 1$ for all k.

19. The simultaneous equations

$$\frac{\partial p}{\partial x} = \frac{\partial q}{\partial t}, \quad \frac{\partial q}{\partial x} = \frac{\partial p}{\partial t},$$

which define $\partial^2 u/\partial x^2 = \partial^2 u/\partial t^2$, are represented by the following finite-difference formulae. Prove that both are stable for $\delta t/\delta x \leqslant 1$.

(a)
$$\frac{1}{2\delta x}(p_{i+1,j} - p_{i-1,j}) = \frac{1}{\delta t}\{q_{i,j+1} - \tfrac{1}{2}(q_{i+1,j} + q_{i-1,j})\},$$

$$\frac{1}{2\delta x}(q_{i+1,j} - q_{i-1,j}) = \frac{1}{\delta t}\{p_{i,j+1} - \tfrac{1}{2}(p_{i+1,j} + p_{i-1,j})\}.$$

(b)
$$\frac{1}{\delta x}(p_{i+\frac{1}{2},j} - p_{i-\frac{1}{2},j}) = \frac{1}{\delta t}(q_{i,j+1} - q_{i,j}),$$

$$\frac{1}{\delta x}(q_{i,j+1} - q_{i-1,j+1}) = \frac{1}{\delta t}(p_{i-\frac{1}{2},j+1} - p_{i-\frac{1}{2},j}).$$

Solution

(a) Substitution of $p_{i,j} = Ae^{\beta i\delta x\sqrt{-1}}\xi^j$ and $q_{i,j} = Be^{\beta i\delta x\sqrt{-1}}\xi^j$ into the equations, and elimination of A/B gives

$$\xi = \cos\beta\,\delta x \pm (\rho\sin\beta\,\delta x)\sqrt{-1} \quad \text{where } \rho = \frac{\delta t}{\delta x}.$$

Hence

$$|\xi|^2 = \cos^2\beta\,\delta x + \rho^2\sin^2\beta\,\delta x \leqslant 1 \text{ for } \rho \leqslant 1.$$

Similarly for (b).

20. Solve the problem in exercise 12 using the finite-difference equations of exercise 19(a), with $\delta x = \delta t = 0\cdot1$.

Solution

$$p_{i,j+1} = \tfrac{1}{2}(p_{i+1,j} + p_{i-1,j} + q_{i+1,j} - q_{i-1,j}),$$
$$q_{i,j+1} = \tfrac{1}{2}(q_{i+1,j} + q_{i-1,j} + p_{i+1,j} - p_{i-1,j}).$$

Along $t = 0$, $p = \partial u/\partial x = \tfrac{1}{2}(1 - 2x)$, $q = \partial u/\partial t = 0$. Along $x = 0$, $u = q = 0$, so $\partial q/\partial t = 0 = \partial p/\partial x$, giving $p_{1,j} = p_{-1,j}$. Put $i = 0$ in the equations above and eliminate $q_{-1,j}$, giving

$$p_{0,j+1} = p_{1,j} + q_{1,j},$$

which is the relationship along the characteristics, since $q_{0,j+1} = 0$. Integrate parallel to Ox using $\delta u = p\,\delta x = \tfrac{1}{2}(p_i + p_{i+1})\delta x$. The values of p, q and u when $t = 0\cdot3$ are as follows.

$x = 0$	0·1	0·2	0·3	0·4	0·5
p	0·2	0·2	0·2	0·2	0·1	0
q	0	−0·1	−0·2	−0·3	−0·3	−0·3
u	0	0·02	0·04	0·06	0·075	0·08

21. The equation $(\partial^2 U/\partial t^2) = (\partial^2 U/\partial x^2)$ is approximated at the point (ih, jk) by the implicit difference scheme

$$\frac{1}{k^2}\delta_t^2 u_{i,j} = \frac{1}{h^2}(\tfrac{1}{4}\delta_x^2 u_{i,j+1} + \tfrac{1}{2}\delta_x^2 u_{i,j} + \tfrac{1}{4}\delta_x^2 u_{i,j-1}).$$

(a) Use the von Neumann method of analysis to prove that the equations are unconditionally stable.

(b) Use the matrix method to establish unconditional stability given that the boundary values are known at $x = 0$ and 1, $0 \leqslant x \leqslant 1$ and that $Nh = 1$.

(c) Prove that the principal part of the local truncation error at the point (ih, jk) is $-\tfrac{1}{12}h^2k^2(1 + 2r^2)(\partial^4 U/\partial x^4)_{i,j}$, where $r = k/h$.

Solution

(a) Equations are

$$-\tfrac{1}{4}r^2 u_{i-1,j+1} + (1 + \tfrac{1}{2}r^2)u_{i,j+1} - \tfrac{1}{4}r^2 u_{i+1,j+1}$$
$$= \tfrac{1}{2}r^2 u_{i-1,j} + (2 - r^2)u_{i,j} + \tfrac{1}{2}r^2 u_{i+1,j} + \tfrac{1}{4}r^2 u_{i-1,j-1}$$
$$+ (-1 - \tfrac{1}{2}r^2)u_{i,j-1} + \tfrac{1}{4}r^2 u_{i+1,j-1}.$$

The error function $E_{i,j} = e^{\sqrt{(-1)}\beta x}\,\xi^j$ is a solution if

$$(1 + r^2 \sin^2 \tfrac{1}{2}\beta h)\xi^2 - (2 - 2r^2 \sin^2 \tfrac{1}{2}\beta h)\xi + (1 + r^2 \sin^2 \tfrac{1}{2}\beta h) = 0.$$

It is easily shown that $|\xi| = 1$.

(b) In matrix form the equations are $\mathbf{u}_{j+1} = \mathbf{A}^{-1}\mathbf{B}\mathbf{u}_j + \mathbf{A}^{-1}\mathbf{C}\mathbf{u}_{j-1} + \mathbf{A}^{-1}\mathbf{b}_j$, where $\mathbf{b}_j$ is a column vector of known constants,

$$\mathbf{A} = (1 + \tfrac{1}{2}r^2)\mathbf{I} - \tfrac{1}{4}r^2\mathbf{E},$$
$$\mathbf{B} = (2 - r^2)\mathbf{I} + \tfrac{1}{2}r^2\mathbf{E},$$
$$\mathbf{C} = (-1 - \tfrac{1}{2}r^2)\mathbf{I} + \tfrac{1}{4}r^2\mathbf{E}$$

and the matrix $\mathbf{E}$ has 1's along each diagonal immediately above and below the main diagonal, zeros elsewhere. A perturbation $\mathbf{e}_0$ of the initial values will satisfy $\mathbf{e}_{j+1} = \mathbf{A}^{-1}\mathbf{B}\mathbf{e}_j + \mathbf{A}^{-1}\mathbf{C}\mathbf{e}_{j-1}$. Hence

$$\begin{bmatrix} \mathbf{e}_{j+1} \\ \mathbf{e}_j \end{bmatrix} = \begin{bmatrix} \mathbf{A}^{-1}\mathbf{B} & \mathbf{A}^{-1}\mathbf{C} \\ \mathbf{I} & \mathbf{O} \end{bmatrix} \begin{bmatrix} \mathbf{e}_j \\ \mathbf{e}_{j-1} \end{bmatrix},$$

i.e., $\mathbf{v}_{j+1} = \mathbf{P}\mathbf{v}_j$. The matrices $\mathbf{A}$, $\mathbf{B}$ and $\mathbf{C}$ have the same system of linearly independent eigenvectors as $\mathbf{E}$. So have $\mathbf{A}^{-1}\mathbf{B}$ and $\mathbf{A}^{-1}\mathbf{C}$. Therefore the eigenvalues λ of $\mathbf{P}$ are given by

$$\det \begin{bmatrix} \alpha_k^{-1}\beta_k - \lambda & \alpha_k^{-1}\gamma_k \\ 1 & -\lambda \end{bmatrix} = 0,$$

$k = 1(1)(N-1)$, where

$$\alpha_k = 1 + r^2 \sin \frac{k\pi}{2N}, \quad \beta_k = 2 - 2r^2 \sin^2 \frac{k\pi}{2N}, \text{ and } \gamma_k = -1 - r^2 \sin^2 \frac{k\pi}{2N}$$

are the eigenvalues of $\mathbf{A}$, $\mathbf{B}$ and $\mathbf{C}$ respectively. It is easily shown that $|\lambda| = 1$.

(c) Expand

$$T_{i,j} = \frac{1}{k^2}\,\delta_t^2 U_{i,j} - \frac{1}{h^2}\,(\tfrac{1}{4}\delta_x^2 U_{i,j+1} + \tfrac{1}{2}\delta_x^2 U_{i,j} + \tfrac{1}{4}\delta_x^2 U_{i,j-1})$$

about the point (i, j).
(Reference Chapter 3, exercise 13 (a).)

22. (a). Verify that the general solution of the equation

$$\partial^2 u/\partial x^2 = \partial^2 u/\partial t^2,$$

namely

$$u = \phi(x + t) + \psi(x - t),$$

is the exact solution of the explicit finite-difference scheme (4.46) for $\delta x = \delta t$.

(b) The solution of the equation $\partial^2 U/\partial x^2 = \partial^2 U/\partial t^2$ satisfies the initial conditions $U = f(x)$ and $\partial U/\partial t = g(x)$ on $t = 0$. When the equation is approximated by the explicit finite-difference scheme (4.46) with $r = 1$, namely,

$$u_{i,j+1} = u_{i+1,j} + u_{i-1,j} - u_{i,j-1},$$

and the derivative condition is approximated by a forward-difference, prove that;

(i) $|e_{i,1}| \leqslant \frac{1}{2} h^2 M_2,$

where $e = U - u$, $\delta t = \delta x = h$ and M_2 is the modulus of the largest value of $\partial^2 U/\partial t^2$ in the first time-interval;

(ii) $e_{i,j+1} = e_{i+1,j} + e_{i-1,j} - e_{i,j-1} + \frac{1}{6} h^4 \eta M_4.$

where M_4 is the modulus of the largest of $\partial^4 U/\partial t^4$ and $\partial^4 U/\partial x^4$ throughout the solution domain and $|\eta| \leqslant 1$.

Hence prove that

$$|e_{i,j}| \leqslant \frac{1}{2} j h^2 M_2 + \frac{1}{12} j(j-1) h^4 M_4, \quad 1 < j < K,$$

and deduce that u converges to U as h tends to zero, where Kh is finite.

Solution

$g_i = (u_{i,1} - u_{i,0})/h$ gives $u_{i,1} = f_i + h g_i$, assuming no initial errors. By Taylor's expansion,

$$U_{i,1} = U_{i,0} + h \partial U_{i,0}/\partial t + \frac{1}{2} h^2 \partial^2 U_{i,\theta}/\partial t^2, \quad (0 < \theta < 1),$$
$$= f_i + h g_i + \frac{1}{2} h^2 \partial^2 U_{i,\theta}/\partial t^2.$$

Hence

$$|e_{i,1}| = |U_{i,1} - u_{i,1}| \leqslant \frac{1}{2} h^2 M_2.$$

Substitution of $u_{i,j} = U_{i,j} - e_{i,j}$ into the finite-difference equation and expansion in terms of $U_{i,j}$ by Taylor's theorem gives

$$e_{i,j+1} = e_{i+1,j} + e_{i-1,j} - e_{i,j-1} + \frac{1}{24}h^4 \left(\frac{\partial^4 U_{i,j+\theta_1}}{\partial t^4} \right.$$

$$\left. + \frac{\partial^4 U_{i,j+\theta_2}}{\partial t^4} - \frac{\partial^4 U_{i+\theta_3,j}}{\partial x^4} - \frac{\partial^4 U_{i+\theta_4,j}}{\partial x^4} \right),$$

where $|\theta_s| < 1$, $(s = 1, 2, 3, 4)$. Hence

$$e_{i,j+1} = e_{i+1,j} + e_{i-1,j} - e_{i,j-1} + \frac{1}{6}h^4 \eta M_4.$$

Draw the straight line characteristics through the point $(i, j+1)$ at $\pm 45°$ to Ox until they meet the line $j = 1$, and mark the points within this triangle contributing terms to $e_{i,j+1}$ when working backwards using the last equation so as to express $e_{i,j+1}$ entirely in terms of errors along $j = 1$. It will be seen there is one point at $(i, j+1)$, two points along $t = jh$, three along $t = (j-1)h$, etc., and $j+1$ along $t = h$. As the $(j+1)$ points along $t = h$ each contribute an error $\leq \frac{1}{2}h^2 M_2$, and the $1 + 2 + \ldots + j = \frac{1}{2}j(j+1)$ points between $t = 2h$ and $(j+1)h$ each contribute an error $\leq \frac{1}{6}h^4 M_4$, it follows that

$$|e_{i,j+1}| \leq \frac{1}{2}(j+1)h^2 M_2 + \frac{1}{12}j(j+1)h^4 M_4.$$

Changing j into $(j-1)$ completes the proof. Since $jh = t$,

$$|e_{i,j}| \leq \frac{1}{2}thM_2 + \frac{1}{12}t^2 h^2 M_4.$$

As h tends to zero this error tends to zero for finite values of t.

Elliptic equations and systematic iterative methods

troduction

ᴌLIPTIC partial differential equations arise usually from equilibrium
steady-state problems and their solutions, in relation to the
lculus of variations, frequently maximize or minimize an integral
presenting the energy of the system. The best known elliptic
uations are Poisson's equation, $\partial^2\phi/\partial x^2 + \partial^2\phi/\partial y^2 = f(x, y)$, often
itten as $\nabla^2\phi = f(x, y)$, and Laplace's equation

$$\partial^2\phi/\partial x^2 + \partial^2\phi/\partial y^2 = \nabla^2\phi = 0.$$

isson's equation, e.g., summarizes the St. Venant theory of tor-
n, the slow motion of incompressible viscous fluid, and the
verse-square law theories of electricity, magnetism and gravitating
atter at points where the charge density, pole strength or mass
nsity respectively, are non-zero. Laplace's equation arises in the
eories associated with the steady flow of heat or electricity in
mogeneous conductors, with the irrotational flow of incompressi-
e fluid, and with potential problems in electricity, magnetism and
avitating matter at points devoid of these entities.
The domain of integration of a two-dimensional elliptic equation
always an area S bounded by a closed curve C. The boundary
ndition usually specifies either the value of the function or the
lue of its normal derivative at every point on C, or a mixture of
th. Unlike hyperbolic equations both conditions cannot be given
bitrarily at any one point. Green's theorem provides an easy proof
this for Poisson's equation by showing that a specified function
lue on C determines a single-valued and differentiable solution
iquely throughout S. From this it follows that the normal deriva-
e at every point on C is uniquely determined. Let ϕ satisfy
$\phi/\partial x^2 + \partial^2\phi/\partial y^2 = f(x, y)$ at every point of S and be specified at
ery point on C. Assume there can be two different solutions ϕ_1
d ϕ_2 inside S. Put $u = \phi_1 - \phi_2$. Then $u = 0$ on C and

$$\partial^2 u/\partial x^2 + \partial^2 u/\partial y^2 = \nabla^2\phi_1 - \nabla^2\phi_2 = f(x, y) - f(x, y) = 0$$

at every point of S. Green's theorem in two dimensions is

$$\int\limits_S \left\{ \left(\frac{\partial u}{\partial x}\right)^2 + \left(\frac{\partial u}{\partial y}\right)^2 \right\} dS = \int\limits_C u\frac{\partial u}{\partial n}\, ds - \int\limits_S u\left(\frac{\partial^2 u}{\partial x^2} + \frac{\partial^2 u}{\partial y^2}\right) dS,$$

where ds is an element of the boundary curve C and dS an eleme
of S. The restrictions on u make the right side zero so $\partial u/\partial x$
$\partial u/\partial y = 0$ because the left side is the sum of squares. Hence u
constant throughout S. Since it is zero on C it must be ze
throughout S, so $\phi_1 = \phi_2$ proving that the solution is unique. As t
solution is known to be continuous the result follows. When $\partial\phi/\partial n$
specified at every point on C the solution is no longer unique. If ϕ
a solution then so is $\phi + A$ where A is any constant. These resu
are valid even when the boundary values are discontinuous, fro
which it follows that discontinuities in boundary values are n
propagated into the solution.

For Laplace's equation it can also be proved that the solutic
throughout S is bounded by the extreme values of ϕ on C and h
no maxima or minima.

The following examples are typical elliptic problems but ha
been worked only to illustrate the use of finite-difference methods.

Example 5.1. *The torsion problem*

The shear stresses and displacements at points within a lor
uniform cylinder in torsion can be calculated from a stress functic
Φ that is constant round the perimeter C of a right cross-section t
the XOY plane, and which satisfies the equation

$$\frac{\partial^2 \Phi}{\partial X^2} + \frac{\partial^2 \Phi}{\partial Y^2} + 2 = 0$$

at every point $P(X, Y)$ of the cross-section.

The problem can be written in non-dimensional form by putting

$$x = X/L, \; y = Y/L \text{ and } \phi = \Phi/L^2,$$

where L is some representative length.

Consider a thick cylindrical tube whose cross-section is symmetr
cal with respect to OX, OY, Fig. 5.1, and such that one quarter of
is the area between the boundaries YFX and $GHKM$. Let
dimensions, in units of L, be as shown; let $\phi = 0$ on YFX and $\phi =$

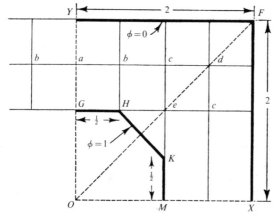

Fig. 5.1

on *GHKM*. Take a square mesh of side $h = \frac{1}{2}$ and denote the different approximate values of ϕ at the mesh points by a, b, c, d, e, taking into account the symmetry of the problem. The simplest finite-difference approximation to the non-dimensional equation

$$\frac{\partial^2 \phi}{\partial x^2} \times \frac{\partial^2 \phi}{\partial y^2} + 2 = 0,$$

is the five-point equation

$$\frac{1}{h^2}(\phi_{i+1,j} - 2\phi_{i,j} + \phi_{i-1,j} + \phi_{i,j+1} - 2\phi_{i,j} + \phi_{i,j-1}) + 2 = 0,$$

where $x = ih$, $y = jh$. With $h = \frac{1}{2}$ we get

$$2(\phi_{i+1,j} + \phi_{i,j+1} + \phi_{i-1,j} + \phi_{i,j-1} - 4\phi_{i,j}) + 1 = 0. \tag{5.1}$$

Application of equation (5.1) to each mesh point leads to

$$2(2b + 1 - 4a) + 1 \quad = 0,$$
$$2(c + a + 1 - 4b) + 1 = 0,$$
$$2(d + b + e - 4c) + 1 = 0,$$
$$2(2c - 4d) + 1 \quad = 0,$$
$$2(2c + 2 - 4e) + 1 \quad = 0.$$

The solution of these five equations for the five pivotal values of ϕ is easily shown to be

$$a = 0{\cdot}737,\; b = 0{\cdot}724,\; c = 0{\cdot}658,\; d = 0{\cdot}454,\; e = 0{\cdot}954.$$

As a solution of the differential equation this is not very accurate because the mesh size is large. In particular the value of e contains a large error because the interior angles at H and K exceed $180°$.

Example 5.2. *Derivative boundary conditions in a heat-conduction problem*

The current in electrical windings gives rise to an internal generation of heat that is dissipated by radiation from the boundaries. The steady temperature distribution in a winding with a rectangular cross-section may therefore be approximated by the solution of the equation for steady heat flow across a rectangle, with internal generation of heat and radiation from its perimeter into the surrounding medium. As the thermal conductivity K parallel to Ox differs from its value λK parallel to Oy the equation for the temperature u is

$$\frac{\partial u^2}{\partial x^2} + \lambda \frac{\partial^2 u}{\partial y^2} = -\frac{q}{K},$$

where q is the number of units of heat generated per unit area per second. This equation can be made non-dimensional as in Chapter 2.

Purely for illustrative purposes let u be the solution of the equation

$$\frac{\partial^2 u}{\partial x^2} + 3 \frac{\partial^2 u}{\partial y^2} = -16 \tag{5.2}$$

inside the square $x = \pm 1$, $y = \pm 1$, Fig. 5.2, satisfy the boundary conditions

$$u = 0 \quad \text{on} \quad x = 1, \quad \frac{\partial u}{\partial y} = -u \quad \text{on} \quad y = 1,$$

and be symmetric with respect to Ox and Oy, i.e.,

$$\frac{\partial u}{\partial x} = 0 \quad \text{on} \quad Oy, \quad \frac{\partial u}{\partial y} = 0 \quad \text{on} \quad Ox.$$

Take a square mesh of side $h = \frac{1}{4}$ and label the mesh points as indicated, equal numbers denoting equal values of u. Denote the mirror image of the point 6 (say) in $y = 1$ by -6, and the value of u

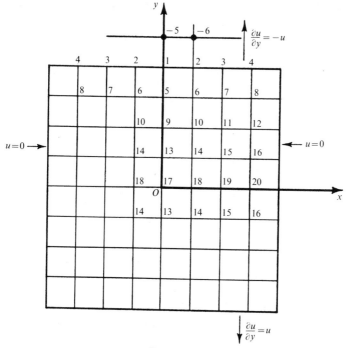

Fig. 5.2

t point 6 by u_6, etc. One finite-difference representation of equation (5.2) is

$$\frac{1}{h^2}(u_{i+1,j} - 2u_{i,j} + u_{i-1,j}) + \frac{3}{h^2}(u_{i,j+1} - 2u_{i,j} + u_{i,j-1}) = -16.$$

When $h = \frac{1}{4}$,

$$u_{i+1,j} + 3u_{i,j+1} + u_{i-1,j} + 3u_{i,j-1} - 8u_{i,j} = -1,$$

nd this equation can be represented very conveniently by the attern of numbers

$$\begin{bmatrix} & 3 & \\ 1 & -8 & 1 \\ & 3 & \end{bmatrix} u = -1. \tag{5.3}$$

t the boundary point 1, on imagining the heat-conducting area to be extended to the first row of external mesh points,

$$u_2 + 3u_{-5} + u_2 + 3u_5 - 8u_1 = -1,$$

and from the boundary condition, using central differences,

$$(u_{-5} - u_5)/2h = 2(u_{-5} - u_5) = -u_1.$$

Elimination of u_{-5} gives the equation

$$-9\tfrac{1}{2}u_1 + 2u_2 + 6u_5 = -1.$$

Similarly, points 2, 3 and 4 yield the equations

$$u_1 - 9\tfrac{1}{2}u_2 + u_3 + 6u_6 = -1,$$

$$u_2 - 9\tfrac{1}{2}u_3 + u_4 + 6u_7 = -1,$$

$$u_3 - 9\tfrac{1}{2}u_4 + 6u_8 = -1.$$

Point 5 gives the equation

$$3u_1 - 8u_5 + 2u_6 + 3u_9 = -1,$$

and point 6 the equation

$$3u_2 + u_5 - 8u_6 + u_7 + 3u_{10} = -1.$$

Similar equations can easily be written down for points 7 to 20, giving twenty linear algebraic equations for twenty unknowns. Their solution is not a trivial computation and would normally be carried out on a digital computer. On writing out the equations in detail it will be found that their matrix form is

$$\mathbf{Au} = \mathbf{b},$$

where $\mathbf{u}$ is a column vector with components $u_1, u_2, \ldots, u_{20}$, $\mathbf{b}$ a column vector with each component -1, and $\mathbf{A}$ a matrix which can be expressed in partitioned form as

$$\begin{bmatrix} (\mathbf{B} - 1\tfrac{1}{2}\mathbf{I}) & 6\mathbf{I} & & & \\ 3\mathbf{I} & \mathbf{B} & 3\mathbf{I} & & \\ & 3\mathbf{I} & \mathbf{B} & 3\mathbf{I} & \\ & & 3\mathbf{I} & \mathbf{B} & 3\mathbf{I} \\ & & & 6\mathbf{I} & \mathbf{B} \end{bmatrix},$$

where

$$\mathbf{B} = \begin{bmatrix} -8 & 2 & & \\ 1 & -8 & 1 & \\ & 1 & -8 & 1 \\ & & 1 & -8 \end{bmatrix}.$$

Their solution is

$$u_1 = 3\cdot067, \quad u_2 = 2\cdot909, \quad u_3 = 2\cdot411, \quad u_4 = 1\cdot496,$$
$$u_5 = 3\cdot720, \quad u_6 = 3\cdot527, \quad u_7 = 2\cdot917, \quad u_8 = 1\cdot801,$$
$$u_9 = 4\cdot169, \quad u_{10} = 3\cdot949, \quad u_{11} = 3\cdot258, \quad u_{12} = 2\cdot000,$$
$$u_{13} = 4\cdot431, \quad u_{14} = 4\cdot195, \quad u_{15} = 3\cdot455, \quad u_{16} = 2\cdot113,$$
$$u_{17} = 4\cdot518, \quad u_{18} = 4\cdot276, \quad u_{19} = 3\cdot520, \quad u_{20} = 2\cdot150.$$

Finite-differences in polar co-ordinates

Finite-difference problems involving circular boundaries can usually be solved more conveniently in polar co-ordinates than Cartesian co-ordinates because they avoid the use of awkward differentiation formulae near the curved boundary.

As an example consider Laplace's equation,

$$\frac{\partial^2 u}{\partial r^2} + \frac{1}{r}\frac{\partial u}{\partial r} + \frac{1}{r^2}\frac{\partial^2 u}{\partial \theta^2} = 0.$$

Define the mesh points in the $r - \theta$ plane by the points of intersection of the circles $r = ih,\ (i = 1, 2, \ldots)$, and the straight lines $= j\,\delta\theta,\ (j = 0, 1, 2, \ldots)$, Fig. 5.3. Laplace's equation at the point

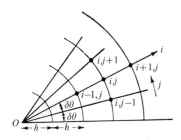

Fig. 5.3

(i, j) may then be approximated by

$$\frac{(u_{i+1,j} - 2u_{i,j} + u_{i-1,j})}{h^2} + \frac{1}{ih} \frac{(u_{i+1,j} - u_{i-1,j})}{2h}$$

$$+ \frac{1}{(ih)^2} \frac{(u_{i,j+1} - 2u_{i,j} + u_{i,j-1})}{(\delta\theta)^2} = 0,$$

giving

$$\left(1 - \frac{1}{2i}\right)u_{i-1,j} + \left(1 + \frac{1}{2i}\right)u_{i+1,j} - 2\left\{1 + \frac{1}{(i\,\delta\theta)^2}\right\}u_{i,j}$$

$$+ \frac{1}{(i\,\delta\theta)^2} u_{i,j-1} + \frac{1}{(i\,\delta\theta)^2} u_{i,j+1} = 0.$$

If these equations are written out in detail for $i = 1, 2, \ldots, n$ and $j = 1, 2, \ldots, m$, and the boundary values are assumed to be known for $i = 0$, $i = (n+1)$, $j = 0$ and $j = (m+1)$, it will be found that their matrix form is

$$\mathbf{Au} = \mathbf{b}$$

where $\mathbf{b}$ is a column vector determined by the boundary values, $\mathbf{u}$ a column vector whose transpose is

$$(u_{1,1}, u_{1,2}, \ldots, u_{1,m}, u_{2,1}, \ldots, u_{2,m}, \ldots, u_{n,1}, u_{n,2}, \ldots, u_{n,m})$$

and $\mathbf{A}$ a matrix which can be written in partitioned form as

$$\begin{bmatrix} \mathbf{B}_1 & (1+\frac{1}{2})\mathbf{I} & & & & \\ (1-\frac{1}{4})\mathbf{I} & \mathbf{B}_2 & (1+\frac{1}{4})\mathbf{I} & & & \\ & (1-\frac{1}{6})\mathbf{I} & \mathbf{B}_3 & & (1+\frac{1}{6})\mathbf{I} & \\ & & & \cdot & & \\ & & & \left(1-\frac{1}{2(n-1)}\right)\mathbf{I} & \mathbf{B}_{n-1} & \left(1+\frac{1}{2(n-1)}\right)\mathbf{I} \\ & & & & \left(1-\frac{1}{2n}\right)\mathbf{I} & \mathbf{B}_n \end{bmatrix}$$

where each **B** and **I** are $m \times m$ matrices and

$$
\mathbf{B}_p =
\begin{bmatrix}
-2\left\{1+\dfrac{1}{(p\,\delta\theta)^2}\right\} & \dfrac{1}{(p\,\delta\theta)^2} & & & & \\[2ex]
\dfrac{1}{(p\,\delta\theta)^2} & -2\left\{1+\dfrac{1}{(p\,\delta\theta)^2}\right\} & \dfrac{1}{(p\,\delta\theta)^2} & & & \\[2ex]
& \dfrac{1}{(p\,\delta\theta)^2} & -2\left\{1+\dfrac{1}{(p\,\delta\theta)^2}\right\} & \dfrac{1}{(p\,\delta\theta)^2} & & \\[2ex]
& & & \ddots & & \\[1ex]
& & & \dfrac{1}{(p\,\delta\theta)^2} & -2\left\{1+\dfrac{1}{(p\,\delta\theta)^2}\right\} & \dfrac{1}{(p\,\delta\theta)^2} \\[2ex]
& & & & \dfrac{1}{(p\,\delta\theta)^2} & -2\left\{1+\dfrac{1}{(p\,\delta\theta)^2}\right\}
\end{bmatrix}
$$

Formulae for derivatives near a curved boundary when using a square mesh

When the boundary is curved and intersects the rectangular mesh at points that are not mesh points the formulae previously used to approximate first- and second-order derivatives at points near the boundary can no longer be used. This section is concerned with the finite-difference approximations to the derivatives at a point such as O, close to the boundary curve C on which the values of u are assumed to be known, Fig. 5.4. Let the mesh be a square of side

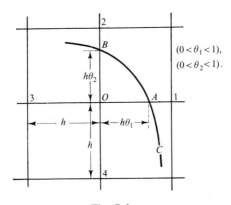

$(0 < \theta_1 < 1),$
$(0 < \theta_2 < 1).$

Fig. 5.4

h. By Taylor's theorem

$$u_A = u_0 + h\theta_1 \frac{\partial u_0}{\partial x} + \tfrac{1}{2}h^2\theta_1^2 \frac{\partial^2 u_0}{\partial x^2} + O(h^3),$$

$$u_3 = u_0 - h\frac{\partial u_0}{\partial x} + \tfrac{1}{2}h^2 \frac{\partial^2 u_0}{\partial x^2} + O(h^3).$$

Elimination of $\partial^2 u_0/\partial x^2$ gives

$$\frac{\partial u_o}{\partial x} = \frac{1}{h}\left\{\frac{1}{\theta_1(1+\theta_1)} u_A - \frac{(1-\theta_1)}{\theta_1} u_0 - \frac{\theta_1}{(1+\theta_1)} u_3\right\},$$

with a leading error of order h^2.

Similarly the elimination of $\partial u_0/\partial x$ leads to

$$\frac{\partial^2 u_0}{\partial x^2} = \frac{1}{h^2}\left\{\frac{2}{\theta_1(1+\theta_1)} u_A + \frac{2}{(1+\theta_1)} u_3 - \frac{2}{\theta_1} u_0\right\},$$

with a leading error of order h, which is of lower accuracy than the usual formula employing only pivotal values.

Hence Poisson's equation $\partial^2 u/\partial x^2 + \partial^2 u/\partial y^2 = f(x, y)$ at the point 0 is approximated by

$$\frac{2u_A}{\theta_1(1+\theta_1)} + \frac{2u_B}{\theta_2(1+\theta_2)} + \frac{2u_3}{(1+\theta_1)} + \frac{2u_4}{(1+\theta_2)} - 2\left(\frac{1}{\theta_1}+\frac{1}{\theta_2}\right)u_0 = h^2 f_0.$$

Finite-difference formulae approximating normal derivatives at points on the boundary *C*, in terms of internal pivotal values, are extremely awkward and can be found in reference 12.

Improvement of the accuracy of solutions

Finer mesh

Although no useful general results concerning the magnitude of the discretization error as a function of the mesh lengths have yet been established, it seems reasonable to assume that this error will usually decrease as the mesh lengths are reduced. One hopes therefore that the sequence of solutions obtained using finer and finer meshes will eventually give a solution that differs from its immediate predecessor by less than some assigned tolerance. With this approach however, the size of the matrix of coefficients increases rapidly and becomes, after a number of refinements, too large for storage in the immediate access store of a computer.

Deferred approach to the limit

This method, suggested by Richardson, is extremely useful when there is a reliable estimate of the discretization error as a function of the mesh length. It is of dubious value however near curved boundaries, near corners with interior angles exceeding 180°, and near boundaries on which specified function values are not smooth.

Let u be the exact solution of the differential equation and u_1, u_2 the approximate solutions at the same mesh-point for mesh lengths h_1 and h_2 respectively. When the leading term in the discretization error is proportional to h^p,

$$u - u_1 = Ah_1^p \quad \text{and} \quad u - u_2 = Ah_2^p.$$

Hence

$$u = \frac{h_2^p u_1 - h_1^p u_2}{h_2^p - h_1^p}.$$

For the five-point formula for Laplace's equation, namely

$$u_{i+1,j} + u_{i-1,j} + u_{i,j+1} + u_{i,j-1} - 4u_{i,j} = 0,$$

the discretization error for a rectangular region with smooth known boundary values is proportional to h^2. In this case, for $h_1 = 2h_2$,

$$u = u_2 + \tfrac{1}{3}(u_2 - u_1).$$

When the value of p is not known an estimate can be found from three approximate solutions at the same grid-point. Taking

$$h_3 = \tfrac{1}{2}h_2 = \tfrac{1}{4}h_1$$

gives

$$\frac{u_2 - u_1}{u_3 - u_2} = 2^p.$$

Deferred-correction method

In this method we calculate an initial approximate solution, difference it to obtain correction terms, add the correction terms to the initial finite-difference equations, then solve the amended equations for a more accurate solution.

As an example consider Laplace's equation. To fourth-order central-differences

$$h^2 \frac{\partial^2 u}{\partial x^2} = \delta_x^2 u - \tfrac{1}{12}\delta_x^4 u,$$

and

$$h^2 \frac{\partial^2 u}{\partial y^2} = \delta_y^2 u - \tfrac{1}{12}\,\delta_y^4 u,$$

where $\delta_x^2 u$ represents the second-order differences obtained by differencing parallel to Ox, i.e.,

$$\delta_x^2 u_{i,j} = \delta_x u_{i+\frac{1}{2},j} - \delta_x u_{i-\frac{1}{2},j} = u_{i+1,j} - 2u_{i,j} + u_{i-1,j}.$$

Hence $h^2(\partial^2 u/\partial x^2 + \partial^2 u/\partial y^2) = 0$ can be approximated by

$$\delta_x^2 u + \delta_y^2 u = \tfrac{1}{12}(\delta_x^4 u + \delta_y^4 u). \tag{5.4}$$

The initial approximation is given by solving $\delta_x^2 u + \delta_y^2 u = 0$, which is, of course, the usual five-point formula

$$u_{i-1,j} + u_{i+1,j} + u_{i,j-1} + u_{i,j+1} - 4u_{i,j} = 0.$$

The numbers $\delta_x^4 u_{i,j}$, $\delta_y^4 u_{i,j}$ are then derived by differencing this solution parallel to Ox and Oy respectively, and used to produce the equations (5.4) for an improved solution.

More accurate finite-difference formulae

Accuracy can also be improved by representing the differential equation by a higher-order finite-difference approximation designed to reduce the truncation error. This increases the number of pivotal values in each difference equation, but for Laplace's and Poisson's equations the increase is not large for truncation errors of order six and four respectively. A number of such formulae have been devised by W. G. Bickley and A. Thom and are given in references 12 and 34. Their mode of derivation is indicated below by the establishment of a nine-point finite-difference approximation to Poisson's equation that has a truncation error of order h^4.

Take a square mesh of side h and label the relative mesh-point positions as in Fig. 5.5.

Denote the Laplacian $\partial^2 u/\partial x^2 + \partial^2 u/\partial y^2$ by $\nabla^2 u$, and let

$$\xi = h\frac{\partial}{\partial x}, \quad \eta = h\frac{\partial}{\partial y}, \quad \mathscr{D}^2 = \frac{\partial^2}{\partial x \partial y},$$

so that

$$\xi^2 + \eta^2 = h^2\nabla^2, \quad \xi\eta = h^2\mathscr{D}^2,$$
$$\xi^4 + \eta^4 = (\xi^2 + \eta^2)^2 - 2\xi^2\eta^2 = h^4(\nabla^4 - 2\mathscr{D}^4).$$

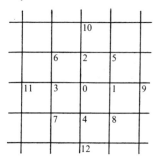

Fig. 5.5

As Taylor's series can be written as

$$u(x+h) = \left(1 + h\frac{d}{dx} + \dots + \frac{h^n}{n!}\frac{d^n}{dx^n} + \dots\right)u(x)$$
$$= (e^{h(d/dx)})u(x),$$

it follows that

$$u_1 = e^{\xi}u_0, \quad u_2 = e^{\eta}u_0, \quad u_3 = e^{-\xi}u_0, \quad u_5 = e^{\xi+\eta}u_0, \text{ etc.}$$

Because Poisson's equation is symmetric in its derivatives, define the following symmetric sums;

$$S_1 = u_1 + u_2 + u_3 + u_4, \quad S_2 = u_5 + u_6 + u_7 + u_8,$$

and

$$S_3 = u_9 + u_{10} + u_{11} + u_{12}.$$

On substituting for $u_1, u_2, \dots$ in terms of u_0 and expanding the exponential operators in powers of ξ and η, it can be shown that

$$S_1 = 4u_0 + h^2\nabla^2 u_0 + \tfrac{1}{12}h^4(\nabla^4 - 2\mathscr{D}^4)u_0 + \tfrac{1}{360}h^6(\nabla^6 - 3\mathscr{D}^2\nabla^2)u_0 + \dots,$$

$$S_2 = 4u_0 + 2h^2\nabla^2 u_0 + \tfrac{1}{6}h^4(\nabla^4 + 4\mathscr{D}^4)u_0 + \tfrac{1}{180}h^6(\nabla^6 + 12\mathscr{D}^4\nabla^2)u_0 + \dots,$$

$$S_3 = 4u_0 + 4h^2\nabla^2 u_0 + \tfrac{4}{3}h^4(\nabla^4 - 2\mathscr{D}^4)u_0 + \tfrac{8}{45}h^6(\nabla^6 - 3\mathscr{D}^4\nabla^2)u_0 + \dots \quad (5.5)$$

Elimination of $\mathscr{D}^4 u_0$ between S_1 and S_2 gives

$$\nabla^2 u_0 = \frac{4S_1 + S_2 - 20u_0}{6h^2} - \tfrac{1}{12}h^2\nabla^4 u_0 + O(h^4).$$

For Poisson's equation the second term on the right is known because $\nabla^2 u = f$, so $\nabla^4 u = \nabla^2 f$. For Laplace's equation the coefficients of h^2 and h^4 both vanish. (See exercise 8.) Hence the nine-point formula $4S_1 + S_2 - 20u_0 = 0$ is a more accurate finite-difference representation of Laplace's equation than $S_1 - 4u_0 = 0$ because its truncation error is of order h^6. (See exercise 9.) The error in $S_1 - 4u_0 = 0$ is of order h^2. The most convenient way of exhibiting this nine-point formula approximating Poisson's equation $\nabla^2 u = f$, is as below.

$$\begin{bmatrix} 1 & 4 & 1 \\ 4 & -20 & 4 \\ 1 & 4 & 1 \end{bmatrix} u = 6h^2 f + \tfrac{1}{2} h^4 \nabla^2 f.$$

Analysis of the discretization error of the five–point difference approximation to Poisson's equation over a rectangle

The following analysis is based on that given in reference 33.

Consider the Dirichlet problem

$$\frac{\partial^2 U}{\partial x^2} + \frac{\partial^2 U}{\partial y^2} = f(x, y), \quad (x, y) \in D, \tag{5.6}$$

$$U(x, y) = g(x, y), \quad (x, y) \in C, \tag{5.7}$$

where $g(x, y)$ is known on the boundary curve C of the rectangular solution domain D defined by $0 < x < a$, $0 < y < b$. (Termed a Dirichlet problem because U is known on C.) Define the mesh points as the points of intersection of the straight lines $x_i = ih$, $i = 0(1)m$ and $y_j = jh$, $j = 0(1)n$, where $mh = a$ and $nh = b$. Denote the set of mesh points in D by D_h and those on C by C_h and let G_h denote the set of all these mesh points, called the union of D_h and C_h and written as $G_h = D_h \cup C_h$, i.e., G_h is the set of all mesh points in $G = D \cup C$.

The five-point difference approximation to equation (5.6) at the point (i, j) is

$$\frac{1}{h^2}(u_{i+1,j} + u_{i-1,j} + u_{i,j+1} + u_{i,j-1} - 4u_{i,j}) = f_{i,j},$$

so the difference equations approximating the problem may be written as

$$Lu_{i,j} = f_{i,j}, \quad (i, j) \in D_h, \tag{5.8}$$

and

$$u_{i,j} = g_{i,j}, \quad (i, j) \in C_h, \tag{5.9}$$

where the five-point difference operator L is defined by

$$Lu_{i,j} = \frac{1}{h^2}(u_{i+1,j} + u_{i-1,j} + u_{i,j+1} + u_{i,j-1} - 4u_{i,j}), \quad (i,j) \in D_h.$$

Our problem is to express, if possible, the discretization error $e_{i,j} = U_{i,j} - u_{i,j}$ at the (i,j)th mesh point of D_h in terms of h. Operating on this error with L leads to

$$\begin{aligned}
Le_{i,j} &= LU_{i,j} - Lu_{i,j}, \quad (i,j) \in D_h, \\
&= LU_{i,j} - f_{i,j}, \text{ by equation (5.8)}, \\
&= T_{i,j}, \quad\quad\quad\quad\quad\quad\quad\quad\quad\quad\quad (5.9a)
\end{aligned}$$

where $T_{i,j}$ is the local truncation error of the difference approximation $Lu - f = 0$ at the point (i,j).

In exercise 6 it is proved that

$$T_{i,j} = \tfrac{1}{12}h^2\left\{\left(\frac{\partial^4 U}{\partial x^4}\right)_{\xi,y_j} + \left(\frac{\partial^4 U}{\partial y^4}\right)_{x_i,\eta}\right\},$$

where $x_i - h < \xi < x_i + h$, $y_j - h < \eta < y_j + h$, and it is assumed that U has continuous derivatives of all orders up to and including the fourth in $G = D \cup C$. If

$$M_4 = \max\left\{\max_G\left|\frac{\partial^4 U}{\partial x^4}\right|, \max_G\left|\frac{\partial^4 U}{\partial y^4}\right|\right\},$$

then

$$\max_{D_h}|T_{i,j}| \leq \tfrac{1}{6}h^2 M_4$$

and hence, by (5.9a),

$$\max_{D_h}|Le_{i,j}| = \max_{D_h}|T_{i,j}| \leq \tfrac{1}{6}h^2 M_4. \quad\quad (5.10)$$

Clearly, we need a result relating $e_{i,j}$ to $Le_{i,j}$. This is provided by a theorem, proved in an addendum to this section, which states that if v is any function defined on the set of mesh points G_h in the rectangular region $0 \leq x \leq a$, $0 \leq y \leq b$, then

$$\max_{D_h}|v| \leq \max_{C_h}|v| + \tfrac{1}{4}(a^2 + b^2)\max_{D_h}|Lv|.$$

Applying this theorem to the discretization error $e_{i,j}$ gives that

$$\max_{D_h}|e_{i,j}| \leqslant \max_{C_h}|e_{i,j}| + \tfrac{1}{4}(a^2+b^2)\max_{D_h}|Le_{i,j}|.$$

But $e_{i,j}=0$ on the boundary because $U_{i,j}=u_{i,j}=g_{i,j}$, $(i,j)\in C_h$, by equations (5.7) and (5.9). Hence, by equation (5.10), we can conclude that

$$\max_{D_h}|e_{i,j}| \leqslant \tfrac{1}{24}(a^2+b^2)h^2 M_4.$$

This is an extremely useful result because it proves that $u_{i,j}$ converges to $U_{i,j}$ as h tends to zero, in spite of the fact that M_4 is unknown for most problems. It also proves that the discretization error is proportional to h^2 so Richardson's 'deferred approach to the limit' method can be used effectively to improve the accuracy of the solution of the difference equations.

Elliptic problems with irregular boundaries

When the closed boundary curve C of a simply-connected open bounded domain D is irregular so that it intersects the sides of the square meshes as in Fig. 5.4, the analysis of the discretization error associated with either the five-point or the nine-point difference approximation to Poisson's equation with known boundary values is much more difficult than for a rectangle. The following result has, however, been obtained for the five-point difference replacement of Laplace's equation. If

$$\frac{\partial^2 U}{\partial x^2} + \frac{\partial^2 U}{\partial y^2} = 0, \quad (x,y)\in D,$$

and

$$U(x,y) = g(x,y), \quad (x,y)\in C,$$

where $g(x,y)$ is continuous and known on C and U has continuous fourth derivatives in $G = D\cup C$, then

$$\max|U(P)-u(P)| \leqslant \tfrac{1}{12}h^2 r^2 M_4 + h^2 M_2, \qquad (5.11)$$

where P is any internal mesh point of D, h is the side of the square mesh, r is the radius of a circle enclosing G, and M_k is defined by

$$M_k = \max\left\{\max_{G}\left|\frac{\partial^k U}{\partial x^k}\right|, \max_{G}\left|\frac{\partial^k U}{\partial y^k}\right|\right\}.$$

For this result to hold it is necessary to approximate Laplace's equation at a point such as O, Fig. 5.4, by the difference equation derived in that section, or to calculate u by interpolating linearly on u_3 and u_A or u_4 and u_B by

$$u_0 = \frac{\theta_1}{1+\theta_1} u_3 + \frac{1}{1+\theta_1} u_A \quad \text{or} \quad u_0 = \frac{\theta_2}{1+\theta_2} u_4 + \frac{1}{1+\theta_2} u_B.$$

Equation (5.11) shows that $u(P)$ converges to $U(P)$ as $h \to 0$ provided M_4 is bounded. This condition is not satisfied if, for example, the boundary contains corners with internal angles in excess of $180°$.

Addendum

If v is any function defined on the set of mesh points $G_h = D_h \cup C_h$ in the rectangular region $0 \le x \le a$, $0 \le y \le b$, then

$$\max_{D_h}|v| \le \max_{C_h}|v| + \tfrac{1}{4}(a^2 + b^2)\max_{D_h}|Lv|,$$

where

$$Lv_{i,j} = \frac{1}{h^2}(v_{i+1,j} + v_{i-1,j} + v_{i,j+1} + v_{i,j-1} - 4v_{i,j}).$$

Proof

Define the function $\phi_{i,j}$ by the equation

$$\phi_{i,j} = \tfrac{1}{4}(x_i^2 + y_j^2) = \tfrac{1}{4}(i^2 + j^2)h^2, \quad (i,j) \in G_h.$$

Clearly

$$0 \le \phi_{i,j} \le \tfrac{1}{4}(a^2 + b^2) \text{ for all } (i,j) \in G_h. \tag{5.12}$$

It also follows that at all points $(i,j) \in D_h$,

$$\phi_{i,j} = \tfrac{1}{4}\{(i+1)^2 + j^2 + (i-1)^2 + j^2 + i^2 + (j+1)^2 + i^2 + (j-1)^2 - 4i^2 - 4j^2\} = 1. \tag{5.13}$$

Now define the functions w^+ and w^- by

$$w^+ = v + N\phi \quad \text{and} \quad w^- = -v + N\phi, \tag{5.14}$$

where

$$N = \max_{D_h}|Lv_{i,j}|.$$

Operating on equations (5.14) with L and using (5.13) leads to

$$Lw_{i,j}^{\pm} = \pm Lv_{i,j} + N, \quad (i, j) \in D_h,$$

$$\geq 0 \text{ in virtue of the definition of } N.$$

In exercise 10 it is proved that if $Lw_{i,j} \geq 0$ for all $(i, j) \in D_h$, then

$$\max_{D_h} w_{i,j} \leq \max_{C_h} w_{i,j}.$$

Applying this result to w^+ and w^- gives that

$$\max_{D_h} w_{i,j}^{\pm} \leq \max_{C_h} w_{i,j}^{\pm} = \max_{C_h}(\pm v_{i,j} + N\phi_{i,j}) \text{ by (5.14).}$$

$$\leq \max_{C_h}(\pm v_{i,j}) + \tfrac{1}{4}(a^2 + b^2)N \text{ by (5.12).}$$

Since $w_{i,j}^{\pm} = \pm v_{i,j} + N\phi_{i,j}$ and $N\phi_{i,j} \geq 0$, it is seen that $\pm v_{i,j} \leq w_{i,j}^{\pm}$ for all $(i, j) \in G_h$. Hence

$$\max_{D_h}(\pm v_{i,j}) \leq \max_{C_h}(\pm v_{i,j}) + \tfrac{1}{4}(a^2 + b^2)N,$$

i.e.,

$$\max_{D_h}|v_{i,j}| \leq \max_{C_h}|v_{i,j}| + \tfrac{1}{4}(a^2 + b^2)\max_{D_h}|Lv_{i,j}|.$$

Comments on the solution of finite-difference equations approximating boundary-value problems

A finite-difference method for approximating a boundary value problem leads to a system of algebraic simultaneous equations. For linear boundary value problems these equations are always linear but their number is generally large and, for this reason, their solution is a major problem in itself. Methods of solution belong essentially to either the class of direct methods or the class of iterative methods.

Direct methods solve the system of equations in a known number of arithmetic operations, and errors in the solution arise entirely from rounding errors introduced during the computation. Basically, these direct methods are elimination methods of which the best known examples are the systematic Gaussian elimination method and the triangular decomposition method which factorizes the matrix $\mathbf{A}$ of the equations $\mathbf{Ax} = \mathbf{b}$ into $\mathbf{A} = \mathbf{LU}$, where $\mathbf{L}$ and $\mathbf{U}$ are lower and upper triangular matrices respectively. In the latter method, once the decomposition has been determined, the solution is calculated from $LUx = b$ by putting $Ux = y$ and then solving $Ly = b$ for y by forward substitution and $Ux = y$ for x by back-substitution. See exercise 19.) With both methods it is usually necessary to employ partial pivoting with scaling to control the growth of rounding errors. The stable non-pivoting elimination procedure given on page 23 for equations with tridiagonal matrices is exceptional. The ideas behind these methods are simple but the details associated with both their efficient computer implementation and the analysis of errors are considerable and can be found in references 36, 38, and 13.

It is of interest to note that when the arithmetic of these methods is carried out on a large modern computer the rounding errors introduced during the computation often have less effect on the solution than the rounding errors in the coefficients and constants of the equations. Say the coefficients and constants of n linear algebraic equations are all less than one in modulus and are known with certainty to only q decimal places, i.e., they contain errors $\leqslant \frac{1}{2}10^{-q}$. Then it has been proved, reference 37, that if the arithmetic of the Gaussian elimination method with partial pivoting is carried out with numbers having more than $(q + \log_{10} n)$ decimal places, i.e., with more than $\log_{10} n$ guarding figures, the solution obtained to q decimal places is the exact solution of a set of n linear equations whose coefficients and constants differ from those of the original equations by less than $\frac{1}{2}10^{-q}$. (In binary arithmetic, the arithmetic most frequently used on a computer, 10^{-q} is replaced by 2^{-q} and $\log_{10} n$ by $\log_2 n$.) In other words, the solution is as accurate as the data warrants. This does not imply that the number of correct significant figures in the solution is the same as in the data. It might be less. If it is, the loss of accuracy occurs *not* through the method of solution but because the equations are ill-conditioned in the sense that small changes in the coefficients produce large changes in the solution. The number of meaningful digits in the solution cannot, in fact, exceed the number of significant figures in the first q decimal places of the smallest pivot.

Although Gaussian elimination and **LU** decomposition are mathematically equivalent, the overall errors of the **LU** method can be made smaller than those of the Gaussian elimination method if inner products are accumulated in double length arithmetic, an operation provided for on some computers. Without double length accumulation the two methods are numerically equivalent as well as mathematically equivalent.

If the accuracy of a computed solution needs to be improved the following method is normally recommended. Denote the computed solution of $\mathbf{Ax} = \mathbf{b}$ by $\mathbf{x}^{(1)}$. Compute the residual $\mathbf{r}^{(1)} = \mathbf{b} - \mathbf{Ax}^{(1)}$ *using double length arithmetic*, i.e., to $2t$ decimal or binary digits instead of the standard t digits. Put $\mathbf{x} = \mathbf{x}^{(1)} + \mathbf{d}$, where $\mathbf{d}$ is the exact correction term to be added to the computed solution. As $\mathbf{A}(\mathbf{x}^{(1)} + \mathbf{d}) = \mathbf{b}$ it follows that $\mathbf{Ad} = \mathbf{b} - \mathbf{Ax}^{(1)} = \mathbf{r}^{(1)}$. The solution of this equation will not be $\mathbf{d}$ exactly but a computed approximation $\mathbf{d}^{(1)}$. Therefore an improved solution will be $\mathbf{x}^{(2)} = \mathbf{x}^{(1)} + \mathbf{d}^{(1)}$, and the procedure can be repeated. In order to keep the solution of the equations to single length arithmetic, i.e., t digits, the residuals could be multiplied by 10^t or 2^t and the corresponding correction terms $\mathbf{d}^{(1)}, \mathbf{d}^{(2)}, \ldots$, multiplied by 10^{-t} or 2^{-t}. (See references 36 and 13.) It should be noted that this method of iterative refinement requires only a small amount of extra arithmetic because the solution of $\mathbf{Ad} = \mathbf{r}^{(1)}$ involves only forward and backward substitutions since $\mathbf{A} = \mathbf{LU}$ has already been found.

If the matrix of coefficients can be entered into the fast store of the computer then direct methods, in general, are quicker and more accurate than iterative methods. (See page 227.) If also the matrix of coefficients has some special property or structure it is usually possible to increase the number of equations that can be solved by very efficient programming. In particular, Martin and Wilkinson, reference 38, have written such a programme for equations with symmetric positive definite matrices, and the NAG library of routines has a programme for banded matrices. Direct methods are certainly preferable to iterative methods when:

(i) Several sets of equations with the same coefficient matrix but different right hand sides have to be solved.

(ii) The matrix is nearly singular. In this case small residuals do not imply small errors in the solution. This is easily seen since $\mathbf{A}^{-1}\mathbf{r}^{(1)} = \mathbf{A}^{-1}\mathbf{b} - \mathbf{x}^{(1)} = \mathbf{x} - \mathbf{x}^{(1)}$. Therefore $(\mathbf{x} - \mathbf{x}^{(1)})$ could have large components when the components of the residual vector are small because some of the elements of $\mathbf{A}^{-1}$ will be large if $\mathbf{A}$ is nearly singular.

The economic use of direct methods has been extended by

Hockney, 1968, reference 17, and George, 1970, references 14 and 15. Stone, 1968, reference 32, has also formulated a very economical iterative method that is more allied to direct methods than to the classical iterative methods and each is dealt with briefly at the end of this book.

As seen in this and earlier chapters, the matrices associated with difference equations approximating partial differential equations are band matrices, i.e., matrices with non-zero elements lying between two sub-diagonals parallel to the main diagonal. They are also usually 'sparse', i.e., the number of zero elements in the matrix is much greater than the number of non-zero elements. For this type of matrix standard Gaussian elimination is inefficient in the sense that it 'fills-in' the zero elements within the band with non-zero numbers that have to be stored in the computer and used at subsequent stages of the elimination process.

With iterative methods however, no arithmetic is associated with zero coefficients so considerably fewer numbers have to be stored in the computer. As a consequence they can be used to solve systems of equations that are too large for the use of direct methods. Programming and data handling are also much simpler than for direct methods. Another advantage, not possessed by direct methods, is their frequent extension to the solution of sets of non-linear equations. The efficient use of iterative methods is very dependent however upon the direct calculation or estimation of the value or values of some acceleration parameter, and upon the coefficient matrix being well-conditioned, otherwise convergence will be slow and the volume of arithmetic enormous. With optimum acceleration parameters the volume of arithmetic of iterative methods for large sets of equations may actually be less than for direct methods.

In this book only the classical Jacobi, Gauss-Seidel, and SOR methods are considered. A far more comprehensive treatment of all present-day iterative methods is given in Young's book, reference 39.

Systematic iterative methods for solving large linear systems of algebraic equations

An iterative method for solving equations is one in which a first approximation is used to calculate a second approximation which in turn is used to calculate a third and so on. The iterative procedure is said to be convergent when the differences between the exact solution and the successive approximations tend to zero as the number of iterations increase. In general, the exact solution is never obtained in a finite number of steps, but this does not matter. What is important is that the successive iterates converge fairly rapidly to values that are correct to a specified accuracy.

As mentioned previously, one would consider using iterative methods when a direct method requires more computer fast storage space than is available and the matrix of coefficients is sparse but well-conditioned, a situation that often arises with the difference equations approximating elliptic boundary value problems.

For simplicity consider the four equations

$$a_{11}x_1 + a_{12}x_2 + a_{13}x_3 + a_{14}x_4 = b_1$$

$$a_{21}x_1 + a_{22}x_2 + a_{23}x_3 + a_{24}x_4 = b_2$$

$$a_{31}x_1 + a_{32}x_2 + a_{33}x_3 + a_{34}x_4 = b_3$$

$$a_{41}x_1 + a_{42}x_2 + a_{43}x_3 + a_{44}x_4 = b_4,$$

(5.15)

where $a_{ii} \neq 0$, $i = 1(1)4$. These may be written as

$$x_1 = \frac{1}{a_{11}}(b_1 - a_{12}x_2 - a_{13}x_3 - a_{14}x_4)$$

$$x_2 = \frac{1}{a_{22}}(b_2 - a_{21}x_1 - a_{23}x_3 - a_{24}x_4)$$

$$x_3 = \frac{1}{a_{33}}(b_3 - a_{31}x_1 - a_{32}x_2 - a_{34}x_4)$$

$$x_4 = \frac{1}{a_{44}}(b_4 - a_{41}x_1 - a_{42}x_2 - a_{43}x_3).$$

(5.16)

Jacobi method

Denote the first approximation to x_i by $x_i^{(1)}$, the second by $x_i^{(2)}$ etc., and assume that n of them have been calculated, i.e., $x_i^{(n)}$ is known for $i = 1(1)4$. Then the Jacobi iterative method expresses the $(n+1)$th iterative values exclusively in terms of the nth iterative values and the iteration corresponding to equations (5.16) is defined

by

$$x_1^{(n+1)} = \frac{1}{a_{11}} \{ b_1 - a_{12} x_2^{(n)} - a_{13} x_3^{(n)} - a_{14} x_4^{(n)} \}$$

$$x_2^{(n+1)} = \frac{1}{a_{22}} \{ b_2 - a_{21} x_1^{(n)} - a_{23} x_3^{(n)} - a_{24} x_4^{(n)} \}$$

(5.17)

$$x_3^{(n+1)} = \frac{1}{a_{33}} \{ b_3 - a_{31} x_1^{(n)} - a_{32} x_2^{(n)} - a_{34} x_4^{(n)} \}$$

$$x_4^{(n+1)} = \frac{1}{a_{44}} \{ b_4 - a_{41} x_1^{(n)} - a_{42} x_2^{(n)} - a_{43} x_3^{(n)} \}.$$

In the general case for m equations,

$$x_i^{(n+1)} = \frac{1}{a_{ii}} \left\{ b_i - \sum_{j=1}^{i-1} a_{ij} x_j^{(n)} - \sum_{j=i+1}^{m} a_{ij} x_j^{(n)} \right\}, \quad i = 1(1)m.$$

Gauss–Seidel method

In this method the $(n+1)$th iterative values are used as soon as they are available and the iteration corresponding to equations (5.16) is defined by

$$x_1^{(n+1)} = \frac{1}{a_{11}} \{ b_1 - a_{12} x_2^{(n)} - a_{13} x_3^{(n)} - a_{14} x_4^{(n)} \}$$

$$x_2^{(n+1)} = \frac{1}{a_{22}} \{ b_2 - a_{21} x_1^{(n+1)} - a_{23} x_3^{(n)} - a_{24} x_4^{(n)} \}$$

(5.18)

$$x_3^{(n+1)} = \frac{1}{a_{33}} \{ b_3 - a_{31} x_1^{(n+1)} - a_{32} x_2^{(n+1)} - a_{34} x_4^{(n)} \}$$

$$x_4^{(n+1)} = \frac{1}{a_{44}} \{ b_4 - a_{41} x_1^{(n+1)} - a_{42} x_2^{(n+1)} - a_{43} x_3^{(n+1)} \}.$$

In the general case for m equations,

$$x_i^{(n+1)} = \frac{1}{a_{ii}} \left\{ b_i - \sum_{j=1}^{i-1} a_{ij} x_j^{(n+1)} - \sum_{j=i+1}^{m} a_{ij} x_j^{(n)} \right\}, \quad i = 1(1)m.$$

Successive over-relaxation method

If the Gauss–Seidel iteration equations are written as

$$x_1^{(n+1)} = x_1^{(n)} + \left[\frac{1}{a_{11}}\{b_1 - a_{11}x_1^{(n)} - a_{12}x_2^{(n)} - a_{13}x_3^{(n)} - a_{14}x_4^{(n)}\}\right]$$

$$x_2^{(n+1)} = x_2^{(n)} + \left[\frac{1}{a_{22}}\{b_2 - a_{21}x_1^{(n+1)} - a_{22}x_2^{(n)} - a_{23}x_3^{(n)} - a_{24}x_4^{(n)}\}\right]$$

$$x_3^{(n+1)} = x_3^{(n)} + \left[\frac{1}{a_{33}}\{b_3 - a_{31}x_1^{(n+1)} - a_{32}x_2^{(n+1)} - a_{33}x_3^{(n)} - a_{34}x_4^{(n)}\}\right] \quad (5.19)$$

$$x_4^{(n+1)} = x_4^{(n)} + \left[\frac{1}{a_{44}}\{b_4 - a_{41}x_1^{(n+1)} - a_{42}x_2^{(n+1)} - a_{43}x_3^{(n+1)} - a_{44}x_4^{(n)}\}\right]$$

it is seen that the expressions in the square brackets are the corrections or changes made to $x_i^{(n)}$, $i = 1(1)4$, by one Gauss–Seidel iteration. (Also called the 'displacements' in many books.) If successive corrections are all one-signed, as they usually are for the approximating difference equations of elliptic problems (see page 241), it would be reasonable to expect convergence to be accelerated if each equation of (5.19) was given a larger correction term than is defined by (5.19). This idea leads to the successive over-relaxation or SOR iteration which is defined by the equations

$$x_1^{(n+1)} = x_1^{(n)} + \frac{\omega}{a_{11}}\{b_1 - a_{11}x_1^{(n)} - a_{12}x_2^{(n)} - a_{13}x_3^{(n)} - a_{14}x_4^{(n)}\} \qquad ($$

$$\cdot \qquad \cdot \qquad \cdot$$
$$\cdot \qquad \cdot \qquad \cdot$$

$$x_4^{(n+1)} = x_4^{(n)} + \frac{\omega}{a_{44}}\{b_1 - a_{41}x_1^{(n+1)} - a_{42}x_2^{(n+1)} - a_{43}x_3^{(n+1)} - a_{44}x_4^{(n)}\}.$$

The factor ω, called the acceleration parameter or relaxation factor, generally lies in the range $1 < \omega < 2$. The determination of the optimum value of ω for maximum rate of convergence will be considered later for a particular class of matrices. The value $\omega = 1$ gives the Gauss–Seidel iteration.

In the general case for m equations the SOR iteration is defined by

$$x_i^{(n+1)} = x_i^{(n)} + \frac{\omega}{a_{ii}}\left\{b_i - \sum_{j=1}^{i-1} a_{ij}x_j^{(n+1)} - \sum_{j=i}^{m} a_{ij}x_j^{(n)}\right\}, \quad i = 1(1)m.$$

This scheme can easily be remembered by noting that it may be written as

$$x_i^{(n+1)} = \frac{\omega}{a_{ii}} \left\{ b_i - \sum_{j=1}^{i-1} a_{ij} x_j^{(n+1)} - \sum_{j=i+1}^{m} a_{ij} x_j^{(n)} \right\} - (\omega - 1) x_i^{(n)}$$

$$= \omega(\text{R.H.S. of the Gauss–Seidel iteration equations})$$

$$-(\omega - 1) x_i^{(n)}. \tag{5.20}$$

Example 5.3

The function U satisfies the equation

$$\frac{\partial U}{\partial t} = \frac{\partial^2 U}{\partial x^2}, \quad 0 < x < 1,$$

the initial condition $U = 1$ when $t = 0$, $0 < x < 1$, and the boundary conditions $U = 0$ at $x = 0$ and 1, $t \geq 0$.

Approximate the differential equation by the Crank–Nicolson equations taking $\delta x = 0.1$ and $r = \delta t/(\delta x)^2 = 1$. Solve them to $4D$ for one time-step by the Jacobi, Gauss–Seidel, and SOR iterative methods, taking $\omega = 1.064$ and the first approximation values as equal to the initial values.

The Crank–Nicolson equations are

$$-u_{i-1,j+1} + 4u_{i,j+1} - u_{i+1,j+1} = u_{i-1,j} + u_{i+1,j}.$$

Denote the unknowns $u_{i,j+1}$, $i = 1(1)9$, by u_i. The equations for the unknowns along the first time-level, taking into account the initial conditions, the boundary conditions and the symmetry about $x = 0.5$ are then

$$4u_1 - u_2 = 1 \quad \text{i.e.,} \quad u_1 = \tfrac{1}{4}(u_2 + 1)$$

$$-u_1 + 4u_2 - u_3 = 2 \qquad u_2 = \tfrac{1}{4}(u_1 + u_3 + 2)$$

$$-u_2 + 4u_3 - u_4 = 2 \qquad u_3 = \tfrac{1}{4}(u_2 + u_4 + 2)$$

$$-u_3 + 4u_4 - u_5 = 2 \qquad u_4 = \tfrac{1}{4}(u_3 + u_5 + 2)$$

$$-2u_4 + 4u_5 = 2 \qquad u_5 = \tfrac{1}{2}(u_4 + 1).$$

The Jacobi iteration equations are

$$u_1^{(n+1)} = \tfrac{1}{4}(u_2^{(n)} + 1),$$
$$u_2^{(n+1)} = \tfrac{1}{4}(u_1^{(n)} + u_3^{(n)} + 2),$$
$$u_3^{(n+1)} = \tfrac{1}{4}(u_2^{(n)} + u_4^{(n)} + 2),$$
$$u_4^{(n+1)} = \tfrac{1}{4}(u_3^{(n)} + u_5^{(n)} + 2)$$

and

$$u_5^{(n+1)} = \tfrac{1}{2}(u_4^{(n)} + 1).$$

Taking the first approximation values $u_i^{(0)}$, $i = 1(1)5$, as equal to the initial values of 1 leads to the first iteration values

$$u_1^{(1)} = \tfrac{1}{4}(u_2^{(0)} + 1) = \tfrac{1}{4}(1 + 1) = 0 \cdot 5,$$
$$u_2^{(1)} = \tfrac{1}{4}(u_1^{(0)} + u_3^{(0)} + 2) = \tfrac{1}{4}(1 + 1 + 2) = 1,$$
$$u_3^{(1)} = \tfrac{1}{4}(u_2^{(0)} + u_4^{(0)} + 2) = \tfrac{1}{4}(1 + 1 + 2) = 1,$$
$$u_4^{(1)} = \tfrac{1}{4}(1 + 1 + 2) = 1$$

and

$$u_5^{(1)} = \tfrac{1}{2}(u_4^{(0)} + 1) = \tfrac{1}{2}(1 + 1) = 1.$$

The solution to $4D$ is obtained after 11 iterations and is displayed in Table 5.1.

TABLE 5.1 (Jacobi)

$u_{i,0} = 0$	1·0	1·0	1·0	1·0	1·0
$n = 0$	0	1·0	1·0	1·0	1·0	1·0
$n = 1$	0	0·5	1·0	1·0	1·0	1·0
$n = 2$	0	0·5	0·875	1·0	1·0	1·0
$n = 3$	0	0·4688	0·875	0·9688	1·0	1·0
$n = 4$	0	0·4688	0·8594	0·9688	0·9922	1·0
$n = 10$	0	0·4641	0·8564	0·9614	0·9890	0·9946
$n = 11$	0	0·4641	0·8564	0·9613	0·9890	0·9945

The Gauss–Seidel iteration equations are

$$u_1^{(n+1)} = \tfrac{1}{4}(u_2^{(n)} + 1),$$
$$u_2^{(n+1)} = \tfrac{1}{4}(u_1^{(n+1)} + u_3^{(n)} + 2),$$
$$u_3^{(n+1)} = \tfrac{1}{4}(u_2^{(n+1)} + u_4^{(n)} + 2),$$
$$u_4^{(n+1)} = \tfrac{1}{4}(u_3^{(n+1)} + u_5^{(n)} + 2),$$

and
$$u_5^{(n+1)} = \tfrac{1}{2}(u_4^{(n+1)} + 1).$$

Therefore
$$u_1^{(1)} = \tfrac{1}{4}(u_2^{(0)} + 1) = \tfrac{1}{4}(1 + 1) = 0\cdot5,$$
$$u_2^{(1)} = \tfrac{1}{4}(u_1^{(1)} + u_3^{(0)} + 2) = \tfrac{1}{4}(0\cdot5 + 1 + 2) = 0\cdot875,$$
$$u_3^{(1)} = \tfrac{1}{4}(u_2^{(1)} + u_4^{(0)} + 2) = \tfrac{1}{4}(0\cdot875 + 1 + 2) = 0\cdot96875,$$
$$u_4^{(1)} = \tfrac{1}{4}(u_3^{(1)} + u_5^{(0)} + 2) = \tfrac{1}{4}(0\cdot96875 + 1 + 2) = 0\cdot99219,$$

and
$$u_5^{(1)} = \tfrac{1}{2}(u_4^{(1)} + 1) = \tfrac{1}{2}(0\cdot99219 + 1) = 0\cdot99609.$$

The solution to $4D$ is obtained after five iterations and is given in Table 5.2.

TABLE 5.2 (Gauss–Seidel)

$i = 0$	1	2	3	4	5
$i_{i,0} = 0$	$1\cdot0$	$1\cdot0$	$1\cdot0$	$1\cdot0$	$1\cdot0$
$n = 0$ 0	$1\cdot0$	$1\cdot0$	$1\cdot0$	$1\cdot0$	$1\cdot0$
$n = 1$ 0	$0\cdot5$	$0\cdot875$	$0\cdot9688$	$0\cdot9922$	$0\cdot9961$
$n = 2$ 0	$0\cdot4688$	$0\cdot8594$	$0\cdot9629$	$0\cdot9898$	$0\cdot9949$
$n = 3$ 0					
$n = 4$ 0					
$n = 5$ 0	$0\cdot4641$	$0\cdot8564$	$0\cdot9613$	$0\cdot9890$	$0\cdot9945$

The SOR iteration equations by (5.20) are
$$u_1^{(n+1)} = \frac{\omega}{4}(u_2^{(n)} + 1) - (\omega - 1)u_1^{(n)} = 0\cdot266(u_2^{(n)} + 1) - 0\cdot064u_1^{(n)},$$
$$u_2^{(n+1)} = 0\cdot266(u_1^{(n+1)} + u_3^{(n)} + 2) - 0\cdot064u_2^{(n)},$$
$$u_3^{(n+1)} = 0\cdot266(u_2^{(n+1)} + u_4^{(n)} + 2) - 0\cdot064u_3^{(n)},$$
$$u_4^{(n+1)} = 0\cdot266(u_3^{(n+1)} + u_5^{(n)} + 2) - 0\cdot064u_4^{(n)},$$
$$u_5^{(n+1)} = 0\cdot532(u_4^{(n+1)} + 1) - 0\cdot064u_5^{(n)}.$$

Therefore
$$u_1^{(1)} = 0\cdot266(u_2^{(0)} + 1) - 0\cdot064u_1^{(0)} = 0\cdot266(1 + 1) - 0\cdot064 = 0\cdot468.$$
$$u_2^{(1)} = 0\cdot266(u_1^{(1)} + u_3^{(0)} + 2) - 0\cdot064u_2^{(0)} = 0\cdot266(0\cdot468 + 3) - 0\cdot064$$
$$= 0\cdot85849.$$
$$u_3^{(1)} = 0\cdot266(0\cdot85849 + 3) - 0\cdot064 = 0\cdot96236.$$
$$u_4^{(1)} = 0\cdot266(0\cdot96236 + 3) - 0\cdot064 = 0\cdot98999.$$
$$u_5^{(1)} = 0\cdot532(0\cdot98999 + 1) - 0\cdot064 = 0\cdot99467.$$

The solution to $4D$ is obtained after three iterations and is shown in Table 5.3.

TABLE 5.3 (S.O.R.)

$i = 0$	1	2	3	4	5
$u_{i,0} = 0$	1·0	1·0	1·0	1·0	1·0
$n = 0 \mid 0$	1·0	1·0	1·0	1·0	1·0
$n = 1 \mid 0$	0·4680	0·8585	0·9624	0·9900	0·9947
$n = 2 \mid 0$	0·4644	0·8566	0·9616	0·9890	0·9945
$n = 3 \mid 0$	0·4641	0·8564	0·9613	0·9890	0·9945

Systematic iterative methods in matrix form

Equations (5.15) in matrix form are

$$\begin{bmatrix} a_{11} & a_{12} & a_{13} & a_{14} \\ a_{21} & a_{22} & a_{23} & a_{24} \\ a_{31} & a_{32} & a_{33} & a_{34} \\ a_{41} & a_{42} & a_{43} & a_{44} \end{bmatrix} \begin{bmatrix} x_1 \\ x_2 \\ x_3 \\ x_4 \end{bmatrix} = \begin{bmatrix} b_1 \\ b_2 \\ b_3 \\ b_4 \end{bmatrix}$$

or, more briefly,

$$\mathbf{A}\mathbf{x} = \mathbf{b}.$$

For our purposes it will be found convenient to express the matrix $\mathbf{A}$ as the sum of its main diagonal elements, its strictly lower triangular elements, and its strictly upper triangular elements and to write it as $\mathbf{A} = \mathbf{D} - \mathbf{L} - \mathbf{U}$, where

$$\mathbf{D} = \begin{bmatrix} a_{11} & 0 & 0 & 0 \\ 0 & a_{22} & 0 & 0 \\ 0 & 0 & a_{33} & 0 \\ 0 & 0 & 0 & a_{44} \end{bmatrix}, -\mathbf{L} = \begin{bmatrix} 0 & 0 & 0 & 0 \\ a_{21} & 0 & 0 & 0 \\ a_{31} & a_{32} & 0 & 0 \\ a_{41} & a_{42} & a_{43} & 0 \end{bmatrix},$$

$$-\mathbf{U} = \begin{bmatrix} 0 & a_{12} & a_{13} & a_{14} \\ 0 & 0 & a_{23} & a_{24} \\ 0 & 0 & 0 & a_{34} \\ 0 & 0 & 0 & 0 \end{bmatrix}.$$

Equations (5.15) can then be written as

$$\mathbf{Dx} = (\mathbf{L} + \mathbf{U})\mathbf{x} + \mathbf{b}$$

and a study of equations (5.17) shows that the corresponding Jacobi iteration can be expressed as

$$\mathbf{Dx}^{(n+1)} = (\mathbf{L} + \mathbf{U})\mathbf{x}^{(n)} + \mathbf{b}, \qquad (5.21)$$

giving

$$\mathbf{x}^{(n+1)} = \mathbf{D}^{-1}(\mathbf{L} + \mathbf{U})\mathbf{x}^{(n)} + \mathbf{D}^{-1}\mathbf{b}.$$

The matrix $\mathbf{D}^{-1}(\mathbf{L} + \mathbf{U})$ is called the point Jacobi iteration matrix. The word 'point' refers to the fact that the algebraic equations approximate a differential equation at a number of mesh points and the iterative procedure expresses the next iterative value at only one mesh point in terms of known iterative values at other mesh points.

Similarly, by equations (5.18), the Gauss–Seidel iteration is defined by

$$\mathbf{Dx}^{(n+1)} = \mathbf{Lx}^{(n+1)} + \mathbf{Ux}^{(n)} + \mathbf{b}. \qquad (5.22)$$

Hence

$$(\mathbf{D} - \mathbf{L})\mathbf{x}^{(n+1)} = \mathbf{Ux}^{(n)} + \mathbf{b},$$

giving that

$$\mathbf{x}^{n+1} = (\mathbf{D} - \mathbf{L})^{-1}\mathbf{Ux}^{(n)} + (\mathbf{D} - \mathbf{L})^{-1}\mathbf{b},$$

which shows that the point Gauss-Seidel iteration matrix is $(\mathbf{D} - \mathbf{L})^{-1}\mathbf{U}$.

The correction or displacement vector $\mathbf{d}^{(n)} = \mathbf{x}^{(n+1)} - \mathbf{x}^{(n)}$ of the SOR method is taken to be ω times the displacement vector $\mathbf{d}_1^{(n)}$ defined by the Gauss–Seidel iteration. By equation (5.22) we have that

$$\mathbf{Dd}_1^{(n)} = \mathbf{D}(\mathbf{x}^{(n+1)} - \mathbf{x}^{(n)}) = \mathbf{Lx}^{(n+1)} + \mathbf{Ux}^{(n)} + \mathbf{b} - \mathbf{Dx}^{(n)}.$$

Hence the SOR iteration, defined by

$$\mathbf{d}^{(n)} = \omega \mathbf{d}_1^{(n)},$$

can be written as

$$\mathbf{x}^{(n+1)} - \mathbf{x}^{(n)} = \omega \mathbf{D}^{-1}(\mathbf{Lx}^{(n+1)} + \mathbf{Ux}^{(n)} + \mathbf{b} - \mathbf{Dx}^{(n)}).$$

Therefore

$$(\mathbf{I} - \omega \mathbf{D}^{-1}\mathbf{L})\mathbf{x}^{(n+1)} = \{(1 - \omega)\mathbf{I} + \omega \mathbf{D}^{-1}\mathbf{U}\}\mathbf{x}^{(n)} + \omega \mathbf{D}^{-1}\mathbf{b}.$$

Hence

$$\mathbf{x}^{(n+1)} = (\mathbf{I} - \omega\mathbf{D}^{-1}\mathbf{L})^{-1}\{(1-\omega)\mathbf{I} + \omega\mathbf{D}^{-1}\mathbf{U}\}\mathbf{x}^{(n)} + (\mathbf{I} - \omega\mathbf{D}^{-1}\mathbf{L})^{-1}\omega\mathbf{D}^{-1}\mathbf{b},$$
$$(5.23)$$

showing that the point SOR iteration matrix is

$$(\mathbf{I} - \omega\mathbf{D}^{-1}\mathbf{L})^{-1}\{(1-\omega)\mathbf{I} + \omega\mathbf{D}^{-1}\mathbf{U}\}.$$

A necessary and sufficient condition for the convergence of iterative methods

Each of the three iterative methods described above can be written as

$$\mathbf{x}^{(n+1)} = \mathbf{G}\mathbf{x}^{(n)} + \mathbf{c}, \qquad (5.24)$$

where $\mathbf{G}$ is the iteration matrix and $\mathbf{c}$ a column vector of known values. This equation was derived from the original equations by rearranging them into the form

$$\mathbf{x} = \mathbf{G}\mathbf{x} + \mathbf{c}, \qquad (5.25)$$

i.e., the unique solution of the m linear equations $\mathbf{Ax} = \mathbf{b}$ is the solution of equation (5.25). The error $\mathbf{e}^{(n)}$ in the nth approximation to the exact solution is defined by $\mathbf{e}^{(n)} = \mathbf{x} - \mathbf{x}^{(n)}$ so it follows by the subtraction of equation (5.24) from equation (5.25) that

$$\mathbf{e}^{(n+1)} = \mathbf{G}\mathbf{e}^{(n)}.$$

Therefore

$$\mathbf{e}^{(n)} = \mathbf{G}\mathbf{e}^{(n-1)} = \mathbf{G}^2\mathbf{e}^{(n-2)} = \ldots = \mathbf{G}^n\mathbf{e}^{(0)}. \qquad (5.26)$$

The sequence of iterative values $\mathbf{x}^{(1)}, \mathbf{x}^{(2)}, \ldots, \mathbf{x}^{(n)}, \ldots$ will converge to $\mathbf{x}$ as n tends to infinity if

$$\lim_{n \to \infty} \mathbf{e}^{(n)} = \mathbf{O}.$$

Since $\mathbf{x}^{(0)}$ and therefore $\mathbf{e}^{(0)}$ is arbitrary it follows that the iteration will converge if and only if

$$\lim_{n \to \infty} \mathbf{G}^n = \mathbf{O}.$$

Assume now that the matrix $\mathbf{G}$ of order m has m linearly independent eigenvectors $\mathbf{v}_s$, $s = 1(1)m$. Then these eigenvectors can be used as a basis for our m-dimensional vector space and the arbitrary error vector $\mathbf{e}^{(0)}$, with its m components, can be expressed uniquely as a linear combination of them, namely,

$$\mathbf{e}^{(0)} = \sum_{s=1}^{m} c_s \mathbf{v}_s,$$

where the c_s are scalars. Hence

$$\mathbf{e}^{(1)} = \mathbf{G}\mathbf{e}^{(0)} = \sum_{s=1}^{m} c_s \mathbf{G}\mathbf{v}_s.$$

But $\mathbf{G}\mathbf{v}_s = \lambda_s \mathbf{v}_s$ by the definition of an eigenvalue, where λ_s is the eigenvalue corresponding to $\mathbf{v}_s$. Hence

$$\mathbf{e}^{(1)} = \sum_{1}^{m} c_s \lambda_s \mathbf{v}_s.$$

Similarly,

$$\mathbf{e}^{(n)} = \sum_{1}^{m} c_s \lambda_s^n \mathbf{v}_s. \tag{5.27}$$

Therefore $\mathbf{e}^{(n)}$ will tend to the null vector as n tends to infinity, for arbitrary $\mathbf{e}^{(0)}$, if and only if $|\lambda_s| < 1$ for all s. In other words, the iteration will converge for arbitrary $\mathbf{x}^{(0)}$ if and only if the spectral radius $\rho(\mathbf{G})$ of $\mathbf{G}$ is less than one.

As a corollary to this result a sufficient condition for convergence is that $\|\mathbf{G}\| < 1$. To prove this we have that $\mathbf{G}\mathbf{v}_s = \lambda_s \mathbf{v}_s$. Hence

$$\|\mathbf{G}\mathbf{v}_s\| = \|\lambda_s \mathbf{v}_s\| = |\lambda_s| \, \|\mathbf{v}_s\|.$$

But for any matrix norm that is compatible with a vector norm $\|\mathbf{v}_s\|$,

$$\|\mathbf{G}\mathbf{v}_s\| \leq \|\mathbf{G}\| \, \|\mathbf{v}_s\|.$$

Therefore

$$|\lambda_s| \, \|\mathbf{v}_s\| \leq \|\mathbf{G}\| \, \|\mathbf{v}_s\|,$$

so

$$|\lambda_s| \leq \|\mathbf{G}\|, \quad s = 1(1)m.$$

It follows from this that a sufficient condition for convergence is that $\|\mathbf{G}\| < 1$. It is not a necessary condition because the norm of $\mathbf{G}$ can exceed one even when $\rho(\mathbf{G}) < 1$.

As an example consider the Jacobi iteration matrix $\mathbf{D}^{-1}(\mathbf{L}+\mathbf{U})$. If the ith equation of $\mathbf{Ax}=\mathbf{b}$ is

$$a_{i1}x_1 + a_{i2}x_2 + \ldots + a_{ii}x_i + \ldots + a_{im}x_m = b_i$$

then the ith row of $\mathbf{D}^{-1}(\mathbf{L}+\mathbf{U})$ is

$$\frac{a_{i1}}{a_{ii}} \quad \frac{a_{i2}}{a_{ii}} \quad . \quad . \quad \frac{a_{i,i-1}}{a_{ii}} \quad 0 \quad \frac{a_{i,i+1}}{a_{ii}} \quad . \quad . \quad \frac{a_{im}}{a_{ii}}.$$

Let the sum of the moduli of these elements be the greatest row sum of the moduli of the elements of the iteration matrix. Then if we take our matrix norm as the infinity norm, the Jacobi iteration will converge if

$$|a_{i1}| + |a_{i2}| + \ldots + |a_{i,i-1}| + 0 + |a_{i,i+1}| + |a_{im}| < |a_{ii}|.$$

This states that the Jacobi method applied to the equations $\mathbf{Ax}=\mathbf{b}$ will converge if $\mathbf{A}$ is a strictly diagonally dominant matrix, i.e., if in each row of $\mathbf{A}$ the modulus of the diagonal element exceeds the sum of the moduli of the off-diagonal elements.

Rate of convergence

Assume that the iteration matrix $\mathbf{G}$ has m linearly independent eigenvectors $\mathbf{v}_s$ corresponding to the eigenvalues λ_s and that $|\lambda_1| > |\lambda_2| \geqslant |\lambda_3| \geqslant \ldots \geqslant |\lambda_m|$. By equation (5.27) the error vector $\mathbf{e}^{(n)}$ can be expressed as

$$\mathbf{e}^{(n)} = \lambda_1^n \left\{ c_1\mathbf{v}_1 + \left(\frac{\lambda_2}{\lambda_1}\right)^n c_2\mathbf{v}_2 + \ldots + \left(\frac{\lambda_m}{\lambda_1}\right)^n c_m\mathbf{v}_m \right\}.$$

For large values of n this shows that

$$\mathbf{e}^{(n)} \simeq \lambda_1^n c_1\mathbf{v}_1.$$

Similarly,

$$\mathbf{e}^{(n+1)} \simeq \lambda_1^{n+1} c_1\mathbf{v}_1,$$

so

$$\mathbf{e}^{(n+1)} \simeq \lambda_1 \mathbf{e}^{(n)}. \tag{5.28}$$

If the ith component of $\mathbf{e}^{(n)}$ is denoted by $e_i^{(n)}$ it is seen that

$$\frac{|e_i^{(n)}|}{|e_i^{(n+1)}|} \simeq \frac{1}{|\lambda_1|} = \frac{1}{\rho(\mathbf{G})}.$$

For simplicity, assume that $e_i^{(n)} = 10^{-2}$ and $e_i^{(n+1)} = 10^{-4}$. Then $e_i^{(n)}/e_i^{(n+1)} = 10^2$ and the logarithm of this to base 10 is 2. More generally, $\log_{10}(1/\rho) = -\log\rho$ gives an indication of the number of decimal digits by which the error is eventually decreased by each convergent iteration. Since, for convergence, $0 < \rho < 1$, the number of decimal digits of accuracy gained per iteration increases as ρ decreases. Alternatively, for large n, $\mathbf{e}^{(n)} \simeq \lambda_1 \mathbf{e}^{(n-1)}$, therefore

$$\mathbf{e}^{(n+p)} \simeq \lambda_1 \mathbf{e}^{(n+p-1)} \simeq \ldots \simeq \lambda_1^p \mathbf{e}^{(n)}, \quad p = 1, 2, \ldots.$$

If we want to reduce the size of the error by 10^{-q}, say, then the number of iterations needed to do this will be the least value of p for which

$$|\lambda_1^p| = \rho^p \leqslant 10^{-q}.$$

Taking logs and remembering that $\log\rho$ is negative for a convergent iteration leads to

$$p \geqslant q/(-\log_{10}\rho),$$

which shows that p decreases as $(-\log\rho)$ increases. Clearly, the number $(-\log\rho)$ provides a measure for the comparison of the rates of convergence of different iterative methods when n is sufficiently large. For this reason $(-\log_e\rho)$ is defined to be *the asymptotic rate of convergence* and is denoted by $R_\infty(\mathbf{G})$.

The *average rate of convergence* $R_n(\mathbf{G})$ after n iterations is defined by

$$R_n(\mathbf{G}) = -\frac{1}{n}\log_e\|\mathbf{G}^n\|_2,$$

where $\|\mathbf{G}^n\|_2$ is the 2-norm or spectral norm of the matrix $\mathbf{G}^n$ which is given by the square root of the spectral radius of $[\mathbf{G}^n]^H[\mathbf{G}^n]$. In reference 39 it is proved that the asymptotic rate of convergence

$$R_\infty(\mathbf{G}) = \lim_{n\to\infty} R_n(\mathbf{G})$$

and that the number of iterations required to reduce the error $\|\mathbf{e}^{(n)}\|$ to $\|\mathbf{e}^{(0)}\|/\alpha$, for sufficiently large n, is $\geqslant -(\log_e\alpha)/R_n$.

Methods for accelerating the convergence of iterative processes

Both of the following methods are applicable to any iterative process when λ_1 is real.

Lyusternik's method (*Reference* 21)

By equation (5.28),

$$\mathbf{e}^{(n+1)} = \lambda_1 \mathbf{e}^{(n)} + \boldsymbol{\delta}^{(n)}, \quad \lambda_1 \text{ real}, \tag{5.29}$$

where the components of $\boldsymbol{\delta}^{(n)}$ are small and $|\lambda_1| < |$. Also, by definition, the exact solution $\mathbf{x}$ and $\mathbf{e}^{(n)}$ are related by

$$\mathbf{x} = \mathbf{x}^{(n)} + \mathbf{e}^{(n)} \tag{5.30}$$
$$= \mathbf{x}^{(n+1)} + \mathbf{e}^{(n+1)}$$
$$= \mathbf{x}^{(n+1)} + \lambda_1 \mathbf{e}^{(n)} + \boldsymbol{\delta}^{(n)}. \tag{5.31}$$

Eliminating $\mathbf{e}^{(n)}$ between (5.30) and (5.31) gives that

$$\mathbf{x} = \frac{\mathbf{x}^{(n+1)} - \lambda_1 \mathbf{x}^{(n)}}{1 - \lambda_1} + \frac{\boldsymbol{\delta}^{(n)}}{1 - \lambda_1},$$

so if $\|\boldsymbol{\delta}^{(n)}\|$ is small compared with $(1 - \lambda_1)$, a good approximation to the solution will be given by

$$\mathbf{x} \simeq \frac{\mathbf{x}^{(n+1)} - \lambda_1 \mathbf{x}^{(n)}}{1 - \lambda_1},$$

which can also be written as

$$\mathbf{x} \simeq \mathbf{x}^{(n)} + \frac{\mathbf{x}^{(n+1)} - \mathbf{x}^{(n)}}{(1 - \lambda_1)}.$$

The last equation shows that small differences between successive iterates do not necessarily imply a close approximation to the solution. If, for example,

$$\max_i |x_i^{(n+1)} - x_i^{(n)}| = \epsilon \quad \text{and} \quad \lambda_1 = 0 \cdot 99$$

then

$$\max_i |x_i^{(n+1)} - x_i^{(n)}| / (1 - \lambda_1) = 100 \, \epsilon,$$

where $x_i^{(n)}$ is the ith component of $\mathbf{x}^{(n)}$. To obtain a solution with a maximum error of ϵ in any component the iteration would have to be continued until

$$\max_i |x_i^{(n+1)} - x_i^{(n)}| \simeq 0 \cdot 01 \, \epsilon.$$

An alternative derivation of this result is given in exercise 12.

For most problems λ_1 will not be known analytically in which case its value must be estimated. One straightforward way of doing this is as follows.

For sufficiently large n,

$$\mathbf{e}^{(n)} \simeq \lambda_1 \mathbf{e}^{(n-1)},$$

therefore

$$\mathbf{e}^{(n+1)} - \mathbf{e}^{(n)} \simeq \lambda_1 (\mathbf{e}^{(n)} - \mathbf{e}^{(n-1)}),$$

i.e.,

$$\{(\mathbf{x} - \mathbf{x}^{(n+1)}) - (\mathbf{x} - \mathbf{x}^{(n)})\} \simeq \lambda_1 \{(\mathbf{x} - \mathbf{x}^{(n)}) - (\mathbf{x} - \mathbf{x}^{(n-1)})\}$$

showing that

$$\mathbf{x}^{(n+1)} - \mathbf{x}^{(n)} \simeq \lambda_1 (\mathbf{x}^{(n)} - \mathbf{x}^{(n-1)}),$$

i.e.,

$$\mathbf{d}^{(n)} \simeq \lambda_1 \mathbf{d}^{(n-1)}. \tag{5.32}$$

Hence

$$\|\mathbf{d}^{(n)}\| \simeq |\lambda_1| \|\mathbf{d}^{(n-1)}\|,$$

so

$$|\lambda_1| = \rho \simeq \frac{\|\mathbf{d}^{(n)}\|}{\|\mathbf{d}^{(n-1)}\|},$$

where the norm of $\mathbf{d}^{(n)}$ can be defined by

$$\|\mathbf{d}^{(n)}\| = \max_i |x_i^{(n+1)} - x_i^{(n)}|, \quad i = 1(1)m,$$

or

$$\|\mathbf{d}^{(n)}\| = |x_1^{(n+1)} - x_1^{(n)}| + |x_2^{(n+1)} - x_2^{(n)}| + \ldots + |x_m^{(n+1)} - x_m^{(n)}|$$

or

$$\|\mathbf{d}^{(n)}\| = \{(x_1^{(n+1)} - x_1^{(n)})^2 + (x_2^{(n+1)} - x_2^{(n)})^2 + \ldots + (x_m^{(n+1)} - x_m^{(n)})^2\}^{\frac{1}{2}}.$$

Equation (5.32) justifies, incidentally, the basis of the SOR iterative method because it proves that when λ_1 is positive the corresponding components of successive correction or displacement vectors are one signed. More refined methods for estimating λ_1 are given in references 39 and 16.

Aitken's method

$$\mathbf{e}^{(n)} \simeq \lambda_1 \mathbf{e}^{(n-1)}, \quad \text{i.e.,} \quad \mathbf{x} - \mathbf{x}^{(n)} \simeq \lambda_1 (\mathbf{x} - \mathbf{x}^{(n-1)})$$

and

$$\mathbf{e}^{(n+1)} \simeq \lambda_1 \mathbf{e}^{(n)}, \quad \text{i.e.,} \quad \mathbf{x} - \mathbf{x}^{(n+1)} \simeq \lambda_1 (\mathbf{x} - \mathbf{x}^{(n)}).$$

Take the ith components of these approximations and eliminate λ_1 by simple division. Then

$$\frac{x_i - x_i^{(n)}}{x_i - x_i^{(n+1)}} \simeq \frac{x_i - x_i^{(n-1)}}{x_i - x_i^{(n)}}$$

and solving this for x_i gives that

$$x_i \simeq \frac{x_i^{(n+1)} x_i^{(n-1)} - (x_i^{(n)})^2}{x_i^{(n+1)} - 2x_i^{(n)} + x_i^{(n-1)}} = x_i^{(n+1)} - \frac{\{x_i^{(n+1)} - x_i^{(n)}\}^2}{x_i^{(n+1)} - 2x_i^{(n)} + x_i^{(n-1)}}.$$

This method of acceleration avoids the explicit estimation of λ_1. It is of interest to note however that the two methods are identical if, in Lyusternick's method, λ_1 is approximated by

$$|x_i^{(n+1)} - x_i^{(n)}|/|x_i^{(n)} - x_i^{(n-1)}| \quad \text{for each } i.$$

As an illustrative example of these methods the following table gives the 9th, 10th and 11th Jacobi iteration values arising from the five-point approximation of $\partial^2 \phi/\partial x^2 + \partial^2 \phi/\partial y^2 + 2 = 0$ over the rectangular domain $0 < x < 2, 0 < y < 1$, where $\phi = 0$ on the boundary. (The torsion problem.) For a square mesh of side $\frac{1}{4}$ there are only eight unknowns because of symmetry. The starting values were zero. The solution ϕ to $4D$ required 65 iterations. Lyusternik's approximation is denoted by L, i.e., $L_i = \phi_i^{(10)} + (\phi_i^{(11)} - \phi_i^{(10)})/(1 - \lambda_1)$, where

$$\lambda_1 = \max_i |\phi_i^{(11)} - \phi_i^{(10)}|/\max_i |\phi_i^{(10)} - \phi_i^{(9)}| = 0 \cdot 8281.$$

Better L values could have been obtained by calculating λ_1 from the average of $|\phi_i^{(11)} - \phi_i^{(10)}|/|\phi_i^{(10)} - \phi_i^{(9)}|$, $i = 1(1)8$. This average is $0 \cdot 8146$ which is closer to the theoretical value of $\frac{1}{2}(\cos \pi/8 + \cos \pi/4) = 0 \cdot 8155$ than is $0 \cdot 8281$. (See page 252). The improved values by Aitken's method are denoted by A.

In practice these methods would be used to improve early iterative values for subsequent iterations.

TABLE 5.4

$\phi^{(9)}$	0·3533	0·4959	0·5486	0·5613	0·4505	0·6430	0·7154	0·7335
$\phi^{(10)}$	0·3616	0·5112	0·5681	0·5827	0·4624	0·6644	0·7434	0·7634
$\phi^{(11)}$	0·3684	0·5235	0·5843	0·5999	0·4791	0·6821	0·7660	0·7880
ϕ	0·3980	0·5782	0·6555	0·6769	0·5138	0·7592	0·8668	0·8968
L	0·4012	0·5828	0·6623	0·6828	0·5177	0·7674	0·8749	0·9071
A	0·4002	0·5733	0·6618	0·6722	0·5093	0·7655	0·8597	0·9070

The optimum acceleration parameter for the SOR method

A problem of paramount importance associated with the **SOR** method is the determination of a suitable value for the acceleration parameter ω. Ideally, we want the optimum value ω_b of ω which minimizes the spectral radius of the SOR iteration matrix and thereby maximizes the rate of convergence of the method. At the present time no formula exists for the determination of ω_b for an arbitrary set of linear equations. Fortunately it can be calculated for many of the difference equations approximating first- and second-order partial differential equations because their matrices are often of a special type called property (A) or 2-cyclic matrices and the significance of this was first revealed by Young in reference 40. He proved that when a 2-cyclic matrix is put into what he termed a consistently ordered form **A**, which can be done by a simple re-ordering of the rows and corresponding columns of the original 2-cyclic matrix, then the eigenvalues λ of the point SOR iteration matrix **H**(ω) associated with **A** are related to the eigenvalues μ of the point Jacobi iteration matrix **B** associated with **A** by the equation

$$(\lambda + \omega - 1)^2 = \lambda \omega^2 \mu^2. \qquad (5.33)$$

From this it can be proved that

$$\omega_b = \frac{2}{1 + \sqrt{\{1 - \rho^2(\mathbf{B})\}}}, \qquad (5.34)$$

where $\rho(\mathbf{B})$ is the spectral radius of the Jacobi iteration matrix.

Varga, reference 35, subsequently generalized these concepts for p-cyclic matrices, $p \geqslant 2$, but in this book only 2-cyclic matrices will be considered because p-cyclic matrices, $p \geqslant 3$, occur very rarely in the mathematics of the physical sciences. Young's definitions of 2-cyclic matrices and consistent ordering will be dealt with later. For our present purposes it is sufficient to know that block tridiagonal matrices **A** of the form

$$\mathbf{A} = \begin{bmatrix} \mathbf{D}_1 & \mathbf{A}_1 & & & & \\ \mathbf{B}_1 & \mathbf{D}_2 & \mathbf{A}_2 & & & \\ & \mathbf{B}_2 & \mathbf{D}_3 & \mathbf{A}_3 & & \\ & & & \cdot & \cdot & \\ & & & \mathbf{B}_{k-2} & \mathbf{D}_{k-1} & \mathbf{A}_{k-1} \\ & & & & \mathbf{B}_{k-1} & \mathbf{D}_k \end{bmatrix}, \quad (5.35)$$

where the $\mathbf{D}_i$, $i = 1(1)k$, are square diagonal matrices not necessarily of the same order, are 2-cyclic and consistently ordered. A special case corresponding to $k = 2$ is

$$\mathbf{A} = \begin{bmatrix} \mathbf{D}_1 & \mathbf{A}_1 \\ \mathbf{B}_1 & \mathbf{D}_2 \end{bmatrix}. \quad (5.36)$$

To avoid interruption of the development of the proof of (5.33) it is convenient at this point to prove the following theorems.

Theorem 1

If the block tridiagonal matrix (5.35) is written as $\mathbf{A} = \mathbf{D} - \mathbf{L} - \mathbf{U}$, where

$$\mathbf{D} = \begin{bmatrix} \mathbf{D}_1 & & & \\ & \mathbf{D}_2 & & \\ & & \cdot & \\ & & & \cdot \\ & & & & \mathbf{D}_k \end{bmatrix}, \quad -\mathbf{L} = \begin{bmatrix} 0 & & & \\ \mathbf{B}_1 & 0 & & \\ & \mathbf{B}_2 & & \\ & & & \\ & & \mathbf{B}_{k-1} & 0 \end{bmatrix}$$

and

$$-\mathbf{U} = \begin{bmatrix} \mathbf{O} & \mathbf{A}_1 & & & \\ & \mathbf{O} & \mathbf{A}_2 & & \\ & & \cdot & \cdot & \\ & & & \mathbf{O} & \mathbf{A}_{k-1} \\ & & & & \mathbf{O} \end{bmatrix}, \quad (5.37)$$

and the matrix $\mathbf{A}(\alpha)$, α any non-zero number, is defined by

$$\mathbf{A}(\alpha) = \mathbf{D} - \alpha\mathbf{L} - \alpha^{-1}\mathbf{L},$$

then

$$\det \mathbf{A}(\alpha) = \det \mathbf{A}.$$

Proof

Let

$$\mathbf{C} = \begin{bmatrix} \mathbf{I}_1 & & & & \\ & \alpha\mathbf{I}_2 & & & \\ & & \alpha^2\mathbf{I}_3 & & \\ & & & \cdot & \\ & & & & \alpha^{k-1}\mathbf{I}_k \end{bmatrix}$$

so that

$$\mathbf{C}^{-1} = \begin{bmatrix} \mathbf{I}_1 & & & & \\ & \alpha^{-1}\mathbf{I}_2 & & & \\ & & \alpha^{-2}\mathbf{I}_3 & & \\ & & & \cdot & \\ & & & & \alpha^{1-k}\mathbf{I}_k \end{bmatrix},$$

where $\mathbf{I}_i$ denotes the unit matrix of the same order as $\mathbf{D}_i$, $i = 1(1)k$.
Then

$$\mathbf{CAC}^{-1} = \begin{bmatrix} \mathbf{D}_1 & \alpha^{-1}\mathbf{A}_1 & & & \\ \alpha\mathbf{B}_1 & \mathbf{D}_2 & \alpha^{-1}\mathbf{A}_2 & & \\ & \cdot & \cdot & \cdot & \\ & & \alpha\mathbf{B}_{k-2} & \mathbf{D}_{k-1} & \alpha^{-1}\mathbf{A}_{k-1} \\ & & & \alpha\mathbf{B}_{k-1} & \mathbf{D}_k \end{bmatrix} = \mathbf{D} - \alpha\mathbf{L} - \alpha^{-1}\mathbf{U} = \mathbf{A}(\alpha).$$

Hence

$$\det \mathbf{A}(\alpha) = \det \mathbf{C} \det \mathbf{A} \det \mathbf{C}^{-1}.$$

But $\det \mathbf{C}^{-1} = 1/\det \mathbf{C}$. Hence the result.
This theorem clearly holds even when each $\mathbf{D}_i$ is a full square matrix
because the diagonal blocks of $\mathbf{CAC}^{-1}$ are independent of α.

Theorem 2

The non-zero eigenvalues of the Jacobi iteration matrix corresponding to the matrix (5.35) occur in pairs $\pm\mu_i$.

Consider the m linear equations

$$\mathbf{Ax} = \mathbf{b}$$

where the $m \times m$ matrix $\mathbf{A}$ has the block tridiagonal form (5.35) and each $\mathbf{D}_i$ is a square diagonal matrix, i.e., $\mathbf{D}$ is an $m \times m$ diagonal matrix. Then by equation (5.21) the Jacobi iteration matrix $\mathbf{B}$ corresponding to $\mathbf{A}$ is $\mathbf{D}^{-1}(\mathbf{L}+\mathbf{U}) = \mathbf{E}+\mathbf{F}$, where $\mathbf{D}^{-1}\mathbf{L} = \mathbf{E}$ and $\mathbf{D}^{-1}\mathbf{U} = \mathbf{F}$ are of the form

$$\mathbf{E} = \begin{bmatrix} 0 & & & & \\ \mathbf{E}_1 & 0 & & & \\ & \mathbf{E}_2 & 0 & & \\ & & \cdot & \cdot & \\ & & & \mathbf{E}_{k-1} & 0 \end{bmatrix} \text{ and } \mathbf{F} = \begin{bmatrix} 0 & \mathbf{F}_1 & & & \\ & 0 & \mathbf{F}_2 & & \\ & & \cdot & \cdot & \\ & & & 0 & \mathbf{F}_{k-1} \\ & & & & 0 \end{bmatrix} \quad (5.37a)$$

The characteristic polynomial of $\mathbf{B}$ is

$$P(\mu) = \det[\mu\mathbf{I} - \mathbf{B}] = \det[\mu\mathbf{I} - \mathbf{E} - \mathbf{F}], \tag{5.38}$$

where $\mu\mathbf{I} - \mathbf{B}$ has the block tridiagonal form of (5.35). Therefore, by theorem 1,

$$\det[\mu\mathbf{I} - \mathbf{E} - \mathbf{F}] = \det[\mu\mathbf{I} - \alpha\mathbf{E} - \alpha^{-1}\mathbf{F}], \tag{5.39}$$

where $\alpha \neq 0$ is any number. With $\alpha = -1$, equation (5.39) gives that

$$\det[\mu\mathbf{I} - \mathbf{E} - \mathbf{F}] = \det[\mu\mathbf{I} + \mathbf{E} + \mathbf{F}] = (-1)^m \det[-\mu\mathbf{I} - \mathbf{E} - \mathbf{F}],$$

i.e.,

$$P(\mu) = (-1)^m P(-\mu).$$

It follows therefore that if m is odd, $P(\mu) = -P(-\mu)$, so $P(\mu)$ is an odd polynomial of μ. Similarly, $P(\mu)$ is an even polynomial when m is even. Hence $P(\mu)$, for some integer $r \geq 0$, will either have the form

$$P(\mu) = \mu^{2r} g(\mu^2), \quad \text{when } m \text{ is even,}$$

or the form $\qquad\qquad\qquad\qquad\qquad\qquad\qquad\qquad (5.39a)$

$$P(\mu) = \mu^{2r+1} f(\mu^2), \quad \text{when } m \text{ is odd,}$$

where $g(x)$ and $f(x)$ denote polynomials in x such that $g(0) \neq 0$ and $f(0) \neq 0$.

By equations (5.39a) the non-zero roots of $P(\mu) = 0$ are given either by $g(\mu^2) = 0$ or by $f(\mu^2) = 0$, thus proving that the non-zero eigenvalues of **B** occur in pairs $\pm\mu_i$. Moreover, for some integer $r \geq 0$, **B** will have $2r$, (m even), or $(2r+1)$, (m odd), eigenvalues equal to zero.

Relationship between the eigenvalues of the Jacobi and SOR point iteration matrices corresponding to a block tridiagonal coefficient matrix

Theorem 3

Let $\mathbf{H}(\omega)$ and **B** represent respectively the SOR and Jacobi point iteration matrices corresponding to the block tridiagonal matrix **A** of equation (5.35). Then if λ is a non-zero eigenvalue of $\mathbf{H}(\omega)$ and μ satisfies the relation

$$(\lambda + \omega - 1)^2 = \lambda \omega^2 \mu^2, \quad \omega \neq 0, \tag{5.40}$$

then μ is an eigenvalue of **B**. Conversely, if μ is an eigenvalue of **B** and satisfies (5.40) then λ is an eigenvalue of $\mathbf{H}((\omega)$.

Proof

By equation (5.23) the point SOR iteration matrix $\mathbf{H}(\omega)$ is

$$\mathbf{H}(\omega) = (\mathbf{I} - \omega\mathbf{E})^{-1}\{(1 - \omega)\mathbf{I} + \omega\mathbf{F}\},$$

where $\mathbf{E} = \mathbf{D}^{-1}\mathbf{L}$ and $\mathbf{F} = \mathbf{D}^{-1}\mathbf{U}$. Therefore the characteristic polynomial of $\mathbf{H}(\omega)$ is

$$Q(\lambda) = \det(\lambda\mathbf{I} - \mathbf{H}(\omega)).$$

As

$$\begin{aligned}
\lambda\mathbf{I} - \mathbf{H}(\omega) &= \lambda(\mathbf{I} - \omega\mathbf{E})^{-1}(\mathbf{I} - \omega\mathbf{E}) - (\mathbf{I} - \omega\mathbf{E})^{-1}\{(1 - \omega)\mathbf{I} + \omega\mathbf{F}\} \\
&= (\mathbf{I} - \omega\mathbf{E})^{-1}\{(\lambda + \omega - 1)\mathbf{I} - \lambda\omega\mathbf{E} - \omega\mathbf{F}\},
\end{aligned}$$

and

$$\det(\mathbf{CD}) = \det\mathbf{C}\det\mathbf{D},$$

it is seen that

$$Q(\lambda) = \det(\mathbf{I} - \omega\mathbf{E})^{-1}\det\{(\lambda + \omega - 1)\mathbf{I} - \lambda\omega\mathbf{E} - \omega\mathbf{F}\}.$$

Now **E** is a strictly lower triangular matrix so $(\mathbf{I} - \omega\mathbf{E})$ is a unit lower triangular matrix, that is, with each diagonal element unity. Since the inverse of a unit lower triangular matrix is another unit lower triangular matrix it follows that $\det(\mathbf{I} - \omega\mathbf{E})^{-1} = 1$. Hence

$$Q(\lambda) = \det\{(\lambda + \omega - 1)\mathbf{I} - \lambda\omega\mathbf{E} - \omega\mathbf{F}\}.$$

This equation for $Q(\lambda)$ holds in fact for any set of equations. When however, it is given that the matrix $\mathbf{A}$ of the equations $\mathbf{Ax} = \mathbf{b}$ i block tridiagonal then it is seen from equation (5.37a) that the matrix $\{(\lambda + \omega - 1)\mathbf{I} - \lambda\omega\mathbf{E} - \omega\mathbf{F}\}$ is also block tridiagonal with each square diagonal block a diagonal matrix. Therefore, by theorem 1,

$$Q(\lambda) = \det\{(\lambda + \omega - 1)\mathbf{I} - \alpha\lambda\omega\mathbf{E} - \alpha^{-1}\omega\mathbf{F}\}, \quad \alpha \neq 0.$$

Put $\alpha = \lambda^{-\frac{1}{2}}$. Then

$$\begin{aligned} Q(\lambda) &= \det\{(\lambda + \omega - 1)\mathbf{I} - \lambda^{\frac{1}{2}}\omega\mathbf{E} - \lambda^{\frac{1}{2}}\omega\mathbf{F}\} \\ &= \det[\omega\lambda^{\frac{1}{2}}\{\omega^{-1}\lambda^{-\frac{1}{2}}(\lambda + \omega - 1)\mathbf{I} - (\mathbf{E} + \mathbf{F})\}] \\ &= \omega^m\lambda^{\frac{1}{2}m}\det\{\omega^{-1}\lambda^{-\frac{1}{2}}(\lambda + \omega - 1)\mathbf{I} - \mathbf{B}\}. \end{aligned}$$

But by equation (5.38),

$$P(\mu) = \det(\mu\mathbf{I} - \mathbf{B}),$$

so

$$Q(\lambda) = \omega^m\lambda^{\frac{1}{2}m}P\{\omega^{-1}\lambda^{-\frac{1}{2}}(\lambda + \omega - 1)\}. \tag{5.41}$$

Now the eigenvalues μ_i of $\mathbf{B}$ are the zeros of the polynomial $P(\mu)$ and the eigenvalues λ_i of $\mathbf{H}(\omega)$ are the zeros of the polynomial $Q(\lambda)$. Hence, by (5.41), it follows that $Q(\lambda) = 0$ if and only i $\omega^{-1}\lambda^{-\frac{1}{2}}(\lambda + \omega - 1)$ is a root of $P(\mu) = 0$, i.e., if and only if

$$\omega^{-1}\lambda^{-\frac{1}{2}}(\lambda + \omega - 1) = \mu.$$

This relation is usually written as

$$(\lambda + \omega - 1)^2 = \lambda\omega^2\mu^2, \quad \omega \neq 0, \quad \lambda \neq 0. \tag{5.43}$$

Additional comments are made on this relationship after the section on consistent ordering.

The Gauss–Seidel point iteration

When $\omega = 1$, equation (5.43) gives that $\lambda = \mu^2$, and $Q(\lambda)$ takes the form

$$Q(\lambda) = \lambda^{\frac{1}{2}m}P(\lambda^{\frac{1}{2}}).$$

But

$$P(\lambda^{\frac{1}{2}}) = \begin{cases} \lambda^r g(\lambda), & \text{if } m \text{ is even,} \\ \lambda^{r+\frac{1}{2}} f(\lambda), & \text{if } m \text{ is odd,} \end{cases}$$

where $g(\lambda)$ and $f(\lambda)$ are the polynomials of (5.39a) with μ^2 replaced by λ. Therefore, to each eigenvalue $\mu_i = 0$ of **B** there corresponds an eigenvalue $\lambda_i = 0$ of **H**(1) and to each non-zero pair of eigenvalues $\pm\mu_i$ of **B** there corresponds the pair $\lambda_i = 0$, μ_i^2 of **H**(1). It follows from this that when the matrix **A** of the equations has the form (5.35) then the Gauss–Seidel method converges if and only if the Jacobi method converges. If both methods converge then

$$\rho\{\mathbf{H}(1)\} = \{\rho(\mathbf{B})\}^2 < 1,$$

and by the definition of the asymptotic rate of convergence R_∞, $R_\infty\{\mathbf{H}(1)\} = 2R_\infty(\mathbf{B})$, i.e., the Gauss–Seidel point iterative method converges twice as fast as the Jacobi point iterative method after a large number of iterations. (See worked example 5.3).

Theoretical determination of the optimum relaxation parameter ω_b

The following results hold when some of the eigenvalues μ_i of the Jacobi iteration matrix are complex (reference 35) but the analysis is difficult so we shall assume each μ_i is real.

Theorem 4

When the matrix **A** is block tridiagonal of the form (5.35), with non-zero diagonal elements, and all the eigenvalues of the Jacobi iteration matrix **B** associated with **A** are real and such that $0 < \rho(\mathbf{B}) < 1$, then

$$\omega_b = \frac{2}{1 + \sqrt{\{1 - \rho^2(\mathbf{B})\}}}$$

and

$$\rho\{\mathbf{H}(\omega_b)\} = \omega_b - 1.$$

Furthermore, the SOR method applied to the equations $\mathbf{Ax} = \mathbf{b}$ converges for all ω in the range $0 < \omega < 2$.

Proof

Write equation (5.43) as

$$\frac{1}{\omega}(\lambda + \omega - 1) = \pm \lambda^{\frac{1}{2}}\mu$$

and let

$$y_1(\lambda) = \frac{1}{\omega}(\lambda + \omega - 1) = \frac{1}{\omega}\lambda + 1 - \frac{1}{\omega}$$

and

$$y_2(\lambda) = \pm \lambda^{\frac{1}{2}}\mu.$$

Then the pair of eigenvalues of $\mathbf{H}(\omega)$ that correspond to the pair of non-zero eigenvalues $\pm\mu$ of $\mathbf{B}$ are the λ values of the points where the straight line

$$y_1 - 1 = \frac{1}{\omega}(\lambda - 1)$$

intersects the parabola

$$y_2^2 = \lambda\mu^2.$$

This straight line passes through the point $(1, 1)$ and its slope $1/\omega$ decreases as ω increases. From Fig. 5.6 it is seen that the largest abscissa of the two points of intersection decreases with increasing ω until the line is a tangent to the parabola. The λ values of the points of intersection are the λ roots of (5.43) which can be rearranged as

$$\lambda^2 + 2\lambda\{(\omega - 1) - \tfrac{1}{2}\omega^2\mu^2\} + (\omega - 1)^2 = 0.$$

For tangency,

$$\{(\omega - 1) - \tfrac{1}{2}\omega^2\mu^2\}^2 - (\omega - 1)^2 = 0,$$

giving

$$\omega = \bar{\omega} = \frac{2\{1 \pm (1 - \mu^2)^{\frac{1}{2}}\}}{\mu^2}. \tag{5.44}$$

The positive sign in (5.44) gives the value of ω corresponding to the other tangent from the point $(1, 1)$ and for which the λ value of the point of contact exceeds 1, i.e., a non-convergent case. The negative sign gives

$$\bar{\omega} = \frac{2}{1 + \sqrt{(1 - \mu^2)}},$$

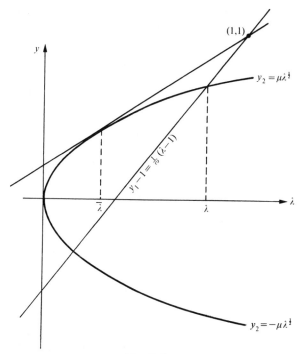

Fig. 5.6

and it is easily shown that the abscissa of the point of tangency is

$$\bar{\lambda} = \bar{\omega} - 1.$$

For values of $\omega > \bar{\omega}$ the λ roots of equation (5.43) are complex, each with modulus $|\lambda| = \omega - 1$ which increases with increasing ω. Since the SOR method diverges for $|\lambda| \ge 1$ it follows that $\omega \ge 2$ gives a divergent iteration, i.e., the iteration is convergent for $0 < \omega < 2$.

Hence we have shown that for a fixed eigenvalue μ of **B** the value of ω which minimizes the corresponding λ eigenvalue of largest modulus is $\bar{\omega}$. Since the maximum rate of convergence of the SOR method is determined by the minimum value of the eigenvalue of largest modulus of $\mathbf{H}(\omega)$, $\lambda = \lambda_1$, say, it follows that

$$\bar{\lambda}_1 = \bar{\omega}_1 - 1,$$

where

$$\bar{\omega}_1 = \frac{2}{1 + \sqrt{(1 - \mu_1^2)}}.$$

and λ_1 corresponds to $\pm\mu_1$. In other words the optimum value of ω for maximum rate of convergence is given by

$$\omega_b = \frac{2}{1 + \sqrt{\{1 - \rho^2(\mathbf{B})\}}},$$

where $\rho(\mathbf{B})$ is the spectral radius of the Jacobi iteration matrix.

The calculation of ω_b

For Poisson's equation over a rectangle of sides ph and qh, with Dirichlet boundary conditions, it is proved in exercise 17 that $\rho(\mathbf{B})$ for the five-point difference approximation, using a square mesh of side h, is

$$\rho(\mathbf{B}) = \frac{1}{2}\left(\cos\frac{\pi}{p} + \cos\frac{\pi}{q}\right).$$

In general, however, either the spectral radius of the Jacobi iteration matrix, or the spectral radius $\rho\{\mathbf{H}(1)\} = \rho^2(\mathbf{B})$ of the Gauss–Seidel iteration matrix must be estimated as on page 241 or by one of the methods given in references 39 and 35.

Re-ordering of equations and unknowns

As will be seen later it is sometimes necessary to re-order a set of equations, together with the unknowns, so that the matrix of the re-ordered set has some desired form. In all that follows the unknown corresponding to the ith mesh point will be denoted by x_i and the subscript will not be changed in any re-ordering. For illustrative purposes, say the equations corresponding to the first, second, third and fourth mesh points are

$$\begin{aligned}
a_{11}x_1 + a_{12}x_2 \phantom{+ a_{23}x_3} + a_{14}x_4 &= b_1, \\
a_{21}x_1 + a_{22}x_2 + a_{23}x_3 \phantom{+ a_{14}x_4} &= b_2, \\
a_{32}x_2 + a_{33}x_3 + a_{34}x_4 &= b_3, \\
a_{41}x_1 + \phantom{a_{32}x_2 +} a_{43}x_3 + a_{44}x_4 &= b_4,
\end{aligned}$$

respectively, which may be written as

$$\begin{bmatrix} a_{11} & a_{12} & 0 & a_{14} \\ a_{21} & a_{22} & a_{23} & 0 \\ 0 & a_{32} & a_{33} & a_{34} \\ a_{41} & 0 & a_{43} & a_{44} \end{bmatrix} \begin{bmatrix} x_1 \\ x_2 \\ x_3 \\ x_4 \end{bmatrix} = \begin{bmatrix} b_1 \\ b_2 \\ b_3 \\ b_4 \end{bmatrix},$$

i.e., as

$$\mathbf{A}\mathbf{x}_A = \mathbf{b}_A.$$

If we now write down the equations at the mesh points four, two, three and one, *in that order*, the column vector of unknowns being in the *same order*, namely, x_4, x_2, x_3, x_1 we obtain

$$\begin{bmatrix} a_{44} & 0 & a_{43} & a_{41} \\ 0 & a_{22} & a_{23} & a_{21} \\ a_{34} & a_{32} & a_{33} & 0 \\ a_{14} & a_{12} & 0 & a_{11} \end{bmatrix} \begin{bmatrix} x_4 \\ x_2 \\ x_3 \\ x_1 \end{bmatrix} = \begin{bmatrix} b_4 \\ b_2 \\ b_3 \\ b_1 \end{bmatrix}, \tag{5.45}$$

which may be represented by

$$\mathbf{C}\mathbf{x}_C = \mathbf{b}_C.$$

It is easily verified that the matrix $\mathbf{C}$ can be derived from matrix $\mathbf{A}$ by interchanging rows one and four of $\mathbf{A}$ to give matrix $\mathbf{B}$ then interchanging columns one and four of $\mathbf{B}$ to give $\mathbf{C}$, or by interchanging columns one and four of $\mathbf{A}$ to give $\mathbf{B}_1$ then rows one and four of $\mathbf{B}_1$ to give $\mathbf{C}$. In matrix algebra, rows one and four of $\mathbf{A}$ can be interchanged by premultiplying $\mathbf{A}$ by the elementary permutation matrix $\mathbf{P}_{14}$, where

$$\mathbf{P}_{14} = \begin{bmatrix} 0 & 0 & 0 & 1 \\ 0 & 1 & 0 & 0 \\ 0 & 0 & 1 & 0 \\ 1 & 0 & 0 & 0 \end{bmatrix}$$

is obtained by interchanging either rows one and four or columns one and four of the unit matrix of order four. Similarly, columns one and four of $\mathbf{B}$ can be interchanged by postmultiplying $\mathbf{B}$ with $\mathbf{P}_{14}$. Hence $\mathbf{C} = \mathbf{P}_{14}\mathbf{A}\mathbf{P}_{14}$. As $\mathbf{x}_C = \mathbf{P}_{14}\mathbf{x}_A$ and $\mathbf{b}_C = \mathbf{P}_{14}\mathbf{b}_A$, the re-ordered equations and unknowns (5.45) may be written as

$$(\mathbf{P}_{14}\mathbf{A}\mathbf{P}_{14})\mathbf{P}_{14}\mathbf{x}_A = \mathbf{P}_{14}\mathbf{b}_A. \tag{5.46}$$

Since $\mathbf{P}_{14} = \mathbf{P}_{41} = \mathbf{P}_{14}^T = \mathbf{P}_{41}^T = \mathbf{P}_{14}^{-1} = \mathbf{P}_{41}^{-1}$, it follows that $\mathbf{C} = \mathbf{P}_{14}\mathbf{A}\mathbf{P}_{14}^T = \mathbf{P}_{14}\mathbf{A}\mathbf{P}_{14}^{-1}$ is an orthogonal similarity transformation of $\mathbf{A}$. Therefore the eigenvalues of $\mathbf{C}$ are those of $\mathbf{A}$. This result would not hold if the unknowns were not ordered in the same way as the equations. It should also be noticed that the coefficient a_{ii} of x_i in the re-ordered equations is a diagonal element of $\mathbf{P}_{14}\mathbf{A}\mathbf{P}_{14}$. Clearly, all possible re-orderings of the equations and unknowns can be obtained by a succession of transformations (5.46). If, for example, the equations and unknowns of (5.46) are to be written down at the mesh points four, three, two and one, in that order, the corresponding matrix equation is

$$(\mathbf{P}_{23}\mathbf{P}_{14}\mathbf{A}\mathbf{P}_{14}\mathbf{P}_{23})\mathbf{P}_{23}\mathbf{P}_{14}\mathbf{x}_A = \mathbf{P}_{23}\mathbf{P}_{14}\mathbf{b}_A,$$

i.e.,

$$(\mathbf{P}\mathbf{A}\mathbf{P}^T)\mathbf{P}\mathbf{x}_A = \mathbf{P}\mathbf{b}_A,$$

where $\mathbf{P} = \mathbf{P}_{23}\mathbf{P}_{14}$ is called a permutation matrix and satisfies $\mathbf{P} = \mathbf{P}^T = \mathbf{P}^{-1}$.

Point iterative methods and re-orderings

In all point iterative methods the equation corresponding to the ith mesh point is 'solved' initially for x_i in terms of the other components of the solution vector $\mathbf{x}$. The iteration then expresses the unknown component $x_i^{(n+1)}$ of the $(n+1)$th iteration vector in terms of known $(n+1)$ and nth iterative values of the remaining components of $\mathbf{x}$.

With the point Jacobi method the $(n+1)$th iterative values are expressed exclusively in terms of nth iterative values so the order in which they are calculated, i.e., the order in which the mesh points are scanned, does not affect the values of successive iterates at a particular mesh point. Mathematically, this obvious result is equivalent to proving that the Jacobi iteration matrices associated with all possible orderings of equations and unknowns have the same set of eigenvalues, and the same set of eigenvectors in the sense that they have the same components but in different orders. The error vectors $\mathbf{x} - \mathbf{x}^{(n)}$ for different orderings will then have the same components but in different orders. This is proved in exercise 18.

With both the Gauss–Seidel and SOR point methods the latest iterates are used as soon as they are available. Any change therefore in the scanning of the mesh points, that is, in the order in which the iteration equations are 'solved' will, in general, affect the rate of convergence of the method. For this reason, any two Gauss–Seidel (or SOR) iteration matrices associated respectively with two different re-orderings of the same set of equations have, in general, different eigenvalues and vectors.

2-cyclic matrices and consistent orderings

The concept of consistent ordering is central to the theory of the SOR iterative method for solving the equations $\mathbf{Ax} = \mathbf{b}$ because at present the calculation of the optimum acceleration parameter is possible only for consistently ordered matrices.

The earliest definition was due to Young, reference 40, and was related to a class of matrices whose members possessed what was termed property (A). (A matrix $\mathbf{A}$ possesses property (A) if there exists a permutation matrix $\mathbf{P}$ such that the similarity transformation $\mathbf{PAP}^T$ is block tridiagonal of the form (5.35)). A number of more general definitions have since been formulated. In particular, Varga, reference 35, extended the concept of consistent ordering to the class of p-cyclic matrices which include property (A) matrices as the special case $p = 2$. All these definitions are dealt with in a book by Young, reference 39, which brings the theory of iterative procedures up to date.

In this book Young's original definitions are used because his concept of an ordering vector enables us to determine very easily a permutation matrix $\mathbf{P}$ that will transform a non-consistently ordered 2-cyclic matrix $\mathbf{A}$ into a consistently ordered 2-cyclic matrix $\mathbf{PAP}^T$. This use of ordering vectors has been extended to p-cyclic matrices in references 39 and 41.

The following presentation of 2-cyclic matrices and consistent ordering is based on N. Papamichael's M. Tech. dissertation, Brunel University, 1970, reference 28.

2-cyclic matrices

The reader is warned that the definition looks difficult. When related to an example however it will be seen to be easy.

Definition 1

The $N \times N$ matrix $\mathbf{A} = (a_{i,j})$ is 2-cyclic if there exist two disjoint subsets S and T of W, the set of the first N positive integers, such that $S \cup T = W$ and if $a_{i,j} \neq 0$ then either $i = j$ or $i \in S$ and $j \in T$ or $i \in T$ and $j \in S$.

Example 5.4

Consider

$$\mathbf{A} = \begin{bmatrix} a_{11} & a_{12} & 0 & 0 & a_{15} \\ a_{21} & a_{22} & a_{23} & 0 & 0 \\ 0 & a_{32} & a_{33} & a_{34} & 0 \\ a_{41} & 0 & a_{43} & a_{44} & 0 \\ a_{51} & 0 & a_{53} & 0 & a_{55} \end{bmatrix}.$$

Mentally exclude the diagonal elements as $i = j$.

In the first row $i = 1$. Let $1 \in S$. Then $j = 2, 5 \in T$.

In the second row $i = 2 \in T$, so $j = 1, 3 \in S$.

The final distribution of the first five positive integers is as below.

	S	T
Row 1	$i = 1$	$j = 2, 5$
Row 2	$j = 1, 3$	$i = 2$
Row 3	$i = 3$	$j = 2, 4$
Row 4	$j = 1, 3$	$i = 4$
Row 5	$j = 1, 3$	$i = 5$

Since the sets $S = \{1, 3\}$ and $T = \{2, 4, 5\}$ are disjoint and $S \cup T = \{1, 2, 3, 4, 5\} = W$, it follows that matrix $\mathbf{A}$ is 2-cyclic. If the matrix was p-cyclic there would be p disjoint non-empty subsets.

Example 5.5

Consider

$$\mathbf{A} = \begin{bmatrix} a_{11} & a_{12} & a_{13} & 0 \\ 0 & a_{22} & a_{23} & a_{24} \\ a_{31} & 0 & a_{33} & a_{34} \\ a_{41} & a_{42} & 0 & a_{44} \end{bmatrix}.$$

As before,

	S	T
Row 1	$i = 1$	$j = 2, 3$
Row 2	$j = 3, 4$	$i = 2$

At the second step of the process the sets S and T cease to be disjoint. Therefore matrix $\mathbf{A}$ is not 2-cyclic.

Theorem 1

A matrix $\mathbf{A}$ is 2-cyclic if and only if there exists a row vector $\boldsymbol{\gamma} = (\gamma_1, \gamma_2, \ldots, \gamma_N)$ with integral components such that if $a_{ij} \neq o$ and $i \neq j$ then $|\gamma_i - \gamma_j| = 1$.

Proof

Assume that matrix A is 2-cyclic. Referring to definition 1 let $\gamma_i = 1$ if $i \in S$ and $\gamma_i = 0$ if $i \in T$. If $a_{ij} \neq 0$ and $i \neq j$ then either $i \in S$ and $j \in T$ and hence $\gamma_i = 1$, $\gamma_j = 0$, or else $i \in T$ and $j \in S$ and hence $\gamma_i = 0$ and $\gamma_j = 1$. In either case $|\gamma_i - \gamma_j| = 1$.

Conversely we must show that if $a_{ij} \neq 0$, $i \neq j$ and $|\gamma_i - \gamma_j| = 1$, then matrix $\mathbf{A}$ is 2-cyclic. Let

(i) S denote a set of integers i for which γ_i is odd,

(ii) T denote a set of integers i for which γ_i is even. For example, if

$$\begin{array}{c|c} \gamma_1 = 3 & \gamma_3 = 4 \\ \gamma_2 = 1 & \gamma_4 = 2 \\ \gamma_5 = 5 & \end{array}$$

then the sets in (i) and (ii) are $S = \{1, 2, 5\}$ and $T = \{3, 4\}$. If $i \in S$, γ_i is odd so it follows that γ_j is even since $|\gamma_i - \gamma_j| = 1$ and the difference of two odd numbers is even. Hence $j \in T$. Similarly $i \in T$ implies that $j \in S$. Hence matrix $\mathbf{A}$ is 2-cyclic.

Definition 2

A vector $\boldsymbol{\gamma}$ with the properties given in theorem 1 is said to be an *ordering vector* for the matrix $\mathbf{A}$.

By this definition an ordering vector for a given 2-cyclic matrix will contain $M \geq 2$ distinct components. The actual numerical value of a component of $\boldsymbol{\gamma}$ is not important. Only the difference between any two components is significant so the components of an ordering vector with M distinct components can always be taken as $0, 1, 2, \ldots, M-1$.

If $\mathbf{A}$ is 2-cyclic then two ordering vectors $\boldsymbol{\gamma}^{(1)}$ and $\boldsymbol{\gamma}^{(2)}$, each with two distinct components, may be obtained by setting

$$\gamma_i^{(1)} = \begin{cases} 0, \text{ if } i \in S \\ 1, \text{ if } i \in T \end{cases}$$

and

$$\gamma_i^{(2)} = \begin{cases} 1, \text{ if } i \in S \\ 0, \text{ if } i \in T \end{cases}$$

Example 5.6

Write down two ordering vectors for the 2-cyclic matrix of example 5.4.

In that example $S = \{1, 3\}$ and $T = \{2, 4, 5\}$. Therefore we can set $\gamma_1 = \gamma_3 = 0$ and $\gamma_2 = \gamma_4 = \gamma_5 = 1$, giving $\boldsymbol{\gamma}^{(1)} = (0, 1, 0, 1, 1)$, or put $\gamma_1 = \gamma_3 = 1$ *and* $\gamma_2 = \gamma_4 = \gamma_5 = 0$ giving $\boldsymbol{\gamma}^{(2)} = (1, 0, 1, 0, 0)$.

In general, for a given 2-cyclic matrix, there exist ordering vectors with $M > 2$ distinct components. Such vectors can be derived from $\boldsymbol{\gamma}^{(1)}$ or $\boldsymbol{\gamma}^{(2)}$ by replacing some of the γ_i by $(\gamma_i + 2)$ but the choice is restricted by conditions specified in reference 41. For small matrices all possible ordering vectors may be obtained by setting $\boldsymbol{\gamma} = (0, \gamma_1, \gamma_2, \ldots, \gamma_{N-1})$ then applying $|\gamma_i - \gamma_j| = 1$ to each $a_{ij} \neq 0$, $i \neq j$. This method applied to example 5.4 yields

$$\boldsymbol{\gamma}^{(1)} = (0, 1, 0, 1, 1), \quad \boldsymbol{\gamma}^{(2)} = (1, 0, 1, 0, 0), \quad \boldsymbol{\gamma}^{(3)} = (1, 0, 1, 2, 2),$$
$$\boldsymbol{\gamma}^{(4)} = (1, 0, 1, 2, 0), \quad \boldsymbol{\gamma}^{(5)} = (1, 0, 1, 0, 2), \quad \boldsymbol{\gamma}^{(6)} = (2, 1, 0, 1, 1),$$
$$\boldsymbol{\gamma}^{(7)} = (1, 2, 1, 0, 2), \quad \boldsymbol{\gamma}^{(8)} = (1, 2, 1, 0, 0), \quad \boldsymbol{\gamma}^{(9)} = (0, 1, 2, 1, 1),$$
$$\boldsymbol{\gamma}^{(10)} = (1, 2, 1, 2, 0).$$

It should be noted that the method indicated gives $\boldsymbol{\gamma}^{(2)}$ as $(0, -1, 0, -1, -1)$ but every component can be made non-negative by adding 1 to each component.

An ordering vector $\boldsymbol{\gamma} = \{\gamma_i\}$ with $M > 2$ distinct components can always be transformed into a vector $\boldsymbol{\gamma}' = \{\gamma_i'\}$ with 2 distinct components by putting $\gamma_i' = \gamma_i \pmod 2$. For example, putting $\gamma_i' = \gamma_i^{(6)} \pmod 2$ transforms $\boldsymbol{\gamma}^{(6)}$ into the vector $(0, 1, 0, 1, 1) = \boldsymbol{\gamma}^{(1)}$.

Consistent orderings

Young defined a consistently ordered matrix as follows.

Definition 3

A 2-cyclic $N \times N$ matrix is consistently ordered if there exists an ordering vector $\boldsymbol{\gamma} = (\gamma_1, \gamma_2, \ldots, \gamma_N)$ such that if $a_{ij} \neq 0$ and $j > i$ then $\gamma_j - \gamma_i = 1$ and if $i > j$ then $\gamma_j - \gamma_i = 1$.

It follows that if the components of γ are in ascending order of magnitude and γ is an ordering vector for the matrix $\mathbf{A}$ then $\mathbf{A}$ is consistently ordered.

The converse is not necessarily true, i.e., a consistently ordered matrix $\mathbf{A}$ might have an ordering vector with successive components not in ascending order of magnitude. (See example 5.8).

Theorem 2

If the matrix $\mathbf{A}$ is 2-cyclic then there exists a permutation matrix $\mathbf{P}$ such that $\mathbf{PAP}^T$ is consistently ordered.

Proof

Let γ be an ordering vector for the matrix $\mathbf{A}$. By interchanging the components of γ it is always possible to obtain a vector γ' with components in ascending order of magnitude.

The matrix $\mathbf{P}$ may be obtained as follows. If the first interchange of the components of γ are γ_i and γ_j, write down the elementary permutation matrix $\mathbf{P}_{ij}$. If the next interchange are the components γ_m and γ_n, premultiply $\mathbf{P}_{ij}$ by $\mathbf{P}_{mn}$. The final matrix $\ldots \mathbf{P}_{mn}\mathbf{P}_{ij}$ given by this process is $\mathbf{P}$ and the final ordering vector γ' is an ordering vector for the consistently ordered matrix $\mathbf{PAP}^T$.

Clearly $\mathbf{P}$ is not unique and the number of consistently ordered matrices $\mathbf{PAP}^T$ that can be obtained this way from γ equals the number of ways γ can be transformed into γ'.

Example 5.7

Consider the matrix $\mathbf{A}$ of example 5.4 for which one ordering vector is $\gamma^{(1)} = (0, 1, 0, 1, 1)$. The vector $\gamma^{(1)'} = (0, 0, 1, 1, 1)$ can be obtained by interchanging elements 2 and 3 of $\gamma^{(1)}$. Therefore $\mathbf{P} = \mathbf{P}_{23}$ and the corresponding consistently ordered matrix $\mathbf{PAP}^T$ is

$$\mathbf{P}_{23}\mathbf{A}\mathbf{P}_{23} = \begin{bmatrix} a_{11} & 0 & \vdots & a_{12} & 0 & a_{15} \\ 0 & a_{33} & \vdots & a_{32} & a_{34} & 0 \\ \cdots & \cdots & & \cdots & \cdots & \cdots \\ a_{21} & a_{23} & \vdots & a_{22} & 0 & 0 \\ a_{41} & a_{43} & \vdots & 0 & a_{44} & 0 \\ a_{51} & a_{53} & \vdots & 0 & 0 & a_{55} \end{bmatrix}$$

(See 'Re-ordering of equations and unknowns'.) Let $\mathbf{P}\mathbf{A}\mathbf{P}^T = \mathbf{C} = [c_{ij}]$, then

$$\mathbf{C} = \begin{bmatrix} c_{11} & 0 & c_{13} & 0 & c_{15} \\ 0 & c_{22} & c_{23} & c_{24} & 0 \\ \hline c_{31} & c_{32} & c_{33} & 0 & 0 \\ c_{41} & c_{42} & 0 & c_{44} & 0 \\ c_{51} & c_{52} & 0 & 0 & c_{55} \end{bmatrix}.$$

As expected, $\mathbf{C}$ is 2-cyclic and consistently ordered with ordering vector $\boldsymbol{\gamma}^{(1)'} = (0, 0, 1, 1, 1)$. Another ordering vector for $\mathbf{A}$ is $\boldsymbol{\gamma}^{(2)} = (1, 0, 1, 0, 0)$. Hence $(1, 0, 1, 0, 0) \rightarrow (0, 0, 1, 0, 1)$. $\mathbf{P}_{15}$. Then $(0, 0, 1, 0, 1) \rightarrow (0, 0, 0, 1, 1)$. $\mathbf{P}_{34}$. Therefore $\mathbf{P} = \mathbf{P}_{34}\mathbf{P}_{15}$ and the corresponding consistently ordered matrix $\mathbf{P}\mathbf{A}\mathbf{P}^T$ is

$$\mathbf{P}_{34}\mathbf{P}_{15}\mathbf{A}\mathbf{P}_{15}\mathbf{P}_{34} = \begin{bmatrix} a_{55} & 0 & 0 & a_{53} & a_{51} \\ 0 & a_{22} & 0 & a_{23} & a_{21} \\ 0 & 0 & a_{44} & a_{43} & a_{41} \\ \hline 0 & a_{32} & a_{34} & a_{33} & 0 \\ a_{15} & a_{12} & 0 & 0 & a_{11} \end{bmatrix}.$$

A third ordering vector is $\boldsymbol{\gamma}^{(3)} = (1, 0, 1, 2, 2)$ for which $\mathbf{P} = \mathbf{P}_{12}$. Therefore a consistent ordering associated with $\boldsymbol{\gamma}^{(3)}$ is

$$\mathbf{P}\mathbf{A}\mathbf{P}^T = \mathbf{P}_{12}\mathbf{A}\mathbf{P}_{12} = \begin{bmatrix} a_{22} & a_{21} & a_{23} & 0 & 0 \\ \hline a_{12} & a_{11} & 0 & 0 & a_{15} \\ a_{32} & 0 & a_{33} & a_{34} & 0 \\ \hline 0 & a_{41} & a_{43} & a_{44} & 0 \\ 0 & a_{51} & a_{53} & 0 & a_{55} \end{bmatrix}.$$

Consistent orderings and block tridiagonal matrices

All the consistently ordered matrices of example 5.7 have the block tridiagonal form

$$\begin{bmatrix} \mathbf{D}_1 & \mathbf{A}_1 & & & \\ \mathbf{B}_1 & \mathbf{D}_2 & \mathbf{A}_2 & & \\ & \cdot & \cdot & \cdot & \\ & & \mathbf{B}_{M-2} & \mathbf{D}_{M-1} & \mathbf{A}_{M-1} \\ & & & \mathbf{B}_{M-1} & \mathbf{D}_M \end{bmatrix}, \quad M \geq 2, \tag{5.47}$$

where each $\mathbf{D}_i$, $i = 1(1)M$, is a square diagonal matrix, the forms associated with $\boldsymbol{\gamma}^{(1)}$ and $\boldsymbol{\gamma}^{(2)}$ corresponding to the special case $M = 2$.

Theorem 3

A 2-cyclic consistently ordered matrix $\mathbf{A}$ has the block tridiagonal form (5.47) if and only if it has an ordering vector with components arranged in ascending order of magnitude.

Proof

Assume that the ordering vector $\boldsymbol{\gamma} = \{\gamma_i\}_{i=1}^N$ with $M \geqslant 2$ distinct components arranged in ascending order of magnitude is an ordering vector for the 2-cyclic consistently ordered $N \times N$ matrix $\mathbf{A}$. Let $\boldsymbol{\gamma}$ have n_0 components equal to 0, n_1 components equal to 1, n_2 components equal to 2, etc., that is,

$$\boldsymbol{\gamma} = (\underbrace{0, 0, \ldots, 0}_{n_0}; \underbrace{1, 1, \ldots, 1}_{n_1}; \underbrace{222 \ldots, 2}_{n_2}; \underbrace{3, \ldots, 3}_{n_3}; \ldots) \quad (5.48)$$

By hypothesis, A is 2-cyclic and consistently ordered so the vector $\boldsymbol{\gamma}$ is such that if $a_{ij} \neq 0$ and $i > j$ then $\gamma_i - \gamma_j = 1$ and if $j > i$ then $\gamma_j - \gamma_i = 1$.

Now consider, for example, the third row of blocks of Fig. 5.7 from left to right. In the first block, $i = (n_0 + n_1 + 1)(1)(n_0 + n_1 + n_2)$, $\gamma_i = 2$, $j = 0(1)n_0$ and $\gamma_j = 0$. Hence $i > j$ but $\gamma_i - \gamma_j = 2$. Therefore $a_{ij} = 0$.

In the second block, $i = (n_0 + n_1 + 1)(1)(n_0 + n_1 + n_2)$, $\gamma_i = 2$, $j = (n_0 + 1)(1)(n_0 + n_1)$ and $\gamma_j = 1$. Hence $i > j$ and $\gamma_i - \gamma_j = 1$ so a_{ij} can be non-zero.

In the third block, irrespective of whether $i > j$ or $j > i$, $\gamma_i = \gamma_j = 2$ so $|\gamma_i - \gamma_j| = 0$, which is contrary to the definition of a 2-cyclic matrix. Hence $a_{ij} = 0$ if $i \neq j$. When $i = j$, a_{ij} need not be zero. Therefore the block on the main diagonal of $\mathbf{A}$ is a square diagonal matrix.

In the fourth block, $j > i$, $\gamma_j = 3$ and $\gamma_i = 2$. Therefore $\gamma_j - \gamma_i = 1$ so a_{ij} can be non-zero.

The same argument applied to blocks $5(1)M$ shows that each of these blocks is a null matrix.

In general, any block which is such that the difference between its row-wise value of γ_1 and its column-wise value of γ_j has a modulus greater that 1 must consist of zero elements. In general, every block on the main diagonal of $\mathbf{A}$ is a diagonal matrix.

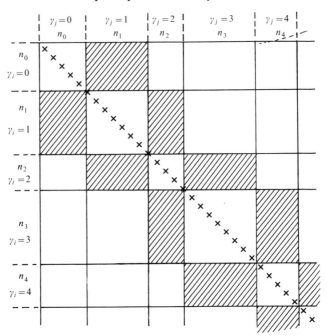

Fig. 5.7

The observations clearly establish matrix **A** as block tridiagonal of the form (5.47).

Conversely, assume that **A** has the block form (5.47) where each $\mathbf{D}_i$ is an $n_{i-1} \times n_{i-1}$ diagonal matrix. To prove that **A** is 2-cyclic and consistently ordered it is sufficient to find an ordering vector for **A** that has components in ascending order of magnitude. Such a vector is given by (5.48).

It follows from the proof of theorem 3 that the consistently ordered block tridiagonal matrix $\mathbf{PAP}^T$ associated with an ordering vector $\boldsymbol{\gamma}$ for **A** is determined by the number of distinct components of $\boldsymbol{\gamma}$ in that $\mathbf{PAP}^T$ has the form (5.47) where:

(i) M is equal to the number of distinct components in $\boldsymbol{\gamma}$.

(ii) Each $\mathbf{D}_i$ is a diagonal matrix of order n_{i-1} where n_{i-1} is the number of components in $\boldsymbol{\gamma}$ equal to $(i-1)$.

In particular, one ordering vector for a 2-cyclic matrix **A** is

$\boldsymbol{\gamma} = \{\gamma_i\}_{i=1}^N$ where

$$\gamma_i = \begin{cases} 0, & \text{if } i \in S \\ 1, & \text{if } i \in T, \end{cases}$$

and S and T are the sets of definition 1. By theorem 3 the 2-cyclic consistently ordered matrix $\mathbf{PAP}^T$ associated with this $\boldsymbol{\gamma}$ will be

$$\begin{bmatrix} \mathbf{D}_1 & \mathbf{A} \\ \mathbf{B} & \mathbf{D}_2 \end{bmatrix}, \tag{5.49}$$

where the $\mathbf{D}_i$ are square diagonal matrices. Hence another definition of 2-cyclic matrices is as follows.

Definition 4

The $N \times N$ matrix $\mathbf{A}$ is 2-cyclic if there exists an $N \times N$ permutation matrix $\mathbf{P}$ such that $\mathbf{PAP}^T$ has the block form (5.49).

As mentioned before, although a block tridiagonal matrix of the form (5.47) is consistently ordered the converse is not necessarily true.

Example 5.8

Consider the matrix

$$\mathbf{A} = \begin{bmatrix} a_{11} & 0 & 0 & a_{14} \\ a_{21} & a_{22} & a_{23} & 0 \\ 0 & a_{32} & a_{33} & 0 \\ a_{41} & 0 & 0 & a_{44} \end{bmatrix}$$

for which one ordering vector is $\boldsymbol{\gamma} = (0, 1, 2, 1)$. $\mathbf{A}$ is not block tridiagonal. Nevertheless it is consistently ordered by definition 3 because

$$\gamma_4 - \gamma_1 = \gamma_2 - \gamma_1 = \gamma_3 - \gamma_2 = 1.$$

Additional comments on consistent ordering and the SOR method

(i) The proof of

$$(\lambda + \omega - 1)^2 = \lambda \omega^2 \mu^2, \quad \omega \neq 0, \quad \lambda \neq 0, \tag{5.50}$$

depended on the matrix $\mathbf{A}$ being block tridiagonal. It has been proved however, reference 39, that this relation holds for any 2-cyclic consistently ordered matrix $\mathbf{A}$ and not just for block tridiagonal matrices. (See exercise 21.)

(ii) If the matrix $\mathbf{A}$ of the equations $\mathbf{A}\mathbf{x} = \mathbf{b}$ is 2-cyclic but inconsistently ordered, its 2-cyclic property implies the existence of a permutation matrix $\mathbf{P}$ such that $\mathbf{P}\mathbf{A}\mathbf{P}^T$ is consistently ordered. The equations and unknowns can then be re-ordered into the form $(\mathbf{P}\mathbf{A}\mathbf{P}^T)\mathbf{P}\mathbf{x} = \mathbf{P}\mathbf{b}$ and this ensures that (5.50) holds. This is the significance of $\mathbf{A}$ being 2-cyclic.

(iii) Relation (5.50) shows that the eigenvalues λ_i of $\mathbf{H}(\omega)$ depend only on the eigenvalues μ_i of the Jacobi iteration matrix $\mathbf{B}$ and the accelerating factor ω. Since the eigenvalues of $\mathbf{B}$ remain unchanged by any transformation of $\mathbf{A}$ to $\mathbf{P}\mathbf{A}\mathbf{P}^T$, (See exercise 18), it follows that the SOR iteration matrices associated with two different orderings $\mathbf{P}_1\mathbf{A}\mathbf{P}_1^T$ and $\mathbf{P}_2\mathbf{A}\mathbf{P}_2^T$ have the same set of eigenvalues. In particular they have the same spectral radii. Therefore the SOR method applied to the equations $(\mathbf{P}_1\mathbf{A}\mathbf{P}_1^T)\mathbf{P}_1\mathbf{x} = \mathbf{P}_1\mathbf{b}$ and $(\mathbf{P}_2\mathbf{A}\mathbf{P}_2^T)\mathbf{P}_2\mathbf{x} = \mathbf{P}_2\mathbf{b}$ gives the same asymptotic rates of convergence because successive error vectors, for sufficiently large n, are related by the equation $\mathbf{e}^{(n+1)} \simeq \lambda_1\mathbf{e}^{(n)}$, where $|\lambda_1| = \rho(\mathbf{H})$.

(iv) If the SOR iteration matrices associated with two different consistent orderings $\mathbf{P}_1\mathbf{A}\mathbf{P}_1^T$ and $\mathbf{P}_2\mathbf{A}\mathbf{P}_2^T$ have also the same set of eigenvectors then the arithmetics of the SOR iterations will be identical because of equation (5.27). This occurs, for example, when two equal components of an ordering vector for a consistently ordered matrix are interchanged. When the matrix $\mathbf{A}$ is block tridiagonal, as in (5.47), the interchange corresponds to the re-ordering of rows and columns within any diagonal submatrix $\mathbf{D}_j$ of $\mathbf{A}$, $j = 1(1)M$.

(v) The concepts of 2-cyclicity and consistent ordering can be generalized to block partitioned matrices. In particular the nine-point finite-difference equations approximating Poisson's equation are 2-cyclic in block form. See reference 39.

Consistent orderings associated with the five-point difference approximation to Poisson's equation

Consider the Dirichlet problem

$$\frac{\partial^2 U}{\partial x^2} + \frac{\partial^2 U}{\partial y^2} = f(x, y), \quad (x, y) \in D,$$

$$U(x, y) = g(x, y), \quad (x, y) \in C, \tag{5.51}$$

where $G = D \cup C$ is the unit square

$$G = \{(x, y) : 0 \leqslant x \leqslant 1, 0 \leqslant y \leqslant 1\}.$$

This notation means that G is the set of points (x, y) such that $0 \leqslant x \leqslant 1$ and $0 \leqslant y \leqslant 1$.) Cover the square G, in the usual manner, by a square mesh of side h, where $Nh = 1$. Then a numerical solution of (5.51) can be obtained by approximating Poisson's equation at the internal mesh points $(x_i, y_j) = (ih, jh)$, $i, j = 1(1)(N-1)$, by the five-point difference equation

$$u_{i-1,j} + u_{i+1,j} + u_{i,j-1} + u_{i,j+1} - 4u_{i,j} = h^2 f_{i,j}. \qquad (5.52)$$

Take $N = 4$ and label the mesh points as shown in Fig. 5.8.

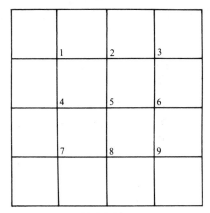

Fig. 5.8

If the difference equations are taken in the natural order of the points, i.e., in the order 1, 2, 3, 4, 5, 6, 7, 8, 9 and the vector of unknowns is ordered in the same way, the matrix of coefficients is

$$\mathbf{A} = \begin{bmatrix} -4 & 1 & 0 & \vdots & 1 & 0 & 0 & \vdots & 0 & 0 & 0 \\ 1 & -4 & 1 & \vdots & 0 & 1 & 0 & \vdots & 0 & 0 & 0 \\ 0 & 1 & -4 & \vdots & 0 & 0 & 1 & \vdots & 0 & 0 & 0 \\ \cdots & & & & & & & & & & \\ 1 & 0 & 0 & \vdots & -4 & 1 & 0 & \vdots & 1 & 0 & 0 \\ 0 & 1 & 0 & \vdots & 1 & -4 & 1 & \vdots & 0 & 1 & 0 \\ 0 & 0 & 1 & \vdots & 0 & 1 & -4 & \vdots & 0 & 0 & 1 \\ \cdots & & & & & & & & & & \\ 0 & 0 & 0 & \vdots & 1 & 0 & 0 & \vdots & -4 & 1 & 0 \\ 0 & 0 & 0 & \vdots & 0 & 1 & 0 & \vdots & 1 & -4 & 1 \\ 0 & 0 & 0 & \vdots & 0 & 0 & 1 & \vdots & 0 & 1 & -4 \end{bmatrix} \begin{matrix} 1 \\ 2 \\ 3 \\ 4 \\ 5 \\ 6 \\ 7 \\ 8 \\ 9 \end{matrix} \qquad (5.53)$$

where the column of numbers on the right-hand side indicates the order of the equations and unknowns. For this matrix the distribution of the elements in the sets S and T of the definition of 2-cyclic matrices is as follows.

	S	T
Row 1	$i = 1$	$j = 2, 4$
Row 2	$j = 1, 3, 5$	$i = 2$
Row 4	$j = 1, 5, 7$	$i = 4$
Row 3	$i = 3$	$j = 2, 6$
Row 5	$i = 5$	$j = 2, 4, 6, 8$
Row 7	$i = 7$	$j = 4, 8$
Row 6	$j = 3, 5, 9$	$i = 6$
Row 8	$j = 5, 7, 9$	$i = 8$
Row 9	$i = 9$	$j = 6, 8$

Thus

$$S = \{1, 3, 5, 7, 9\}, \quad T = \{2, 4, 6, 8\},$$

$S \cup T = W$ and $S \cap T = \phi$. Therefore matrix **A** is 2-cyclic.

It follows immediately that two possible ordering vectors for **A** are

$$\boldsymbol{\gamma}^{(1)} = (0, 1, 0, 1, 0, 1, 0, 1, 0)$$

and

$$\boldsymbol{\gamma}^{(2)} = (1, 0, 1, 0, 1, 0, 1, 0, 1).$$

A third ordering vector which can either be obtained by means of a theorem given by Papamichael and Smith, reference 14, or as indicated earlier, is

$$\boldsymbol{\gamma}^{(3)} = (0, 1, 2, 1, 2, 3, 2, 3, 4).$$

Although the components of $\boldsymbol{\gamma}^{(3)}$ are not in ascending order of magnitude it is easily checked that matrix **A** is consistently ordered with respect to $\boldsymbol{\gamma}^{(3)}$. Therefore matrix **A** is consistently ordered but does not have the block tridiagonal form of (5.47) because each diagonal block is not a diagonal submatrix.

A block tridiagonal consistent ordering of **A**, associated with $\boldsymbol{\gamma}^{(1)}$, may be obtained by scanning the points in the order 1, 3, 5, 7, 9, 2,

4, 6, 8 and gives the matrix

$$\mathbf{A}_1 = \begin{bmatrix} -4 & 0 & 0 & 0 & 0 & 1 & 1 & 0 & 0 \\ 0 & -4 & 0 & 0 & 0 & 1 & 0 & 1 & 0 \\ 0 & 0 & -4 & 0 & 0 & 1 & 1 & 1 & 1 \\ 0 & 0 & 0 & -4 & 0 & 0 & 1 & 0 & 1 \\ 0 & 0 & 0 & 0 & -4 & 0 & 0 & 1 & 1 \\ 1 & 1 & 1 & 0 & 0 & -4 & 0 & 0 & 0 \\ 1 & 0 & 1 & 1 & 0 & 0 & -4 & 0 & 0 \\ 0 & 1 & 1 & 0 & 1 & 0 & 0 & -4 & 0 \\ 0 & 0 & 1 & 1 & 1 & 0 & 0 & 0 & -4 \end{bmatrix} \begin{matrix} 1 \\ 3 \\ 5 \\ 7 \\ 9 \\ 2 \\ 4 \\ 6 \\ 8 \end{matrix} \quad (5.54)$$

Similarly, a block tridiagonal consistent ordering associated with $\boldsymbol{\gamma}^{(2)}$ may be obtained by scanning the points in the order 2, 4, 6, 8, 1, 3, 5, 7, 9. It is seen therefore that consistent orderings of $\mathbf{A}$ can be obtained by labelling the mesh points black and white, as on a chessboard, and taking all the 'white' equations before the 'black' equations, or vice-versa, the unknowns being ordered in the same manner as the equations.

Since

$$\mathbf{P}_{67}\mathbf{P}_{34}[\boldsymbol{\gamma}^{(3)}]^T = (0, 1, 1, 2, 2, 3, 3, 4)^T,$$

a block tridiagonal consistent ordering of $\mathbf{A}$ associated with $\boldsymbol{\gamma}^{(3)}$ may be obtained by scanning the points in the order 1, 2, 4, 3, 5, 7, 6, 8, 9 and gives the matrix

$$\mathbf{A}_2 = \begin{bmatrix} -4 & 1 & 1 & 0 & 0 & 0 & 0 & 0 & 0 \\ 1 & -4 & 0 & 1 & 1 & 0 & 0 & 0 & 0 \\ 1 & 0 & -4 & 0 & 1 & 1 & 0 & 0 & 0 \\ 0 & 1 & 0 & -4 & 0 & 0 & 1 & 0 & 0 \\ 0 & 1 & 1 & 0 & -4 & 0 & 1 & 1 & 0 \\ 0 & 0 & 1 & 0 & 0 & -4 & 0 & 1 & 0 \\ 0 & 0 & 0 & 1 & 1 & 0 & -4 & 0 & 1 \\ 0 & 0 & 0 & 0 & 1 & 1 & 0 & -4 & 1 \\ 0 & 0 & 0 & 0 & 0 & 0 & 1 & 1 & -4 \end{bmatrix} \begin{matrix} 1 \\ 2 \\ 4 \\ 3 \\ 5 \\ 7 \\ 6 \\ 8 \\ 9 \end{matrix}$$

Another block tridiagonal consistent ordering of **A**, associated with $\gamma^{(3)}$, is obtained by scanning the points in the order 1, 4, 2, 7, 5, 3, 8, 6, 9. Hence a consistent ordering of **A** can be obtained by scanning the mesh points on successive diagonals in the same direction.

The matrix obtained by scanning successive mesh lines in opposite directions, i.e., by scanning the points in the order 1, 2, 3, 6, 5, 4, 7, 8, 9, is not consistently ordered.

Stone's strongly implicit iterative method

The point SOR method is fully explicit, the pth equation being used to calculate the next iterative value of only the component u_p. A strongly implicit method calculates all of the next iterative values by a direct elimination method. Naturally one hopes that the successive vectors of iterative values will approximate the exact solution vector very closely after only a few iterations. Let the function U satisfy the elliptic differential equation

$$a_1(x, y)\frac{\partial^2 U}{\partial x^2} + a_2(x, y)\frac{\partial^2 U}{\partial y^2} + a_3(x, y)\frac{\partial U}{\partial x}$$
$$+ a_4(x, y)\frac{\partial U}{\partial y} + a_5(x, y)U = Q(x, y) \quad (5.55)$$

at every interior point of the rectangle

$$S = \{(x, y) : 0 \leq x \leq a, 0 \leq y \leq b\}$$

and have known values on its boundary. Cover the solution domain with rectangular meshes of sides h and k such that $(N+1)h = a$, $(M+1)k = b$, and label the internal mesh points as shown in Fig. 5.9 so as to avoid double subscript notation. Then a five-point difference approximation to (5.55) at the pth mesh point can be written as

$$B_p u_{p-N} + D_p u_{p-1} + E_p u_p + F_p u_{p+1} + H_p u_{p+N} = q_p. \quad (5.56)$$

When the equations at the mesh points 1, 2, 3, ..., p, $p+1$, ..., MN, in that order, are expressed as $\mathbf{Au} = \mathbf{q}$, the matrix **A** is as shown in Fig. 5.10, where all the coefficients B_p, $p = (N+1)(1)MN$, lie on the diagonal labelled B, all the coefficients D_p, $p = 2(1)MN$, lie on the diagonal labelled D, etc. The F and D diagonals are immediately above and below the E diagonal. Thus

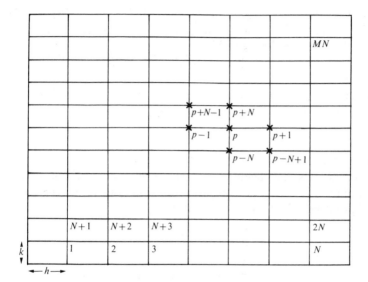

Fig. 5.9

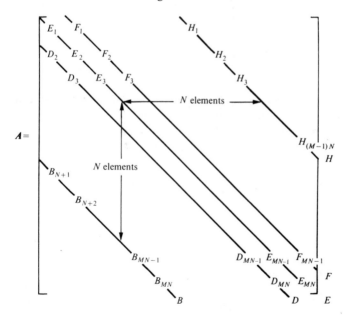

Fig. 5.10

matrix **A** is of bandwidth $2N$. For large N the standard **LU** decomposition 'fills-in' a very large number of zeros, the zeros between the B and H diagonals.

Stone's idea was to modify the matrix **A** by the addition of a 'small' matrix **N** so that:

(i) The factorization of $(\mathbf{A}+\mathbf{N})$ into the product $\bar{\mathbf{L}}\bar{\mathbf{U}}$ involves much less arithmetic than the standard **LU** decomposition of **A**.

(ii) The elements of $\bar{\mathbf{L}}$ and $\bar{\mathbf{U}}$ are easily calculated.

(iii) $\|\mathbf{N}\| \ll \|\mathbf{A}\|$.

Assuming this has been done, the development of an iterative procedure to calculate the solution of $\mathbf{Au}=\mathbf{q}$ is then quite straightforward because the equation can be written as

$$(\mathbf{A}+\mathbf{N})\mathbf{u} = (\mathbf{A}+\mathbf{N})\mathbf{u} + (\mathbf{q}-\mathbf{Au})$$

which suggests the iterative procedure

$$(\mathbf{A}+\mathbf{N})\mathbf{u}^{(n+1)} = (\mathbf{A}+\mathbf{N})\mathbf{u}^{(n)} + (\mathbf{q}-\mathbf{Au}^{(n)}). \qquad (5.57)$$

When the right-hand side is known, equation (5.57) gives an efficient method for solving *directly* for $\mathbf{u}^{(n+1)}$ because the factorization of $(\mathbf{A}+\mathbf{N})$ into $\bar{\mathbf{L}}\bar{\mathbf{U}}$ is efficient. If $\|\mathbf{N}\| \ll \|\mathbf{A}\|$ one would intuitively expect a rapid rate of convergence.

To achieve his objectives Stone decided that $\bar{\mathbf{L}}$ and $\bar{\mathbf{U}}$ would each have only three non-zero elements per row as illustrated in Fig. 5.11. The non-zero elements of $\bar{\mathbf{L}}$ lie on the diagonals b, c and d, corresponding to the diagonals B, D and E of **A** respectively, and the non-zero elements of $\bar{\mathbf{U}}$ lie on the diagonals $1,e$ and f, corresponding to the diagonals E, F and H of **A** respectively.

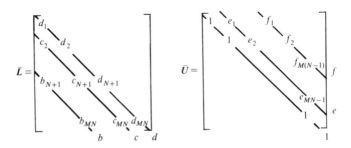

Fig. 5.11

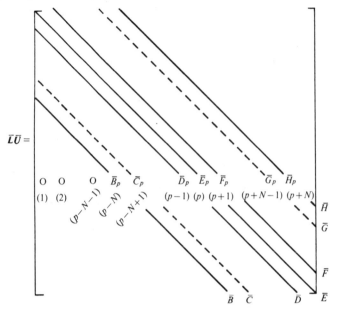

$\bar{L}\bar{U} =$

O O   O $\bar{B}_p$ $\bar{C}_p$ $\bar{D}_p$ $\bar{E}_p$ $\bar{F}_p$ $\bar{G}_p$ $\bar{H}_p$

(1) (2) $(p-N-1)$ $(p-N)$ $(p-N+1)$ $(p-1)$ (p) $(p+1)$ $(p+N-1)$ $(p+N)$

$\bar{B}$ $\bar{C}$ $\bar{D}$ $\bar{E}$

$\bar{H}$ $\bar{G}$ $\bar{F}$

(The bracketed numbers (1), (2), ..., (p),... indicate the position of the first, second,..., pth,... elements along the pth row.)

Fig. 5.12

The product $\bar{L}\bar{U}$ has seven non-zero elements per row, Fig. 5.12, which lie along the diagonals $\bar{B}$, $\bar{C}$, $\bar{D}$, $\bar{E}$, $\bar{F}$ and $\bar{H}$. The diagonals $\bar{B}$, $\bar{D}$, $\bar{E}$, $\bar{F}$ and $\bar{H}$ correspond to the diagonals B, D, E, F and H of Fig. 5.10, and the additional diagonals $\bar{C}$ and $\bar{G}$ are immediately next to the $\bar{B}$ and $\bar{H}$ diagonals.

By Fig. 5.12 it is seen that the pth equation of

$$\bar{L}\bar{U}\mathbf{u} = \mathbf{q}\{ = (\mathbf{A}+\mathbf{N})\mathbf{u}\}$$

is

$$(\bar{B}_p u_{p-N} + \bar{D}_p u_{p-1} + \bar{E}_p u_p + \bar{F}_p u_{p+1} + \bar{H}_p u_{p+N})$$
$$+ (\bar{C}_p u_{p-N+1} + \bar{G}_p u_{p+N-1}) = q_p. \quad (5.58)$$

A comparison of equations (5.58) and (5.56) shows that the first five terms of (5.58) will coincide with those of (5.56) if

$$\bar{B}_p = B_p, \quad \bar{D}_p = D_p, \quad \bar{E}_p = E_p, \quad \bar{F}_p = F_p \text{ and } \bar{H}_p = H_p. \quad (5.59)$$

The matrix $\mathbf{N}$ would then consist of the diagonals $\bar{C}$ and $\bar{G}$. In this case the unknown elements b, c, d, e and f of $\bar{\mathbf{L}}$ and $\bar{\mathbf{U}}$, and the unknown elements $\bar{C}$ and $\bar{G}$ of $\mathbf{N}$, would be found from equations (5.59) and the relationships between the elements of $\bar{\mathbf{L}}\bar{\mathbf{U}}$ and those of $\bar{\mathbf{L}}$ and $\bar{\mathbf{U}}$ which are,

$$\bar{B}_p = b_p \tag{5.60a}$$

$$\bar{D}_p = c_p \tag{5.60b}$$

$$\bar{E}_p = b_p f_{p-N} + c_p e_{p-1} + d_p \tag{5.60c}$$

$$\bar{F}_p = d_p e_p \tag{5.60d}$$

$$\bar{H}_p = d_p f_p \tag{5.60e}$$

$$\bar{C}_p = b_p e_{p-N} \text{ and } \bar{G}_p = c_p f_{p-1}. \tag{5.60f}$$

Equations (5.59) and (5.60a) to (5.60e) would give b, c, d, e and f. Equations (5.60f) would then give $\bar{C}_p$ and $\bar{G}_p$. Stone found, however, that this choice of $\mathbf{N}$ did not give a rapidly convergent iteration.

Considering equation (5.58), in which the term $(\bar{C}_p u_{p-N+1} + \bar{G}_p u_{p+N-1})$ is the pth component of $\mathbf{Nu}$ for the $\mathbf{N}$ considered above, Stone then decided to diminish the magnitude of this term by subtracting from it a closely equivalent expression. By Taylor expansions it is easily shown that

$$u_{p-N+1} = -u_p + u_{p+1} + u_{p-N} + O(hk)$$

and

$$u_{p+N-1} = -u_p + u_{p+N} + u_{p-1} + O(hk).$$

Therefore $u_{p-N+1} - (-u_p + u_{p+1} + u_{p-N}) = O(hk)$.

At this stage Stone also introduced an acceleration parameter α, $0 < \alpha < 1$, and defined the pth component of $\mathbf{Nu}$ to be

$$\bar{C}_p\{u_{p-N+1} - \alpha(-u_p + u_{p+1} + u_{p-N})\}$$

$$+ \bar{G}_p\{u_{p+N-1} - \alpha(-u_p + u_{p+N} + u_{p-1})\}. \tag{5.61}$$

Hence the pth equation of $(\mathbf{A} + \mathbf{N})\mathbf{u} = \mathbf{q}$ is, by equations (5.56) and (5.61),

$$(B_p - \alpha\bar{C}_p)u_{p-N} + (D_p - \alpha\bar{G}_p)u_{p-1} + \{E_p + \alpha(\bar{C}_p + \bar{G}_p)\}u_p$$

$$+ (F_p - \alpha\bar{C}_p)u_{p+1} + (H_p - \alpha\bar{G}_p)u_{p+N} + \bar{C}_p u_{p-N+1} + \bar{G}_p u_{p+N-1} = q_p$$

The relationships between the elements of $\bar{\mathbf{L}}\bar{\mathbf{U}} = \mathbf{A} + \mathbf{N}$ and those of

$\bar{\mathbf{L}}$ and $\bar{\mathbf{U}}$ are then

$$B_p - \alpha \bar{C}_p = b_p \tag{5.62a}$$

$$D_p - \alpha \bar{G}_p = c_p \tag{5.62b}$$

$$E_p + \alpha(\bar{C}_p + \bar{G}_p) = b_p f_{p-N} + c_p e_{p-1} + d_p \tag{5.62c}$$

$$F_p - \alpha \bar{C}_p = d_p e_p \tag{5.62d}$$

$$H_p - \alpha \bar{G}_p = d_p f_p \tag{5.62e}$$

$$\bar{C}_p = b_p e_{p-N} \quad \text{and} \quad \bar{G}_p = c_p f_{p-1}. \tag{5.62f}$$

After substituting for $\bar{C}_p$ and $\bar{G}_p$ from (5.62f), the five equations (5.62a) to (5.62e) can be written more usefully as

$$b_p = \frac{B_p}{(1 + \alpha e_{p-N})} \tag{5.63a}$$

$$c_p = \frac{D_p}{(1 + \alpha f_{p-1})} \tag{5.63b}$$

$$d_p = E_p + \alpha(b_p e_{p-N} + c_p f_{p-1}) - b_p f_{p-N} - c_p e_{p-1} \tag{5.63c}$$

$$e_p = \frac{(F_p - \alpha b_p e_{p-N})}{d_p} \tag{5.63d}$$

$$f_p = \frac{(H_p - \alpha c_p f_{p-1})}{d_p}, \tag{5.63e}$$

and give the coefficients in $\bar{\mathbf{L}}$ and $\bar{\mathbf{U}}$ sequentially for $p = 1, 2, \ldots$, MN. For example, if $N = 3$ and $M = 4$, then

$$B_1 = 0 = B_2 = B_3 = D_1 = F_{12} = H_{12} = H_{11} = H_{10},$$

$$c_1 = 0 = b_1 = b_2 = b_3 = e_{12} = f_{12} = f_{11} = f_{10},$$

and every letter with a zero or negative subscript will also be zero. Hence equations (5.63a) to (5.63e) give that

$$b_1 = 0, \quad c_1 = 0, \quad d_1 = E_1, \quad e_1 = F_1/d_1 = F_1/E_1,$$

$$f_1 = H_1/d_1 = H_1/E_1,$$

and

$$b_2 = 0, \quad c_2 = \frac{D_2}{1 + \alpha f_1} = \frac{D_2 E_1}{(E_1 + \alpha H_1)},$$

$$d_2 = E_2 + \alpha c_2 f_1 - c_2 e_1 = \text{etc.}$$

The implementation of the iteration

In practice, the iteration defined by equation (5.57), namely

$$(\mathbf{A}+\mathbf{N})\mathbf{u}^{(n+1)} = (\mathbf{A}+\mathbf{N})\mathbf{u}^{(n)} + (\mathbf{q}-\mathbf{A}\mathbf{u}^{(n)}),$$

where $(\mathbf{A}+\mathbf{N}) = \bar{\mathbf{L}}\bar{\mathbf{U}}$, is dealt with as follows.

Let $\mathbf{d}^{(n)} = \mathbf{u}^{(n+1)} - \mathbf{u}^{(n)}$ and $\mathbf{R}^{(n)} = \mathbf{q}-\mathbf{A}\mathbf{u}^{(n)}$.

Then by (5.57) a complete cycle of the iteration consists of the solution of

$$\bar{\mathbf{L}}\bar{\mathbf{U}}\mathbf{d}^{(n)} = \mathbf{R}^{(n)}, \tag{5.64}$$

followed by

$$\mathbf{u}^{(n+1)} = \mathbf{u}^{(n)} + \mathbf{d}^{(n)},$$

which is the iterative refinement described on page 226. Equation (5.64) would, of course, be solved by the forward and backward substitutions

$$\bar{\mathbf{L}}\mathbf{y}^{(n)} = \mathbf{R}^{(n)}$$

and

$$\bar{\mathbf{U}}\mathbf{d}^{(n)} = \mathbf{y}^{(n)}.$$

An additional acceleration parameter ω can also be introduced into the procedure by replacing (5.64) with

$$\bar{\mathbf{L}}\bar{\mathbf{U}}\mathbf{d}^{(n)} = \omega\mathbf{R}^{(n)},$$

as in reference 10. Further details concerning the calculation of α and the solution of the equations are given in Stone's paper, reference 32. His results indicate that the method is economical arithmetically in relation to older methods and that its rate of convergence is much less sensitive to the choice of iteration parameters than are the SOR and ADI methods.

Two recent direct methods

A method for 'variables separable' equations

The following method which depends upon the differential equation being 'variables separable', although that is not immediately obvious, was first proposed by Hockney, 1966, reference 17, who considered the problem

$$\frac{\partial^2 U}{\partial x^2} + \frac{\partial^2 U}{\partial y^2} = g(x, y) \quad (x, y) \in D,$$

$$U = 0, \qquad (x, y) \in C,$$

where C is the boundary of the rectangular domain $D = \{(x, y) : 0 < x < a, \ 0 < y < b\}$. Using Fig. 5.9 and a square mesh, the five-point difference equations approximating this problem may be written in partitioned form as

$$\begin{bmatrix} \mathbf{B} & \mathbf{I} & & & \\ \mathbf{I} & \mathbf{B} & \mathbf{I} & & \\ & \mathbf{I} & \mathbf{B} & \mathbf{I} & \\ & & \cdot & \cdot & \cdot \\ & & & \mathbf{I} & \mathbf{B} \end{bmatrix} \begin{bmatrix} \mathbf{u}_1 \\ \mathbf{u}_2 \\ \mathbf{u}_3 \\ \vdots \\ \mathbf{u}_M \end{bmatrix} = \begin{bmatrix} \mathbf{b}_1 \\ \mathbf{b}_2 \\ \mathbf{b}_3 \\ \vdots \\ \mathbf{b}_M \end{bmatrix} \tag{5.65}$$

where $\mathbf{u}_r$ is the vector of mesh values along $y = rh$, $r = 1(1)M$, $\mathbf{b}_r$ is a known vector corresponding to $\mathbf{u}_r$ and the $N \times N$ matrix $\mathbf{B}$ is

$$\begin{bmatrix} -4 & 1 & & \\ 1 & -4 & 1 & \\ & \cdot & \cdot & \cdot \\ & & 1 & -4 \end{bmatrix},$$

where N is the number of mesh points along a row parallel to Ox. By (5.65)

$$\mathbf{B}\mathbf{u}_1 + \mathbf{u}_2 = \mathbf{b}_1$$

$$\mathbf{u}_1 + \mathbf{B}\mathbf{u}_2 + \mathbf{u}_3 = \mathbf{b}_2 \tag{5.66}$$

$$\cdot \quad \cdot \quad \cdot$$

$$\mathbf{u}_{M-1} + \mathbf{B}\mathbf{u}_M = \mathbf{b}_M.$$

Let $\mathbf{q}_r$ be an eigenvector of $\mathbf{B}$ corresponding to the eigenvalue λ_r. Then

$$\mathbf{B}\mathbf{q}_r = \lambda_r \mathbf{q}_r, \quad r = 1(1)M,$$

and this set of equations can be written as

$$\mathbf{B}\mathbf{Q} = \mathbf{Q} \operatorname{diag}(\lambda_1, \lambda_2, \ldots, \lambda_M),$$

where $\mathbf{Q}$ is the modal matrix $[\mathbf{q}_1 \ \mathbf{q}_2 \ldots \mathbf{q}_M]$. But $\mathbf{B}$ is symmetric. therefore the eigenvectors $\mathbf{q}_r$, $r = 1(1)M$, can be normalized so that $\mathbf{Q}^T\mathbf{Q} = \mathbf{I}$. Hence $\mathbf{Q}^T\mathbf{B}\mathbf{Q} = \operatorname{diag}(\lambda_1, \lambda_2, \ldots, \lambda_M) = \mathbf{\Lambda}$, say.

Let

$$\bar{\mathbf{u}}_r = \mathbf{Q}^T \mathbf{u}_r \text{ and } \bar{\mathbf{b}}_r = \mathbf{Q}^T \mathbf{b}_r, \tag{5.67}$$

from which it follows that

$$\mathbf{u}_r = \mathbf{Q}\bar{\mathbf{u}}_r \text{ and } \mathbf{b}_r = \mathbf{Q}\bar{\mathbf{b}}_r. \tag{5.68}$$

Substituting from (5.68) into (5.66) and premultiplying throughout with $\mathbf{Q}^T$ leads to the equations

$$\begin{aligned}
\boldsymbol{\Lambda}\bar{\mathbf{u}}_1 + \bar{\mathbf{u}}_2 &= \bar{\mathbf{b}}_1 \\
\bar{\mathbf{u}}_1 + \boldsymbol{\Lambda}\bar{\mathbf{u}}_2 + \bar{\mathbf{u}}_3 &= \bar{\mathbf{b}}_2 \\
\cdot \quad \cdot \quad \cdot \\
\bar{\mathbf{u}}_{M-1} + \boldsymbol{\Lambda}\bar{\mathbf{u}}_M &= \bar{\mathbf{b}}_M.
\end{aligned} \tag{5.69}$$

Denote the ith components of $\bar{\mathbf{u}}_r$ and $\bar{\mathbf{b}}_r$ by $\bar{u}_{i,r}$ and $\bar{b}_{i,r}$ respectively and select the ith row of each of the equations (5.69). This gives the tridiagonal system of equations

$$\begin{aligned}
\lambda_i \bar{u}_{i,1} + \bar{u}_{i,2} &= \bar{b}_{i,1} \\
\bar{u}_{i,1} + \lambda_i \bar{u}_{i,2} + \bar{u}_{i,3} &= \bar{b}_{i,2} \\
\bar{u}_{i,2} + \lambda_i \bar{u}_{i,3} + \bar{u}_{i,4} &= \bar{b}_{i,3} \\
\cdot \quad \cdot \quad \cdot \\
\bar{u}_{i,M-1} + \lambda_i \bar{u}_{i,M} &= \bar{b}_{i,M},
\end{aligned} \tag{5.70}$$

for $\bar{u}_{i,r}$, $r = 1(1)M$. All the components of $\bar{\mathbf{u}}_r$, $r = 1(1)M$ can clearly be found by solving N such sets of equations for $\bar{u}_{i,r}$, $i = 1(1)N$. The procedure is therefore:

(i) Calculate the eigenvalues and eigenvectors of $\mathbf{B}$. (These are well known for the problem considered. See page 113.)

(ii) Compute $\bar{\mathbf{b}}_r = \mathbf{Q}^T \mathbf{b}_r$.

(iii) Solve equations (5.70), which is easily done.

(iv) Calculate $\mathbf{u}_r = \mathbf{Q}_r \bar{\mathbf{u}}_r$.

This method has been extended to more general self-adjoint 'variables separable' elliptic equations, to problems with derivative boundary conditions, and with irregular boundaries, see references 4 and 5, but research on the method is still relatively recent.

George's dissection method

As mentioned previously the standard Gauss elimination method or solving equations with a large band-width coefficient matrix **A** is nefficient in the sense that zero elements within the band are replaced by non-zero elements that have to be stored in the computer and used at subsequent stages of the elimination.

George, 1973, reference 15, by a combination of analysis, graph theory and intuition, formulated an ordering of the equations that gave substantial reductions in the 'fill-up', the computer storage required and in the volume of the arithmetic of the elimination process. His ordering has since been proved to be virtually optimal. For the five-point difference approximation of a Dirichlet elliptic problem defined over a rectangular region, and with a mesh giving $N \times N$ equations $\mathbf{Au} = \mathbf{b}$, the number of non-zero elements in the final upper triangular matrix **U** of $\mathbf{A} = \mathbf{LU}$ is $O(N^2 \log_2 N)$ and the volume of associated arithmetic is $O(N^3)$. The corresponding figures for the natural reading order of the mesh points is $O(N^3)$ and $O(N^4)$. For large N the savings in storage and effort are clearly considerable. A simpler but less efficient ordering has also been given by George (1972) in reference 14.

EXERCISES AND SOLUTIONS

1. The function ϕ satisfies the equation

$$\frac{\partial^2 \phi}{\partial x^2} + \frac{\partial^2 \phi}{\partial y^2} + 2 = 0$$

at every point inside the square bounded by the straight lines $x = \pm 1$, $y = \pm 1$, and is zero on the boundary. Calculate a finite-difference solution using a square mesh of side $\frac{1}{2}$. (The non-dimensional form of the torsion problem for a solid elastic cylinder with a square cross-section.)

Assuming the discretization error is proportional to h^2 calculate an improved value of ϕ at the point $(0, 0)$. (The analytical solution value is $0 \cdot 589$.)

Solution

Because of the symmetry there are only three unknowns; ϕ_1 at $(0, 0)$, ϕ_2 at $(\frac{1}{2}, 0)$, ϕ_3 at $(\frac{1}{2}, \frac{1}{2})$. The equations are $8\phi_2 - 8\phi_1 + 1 = 0$, $4\phi_3 + 2\phi_1 - 8\phi_2 + 1 = 0$ and $4\phi_2 - 8\phi_3 + 1 = 0$, giving $\phi_1 = 0 \cdot 562$, $\phi_2 = 0 \cdot 438$ and $\phi_3 = 0 \cdot 344$ to 3D.

A coarse mesh of side $h = 1$ gives the finite-difference equation $-4\phi_1 + 2 = 0$, so $\phi_1 = 0 \cdot 5$. Hence the 'deferred approach to the limit' method gives an improved value of $\phi_1 = 0 \cdot 562 + \frac{1}{3}(0 \cdot 062) = 0 \cdot 583$, which is very close to the exact value of $0 \cdot 589$ in spite of the crude mesh $h = 1$.

2. The function u satisfies the equation

$$\frac{\partial^2 u}{\partial x^2} + \frac{\partial^2 u}{\partial y^2} - 32u = 0$$

at every point inside the square $x = \pm 1$, $y = \pm 1$, and is subject to the boundary conditions

(i) $u = 0$ on $y = 1$, $\quad -1 \leqslant x \leqslant 1$,

(ii) $u = 1$ on $y = -1$, $\quad -1 \leqslant x \leqslant 1$,

(iii) $\partial u/\partial x = -\frac{1}{2}u$ in $x = 1$, $\quad -1 < y < 1$,

(iv) $\partial u/\partial x = \frac{1}{2}u$ on $x = -1$, $\quad -1 < y < 1$.

Take a square mesh of side $\frac{1}{4}$ and label the points with co-ordinates $(0, \frac{3}{4})$, $(\frac{1}{4}, \frac{3}{4})$, $(\frac{1}{2}, \frac{3}{4})$, $(\frac{3}{4}, \frac{3}{4})$, $(1, \frac{3}{4})$, $(0, \frac{1}{2})$, $(\frac{1}{4}, \frac{1}{2})$, ... as 1, 2, 3, 4, 5, 6, 7, ... etc. (similar to Fig. 5.2). Using the simplest central-difference formulae, show that the thirty-five finite-difference equations approximating this problem can be written in matrix form as

$$\mathbf{Au} = \mathbf{b},$$

where $\mathbf{u}$ is a column vector whose transpose is $(u_1, u_2, u_3, \ldots, u_{34}, u_{35})$, $\mathbf{b}$ a (35×1) column vector whose transpose is $(0, 0, \ldots, 0, -1, -1, -1, -1, -1)$, and $\mathbf{A}$ a matrix which can be written in partitioned form as

$$\begin{bmatrix} \mathbf{B} & \mathbf{I} & & & & & \\ \mathbf{I} & \mathbf{B} & \mathbf{I} & & & & \\ & \mathbf{I} & \mathbf{B} & \mathbf{I} & & & \\ & & \mathbf{I} & \mathbf{B} & \mathbf{I} & & \\ & & & \mathbf{I} & \mathbf{B} & \mathbf{I} & \\ & & & & \mathbf{I} & \mathbf{B} & \mathbf{I} \\ & & & & & \mathbf{I} & \mathbf{B} \end{bmatrix},$$

where

$$B = \begin{bmatrix} -6 & 2 & & & \\ 1 & -6 & 1 & & \\ & 1 & -6 & 1 & \\ & & 1 & -6 & 1 \\ & & & 2 & -6\frac{1}{4} \end{bmatrix}$$

and

$$I = \begin{bmatrix} 1 & & & & \\ & 1 & & & \\ & & 1 & & \\ & & & 1 & \\ & & & & 1 \end{bmatrix}$$

(This is a special case, in non-dimensional form, of the equation $\partial^2 u/\partial x^2 + \partial^2 u/\partial y^2 - 2H(u - u_0)/KD = 0$, which determines the steady temperature at points on a thin flat plate radiating heat from its surface into a medium at temperature u_0. D represents its thickness, K its thermal conductivity and H its surface conductance.)

Solution

Proceed as in worked example 5.2, after noting that the problem is symmetric with respect to $x = 0$.

3. The slow steady motion of viscous fluid through a cylindrical tube, whose cross section is the area S bounded by the closed curve C, can be calculated from a function ψ that satisfies Laplace's equation at all points of S and equals $\frac{1}{2}r^2$ on the curve C, where (r, θ) are the polar co-ordinates of a point in the plane of S. Calculate a finite-difference solution for flow through a circular sector bounded by the lines $\theta = 0$, $\theta = 0\cdot8$ radians, and the circle $r = 1$, at the mesh points defined by $r = \frac{1}{3}i$, $(i = 1, 2)$, $\theta = 0\cdot2j$, $(j = 1, 2, 3)$.

Solution

The problem is symmetrical about $\theta = 0\cdot4$ so there are four unknowns, ψ_1 at $(\frac{1}{3}, 0\cdot2)$, ψ_2 at $(\frac{1}{3}, 0\cdot4)$, ψ_3 at $(\frac{2}{3}, 0\cdot2)$ and ψ_4 at

$(\frac{2}{3}, 0\cdot4)$. The boundary values, and the polar co-ordinate finite-difference form of Laplace's equation, give the equations

$$-52\psi_1 + 25\psi_2 + 1\tfrac{1}{2}\psi_3 + 1\tfrac{7}{18} = 0,$$
$$50\psi_1 - 52\psi_2 + 1\tfrac{1}{2}\psi_4 = 0,$$
$$\tfrac{3}{4}\psi_1 - 14\tfrac{1}{2}\psi_3 + 6\tfrac{1}{4}\psi_4 + 2\tfrac{1}{72} = 0,$$
$$\tfrac{3}{4}\psi_2 + 12\tfrac{1}{2}\psi_3 - 14\tfrac{1}{2}\psi_4 + \tfrac{5}{8} = 0.$$

Their solution is

$$\psi_1 = 0\cdot0705, \quad \psi_2 = 0\cdot0756, \quad \psi_3 = 0\cdot2591, \quad \psi_4 = 0\cdot2704.$$

4. Derive the finite-difference form of Laplace's equation in polar co-ordinates as given earlier in the chapter. The function u satisfies Laplace's equation at every point within the circular sector bounded by the straight lines $\theta = 0$, $\theta = (m+1)\delta\theta$, and the circle $r = n\delta r$, where m and n are integers. It also has known values along these radii, and a known normal derivative, $\partial u/\partial r$, on the circular boundary. When this known derivative is represented by the central-difference formula $(u_{n+1,j} - u_{n-1,j})/2\delta r$, show that the finite-difference equations can be written as

$$
\begin{bmatrix}
\mathbf{B}_1 & (1+\tfrac{1}{2})\mathbf{I} & & & \\
(1-\tfrac{1}{4})\mathbf{I} & \mathbf{B}_2 & (1+\tfrac{1}{4})\mathbf{I} & & \\
& (1-\tfrac{1}{6})\mathbf{I} & \mathbf{B}_3 & (1+\tfrac{1}{6})\mathbf{I} & \\
& & \left(1-\dfrac{1}{2(n-1)}\right)\mathbf{I} & \mathbf{B}_{n-1} & \left(1+\dfrac{1}{2(n-1)}\right)\mathbf{I} \\
& & & 2\mathbf{I} & \mathbf{B}_n
\end{bmatrix}
\times
\begin{bmatrix}
\mathbf{u}_1 \\
\mathbf{u}_2 \\
\mathbf{u}_3 \\
\vdots \\
\mathbf{u}_n
\end{bmatrix}
=
\begin{bmatrix}
\mathbf{b}_1 \\
\mathbf{b}_2 \\
\mathbf{b}_3 \\
\vdots \\
\mathbf{b}_n
\end{bmatrix},
$$

where $\mathbf{u}_r$ is a column vector with components $(u_{r,1}, u_{r,2}, \ldots, u_{r,m})$,

$\mathbf{b}_1$, $\mathbf{b}_2$, ..., are known column vectors determined by the boundary conditions, $\mathbf{I}$ is the unit matrix of order m, and $\mathbf{B}_p$ the same square matrix of order m as given earlier in this chapter in the section on finite-differences in polar co-ordinates.

Solution

Write out the finite-difference equations approximating Laplace's equation for $i = 1(1)(n-1)$ and $j = 1(1)m$. Obtain the last block of m equations by eliminating the 'fictitious' value $u_{n+1,j}$ between the boundary derivative formula and the finite-difference equation with $i = n$.

5. The solution domain for the two-dimensional Laplace equation $\nabla^2 \phi = 0$ is the area bounded by the closed curve C on which the values of ϕ are known. Derive the formula for the residual at the mesh point O when the curve C intersects both of the perpendicular mesh lines of length h that pass through O.

A semi-circular lamina of radius $2h$, and uniform conductivity, has its diameter kept at a temperature of $0°$ and its circumference at a temperature of $100°$. Calculate a finite-difference solution to the steady-state temperatures at the nodal points of a square mesh of side h.

Solution

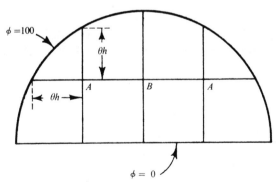

Fig. 5.13

$\theta = \sqrt{3} - 1.$

Hence

$$R_A = \frac{400}{(\sqrt{3}-1)\sqrt{3}} + \frac{2\phi_B}{\sqrt{3}} - \frac{4\phi_A}{\sqrt{3}-1},$$

$$R_B = 2\phi_A + 100 - 4\phi_B.$$

$R_A = R_B = 0$ gives $\phi_A = 70 \cdot 5°$, $\phi_B = 60 \cdot 2°$.

Although this coarse mesh is used only to provide a simple exercise it does, in fact, give the temperatures correct to within almost 1°. The analytical solution to this problem by the method of separation of the variables, and in terms of polar co-ordinates (r, θ), is

$$\phi = \frac{400}{\pi} \sum_{n=1}^{\infty} \frac{1}{(2n-1)} \left(\frac{r}{2h}\right)^{2n-1} \sin(2n-1)\theta,$$

and gives $\phi_A = 70 \cdot 5°$, $\phi_B = 59°$.

6. Prove that the truncation error of the five-point finite-difference formula approximating Laplace's equation at the point (x_i, y_j), for a square mesh of side h, can be written as

$$\tfrac{1}{12}h^2 \left\{ \frac{\partial^4}{\partial x^4} u(\xi, y_j) + \frac{\partial^4}{\partial y^4} u(x_i, \eta) \right\},$$

where $x_i - h < \xi < x_i + h$, $y_j - h < \eta < y_j + h$, and it is assumed that the first, second, third and fourth-order partial derivatives of u with respect to x and y are continuous throughout these intervals respectively.

Solution

$$T_{i,j} = \frac{1}{h^2} (u_{i+1,j} + u_{i,j+1} + u_{i-1,j} + u_{i,j-1} - 4u_{i,j}) - \left(\frac{\partial^2 u_{i,j}}{\partial x^2} + \frac{\partial^2 u_{i,j}}{\partial y^2}\right).$$

By Taylor's expansion,

$$u_{i+1,j} = u_{i,j} + h\frac{\partial u_{i,j}}{\partial x} + \tfrac{1}{2}h^2\frac{\partial^2 u_{i,j}}{\partial x^2} + \tfrac{1}{6}h^3\frac{\partial^3 u_{i,j}}{\partial x^3} + \tfrac{1}{24}h^4\frac{\partial^4}{\partial x^4} u(\xi_1, y_j),$$

where $x_i < \xi_1 < x_i + h$, etc.

7. The elliptic equation

$$\frac{\partial^2 u}{\partial x^2} + \frac{\partial^2 u}{\partial y^2} + d\frac{\partial u}{\partial x} + e\frac{\partial u}{\partial y} + fu = 0$$

is satisfied by the function u at every point of an area S. Prove that a non-constant value of u cannot assume a positive maximum or a negative minimum inside S when f is negative.

Solution

Assume u has a positive maximum at the point P in S. Then at P, $\partial u/\partial x = \partial u/\partial y = 0$, $\partial^2 u/\partial x^2 \leq 0$ and $\partial^2 u/\partial y^2 \leq 0$. Hence the left side of the equation is negative, in which case u cannot be a solution. Similarly for u a negative minimum. This result also holds for $f = 0$, but the proof is more difficult.

8. Derive equations (5.5) for S_1, S_2 and S_3. Deduce that,

$$(a) \quad \nabla^2 u_0 = \frac{4S_1 + S_2 - 20u_0}{6h^2} - \tfrac{1}{12}h^2\nabla^4 u_0$$

$$- \tfrac{1}{180}h^4(\tfrac{1}{2}\nabla^6 u_0 + \mathcal{D}^4\nabla^2 u_0) + O(h^6);$$

$$(b) \quad \nabla^4 u_0 = \frac{1}{h^4}(S_3 + 2S_2 - 8S_1 + 20u_0) + O(h^2).$$

9. Use exercise $8(a)$ to show that when f is a constant, Poisson's equation,

$$\nabla^2 u + f = 0,$$

can be represented at the central point point '0', Fig. 5.5, by the ninepoint finite-difference equation

$$\begin{bmatrix} 1 & 4 & 1 \\ 4 & -20 & 4 \\ 1 & 4 & 1 \end{bmatrix} u + 6h^2 f = 0,$$

and that the truncation error is of order h^6.

Use this result to solve exercise 1, the torsion problem for a square section.

Solution

In terms of the notation of exercise 1 the equations are

$$\phi_1 + 8\phi_2 - 20\phi_3 + 3 = 0,$$
$$4\phi_1 - 18\phi_2 + 8\phi_3 + 3 = 0,$$
$$-20\phi_1 + 16\phi_2 + 4\phi_3 + 3 = 0.$$

Hence $\phi_1 = 0\cdot590$, $\phi_2 = 0\cdot459$, $\phi_3 = 0\cdot363$ to 3D.

As the analytical value of ϕ_1 is $0\cdot589$ it is seen that the nine-point formula gives a very accurate solution in this case.

10. The simply connected open bounded domain D with closed boundary curve C is covered with a square mesh defined by the lines $x_i = ih$, $y_j = jh$, $i, j = 0, \pm1, \pm2, \ldots$. The set of mesh points interior to D is denoted by D_h and the set on C by C_h. The function $w_{i,j}$ defined on $D_h \cup C_h$ is such that $Lw_{i,j} \geq 0$ for all $(i, j) \in D_h$, where $Lw_{i,j} = (w_{i+1,j} + w_{i-1,j} + w_{i,j+1} + w_{i,j-1} - 4w_{i,j})/h^2$.

Prove that $\max\limits_{D_h} w_{i,j} \leq \max\limits_{C_h} w_{i,j}$, i.e., that the maximum of $w_{i,j}$ is on C_h.

Solution

The proof is by contradiction. Using the numbering of Fig. 5.5 with P replacing 0, $P \in D_h$, $Lw_p = (w_1 + w_2 + w_3 + w_4 - 4w_P)/h^2 \geq 0$ by hypothesis. Therefore $w_P \leq \frac{1}{4}(w_1 + w_2 + w_3 + w_4)$. Assume now that $w_{i,j}$ has a maximum value M at P, i.e., $M = w_p \geq w_{i,j}$, $(i, j) \in D_h$ and $M > w_{i,j}$, $(i, j) \in C_h$. In this case the preceding inequality can hold only as an equality with $w_1 = w_2 = w_3 = w_4 = M$. By choosing P as the point 1 and repeating the argument, etc., it follows that $w_{i,j} = M$ at all points of D_h and C_h. This contradicts the assumption $w_{i,j} < M$ on C_h.

11. The function U satisfies Laplace's equation at the points of the square $0 < x < 4$, $0 < y < 4$ and the Dirichlet boundary conditions:

  (i) $U = 0$ along $x = 0$ and $y = 0$,
  (ii) $U = x^3$ along $y = 4$, $0 \leq x \leq 4$,
  (iii) U linear and continuous along $x = 4$, $0 \leq y \leq 4$.

Use a square mesh of side 1 and write down the five-point difference equations approximating this problem at the nine internal mesh points. Either write and run a programme to solve these equations by the SOR method or carry out two SOR iterations using a hand calculator.

Solution

By page 252,

$$\rho(\mathbf{B}) = \frac{1}{2}\left(\cos\frac{\pi}{4} + \cos\frac{\pi}{4}\right) = \frac{1}{\sqrt{2}}.$$

Hence $\omega = 2/\{1 + \sqrt{(1-\rho^2)}\} = 1\cdot1716$. Labelling the mesh points 1, 2, 3, ..., 9 from left to right and from $y = 1(1)3$, a consistent ordering is obtained by taking the equations in the same order. The corresponding SOR equations are

$$u_1^{(n+1)} = 0\cdot2929(u_2^{(n)} + u_4^{(n)}) - 0\cdot1716\,u_1^{(n)}$$
$$u_2^{(n+1)} = 0\cdot2929(u_1^{(n+1)} + u_3^{(n)} + u_5^{(n)}) - 0\cdot1716\,u_2^{(n)}$$
$$u_3^{(n+1)} = 0\cdot2929(u_2^{(n+1)} + u_6^{(n)} + 16) - 0\cdot1716\,u_3^{(n)}$$
$$\cdots\cdots\cdots\cdots$$
$$u_9^{(n+1)} = 0\cdot2929(u_6^{(n+1)} + u_8^{(n+1)} + 75) - 0\cdot1716\,u_9^{(n)}.$$

Taking all initial values as zero the solution to $4D$ requires 11 iterations and $(u_1, u_2, \ldots, u_9)$ is $(2\cdot6071, 5\cdot9643, 10\cdot5, 4\cdot4643, 10\cdot75, 20\cdot0357, 4\cdot5, 12\cdot5357, 26\cdot8929)$. The values of the second iteration are $(0, 1\cdot3726, 7\cdot4316, 0\cdot0858, 4\cdot2860, 18\cdot5255, 0\cdot9792, 11\cdot0331, 26\cdot1934)$.

12. (i) Deduce that the eigenvalue λ_1 of largest modulus of a real $m \times m$ iteration matrix $\mathbf{G}$ is real if $|\lambda_1| > |\lambda_i|$, $i = 2(1)m$.

(ii) Assuming that the conditions of (i) hold, show that $\mathbf{d}^{(n)} \simeq \lambda_1 \mathbf{d}^{(n-1)}$, for sufficiently large n, where the displacement vector $\mathbf{d}^{(n)} = \mathbf{x}^{(n+1)} - \mathbf{x}^{(n)}$.

(iii) Assuming that $1 > |\lambda_1| > |\lambda_i|$, $i = 2(1)m$, derive Lyusternik's acceleration process $\mathbf{x} \simeq \mathbf{x}^{(n)} + \mathbf{d}^{(n)}/(1 - \lambda_1)$ by writing the solution $\mathbf{x}$ as the limiting value of the infinite series

$$\mathbf{x} = \mathbf{x}^{(n)} + \mathbf{d}^{(n)} + \mathbf{d}^{(n+1)} + \ldots$$

Solution

(i) The elements of **G** are real so complex roots occur in conjugate pairs. Hence λ_1 is real.

(ii) As per text.

(iii) By (ii), $\mathbf{d}^{(n+1)} \simeq \lambda_1 \mathbf{d}^{(n)}$, $\mathbf{d}^{(n+2)} \simeq \lambda_1^2 \mathbf{d}^{(n)}$, etc. Hence

$$\mathbf{x} \simeq \mathbf{x}^{(n)} + (1 + \lambda_1 + \lambda_1^2 + \ldots)\mathbf{d}^{(n)}$$

$$= \mathbf{x}^{(n)} + \frac{\mathbf{d}^{(n)}}{(1 - \lambda_1)}, \quad \text{since} \quad 0 < \lambda_1 < 1.$$

13. The following numbers to $5D$ are the third components of the fifth, sixth and seventh iteration vectors respectively of a Jacobi iterative solution.

$$0 \cdot 41504, \quad 0 \cdot 45874, \quad 0 \cdot 49500.$$

Use Aitken's method to calculate an improved value for this component. Verify that Lyusternik's method gives the same value to $4D$ when λ_1 is approximated by $d_3^{(6)}/d_3^{(5)}$.

Solution

$x_3 \simeq 0 \cdot 495 - (0 \cdot 03626)^2/(-0 \cdot 00744) = 0 \cdot 67172.$

$d_3^{(5)} = 0 \cdot 0437$, $d_3^{(6)} = 0 \cdot 03626$, $\lambda_1 = 0 \cdot 8297$. By Lyusternik's method $x_3 \simeq x_3^{(6)} + d_3^{(6)}/(1 - \lambda_1)$ gives $x_3 \simeq 0 \cdot 6717$ to $4D$.

(The problem from which these figures came required 65 iterations for $4D$ accuracy. $x_3 = 0 \cdot 6555$.)

14. The Crank–Nicolson equations approximating $\partial U/\partial t = \partial^2 U/\partial x^2$, where U is known at $x = 0$ and $x = Nh$, may be written as

$$u_i = \rho(u_{i-1} + u_{i+1}) + c_i, \quad i = 1(1)(N-1),$$

where u_i denotes the unknown $u_{i,j+1}$, c_i denotes a known number for all $i = 1(1)(N-1)$, $\rho = r/2(1 + r)$ and $r = k/h^2$. Prove that the Gauss-Seidel iterative procedure for solving these equations converges for all positive values of r, it being assumed that the $(n+1)$th iterative values are calculated systematically from $i = 1(1)(N-1)$.

Solution

The iteration is defined by the equation

$$u_i^{(n+1)} = \rho(u_{i-1}^{(n+1)} + u_{i+1}^{(n)}) + c_i.$$

The simplest way to obtain the Gauss-Seidel iteration matrix is to substitute for $u_{i-1}^{(n+1)}$ in terms of $u_{i-1}^{(n)}$ from the preceding equation. Then

$$u_1^{(n+1)} = \rho u_2^{(n)} + (\rho u_0 + c_1) = \rho u_2^{(n)} + c_1',$$

where c_1' is known because u_0 is a boundary value.

$$\begin{aligned}
u_2^{(n+1)} &= \rho u_3^{(n)} + \rho u_1^{(n+1)} + c_2 \\
&= \rho u_3^{(n)} + \rho^2 u_2^{(n)} + (\rho c_1' + c_2) \\
&= \rho u_3^{(n)} + \rho^2 u_2^{(n)} + c_1', \qquad \text{etc.}
\end{aligned}$$

In matrix form,

$$\mathbf{u}^{(n+1)} = \begin{bmatrix}
0 & \rho & 0 & & & \\
0 & \rho^2 & \rho & 0 & & \\
0 & \rho^3 & \rho^2 & \rho & 0 & \\
\cdot & & & & & \\
\cdot & & & & & \\
0 & \rho^{N-1} & \rho^{N-2} & & & \rho^2
\end{bmatrix} \mathbf{u}^{(n)} + \mathbf{c}$$

The second column gives the largest column sum of the elements of the iteration matrix. If λ_i represents an eigenvalue of the matrix then by Gerschgorin's first theorem,

$$\max_i |\lambda_i| \leqslant \rho + \rho^2 + \ldots + \rho^{N-1} < \frac{\rho}{1-\rho} = \frac{r}{2+r} < 1.$$

15. For the equations

$$\begin{aligned}
x_1 + 2x_2 + 4x_3 &= 1, \\
\tfrac{1}{8}x_1 + x_2 + x_3 &= 3
\end{aligned}$$

and

$$-x_1 + 4x_2 + x_3 = 7,$$

prove that the Jacobi iteration is convergent but that the Gauss-Seidel iteration is divergent.

Solution

The Jacobi iteration matrix is

$$\begin{bmatrix} 0 & -2 & -4 \\ -\frac{1}{8} & 0 & -1 \\ 1 & -4 & 0 \end{bmatrix}$$

Its eigenvalues μ are given by

$$\det\begin{bmatrix} -\mu & -2 & -4 \\ -\frac{1}{8} & -\mu & -1 \\ 1 & -4 & -\mu \end{bmatrix} = 0 = -\mu(\mu^2 - \tfrac{1}{4}).$$

Hence $\mu = 0$, $\pm\frac{1}{2}$, so the iteration is convergent.

The eigenvalues λ of the Gauss–Seidel iteration matrix can be calculated either as in exercise 16 or by substituting for $x_i^{(n+1)}$ in terms of $x_i^{(n)}$ from preceding Gauss–Seidel iteration equations. The Gauss–Seidel iteration equations are

$$x_1^{(n+1)} = -2x_2^{(n)} - 4x_3^{(n)} + 1,$$
$$x_2^{(n+1)} = -\tfrac{1}{8}x_1^{(n+1)} - x_3^{(n)} + 3,$$
$$x_3^{(n+1)} = x_1^{(n+1)} - 4x_2^{(n+1)} + 7.$$

By the first two equations,

$$x_2^{(n+1)} = -\tfrac{1}{8}(-2x_2^{(n)} - 4x_3^{(n)} + 1) - x_3^{(n)} + 3, \text{ etc.,}$$

leading finally to the Gauss–Seidel iteration matrix

$$\begin{bmatrix} 0 & -2 & -4 \\ 0 & \frac{1}{4} & -\frac{1}{2} \\ 0 & -3 & -2 \end{bmatrix}.$$

Its eigenvalues λ satisfy $\lambda(\lambda^2 + 1\tfrac{3}{4}\lambda - 2) = 0$, so $\lambda = 0$, $0\cdot788$ and $-2\cdot538$.

16. A function satisfies Laplace's equation at every point inside the square bounded by the straight lines $x = \pm 1$, $y = \pm 1$, has known boundary values, and is symmetrical with respect to Ox and Oy. Write down the simplest finite-difference equations giving an approximate solution at the nodal points of a square of side $\frac{1}{2}$. Prove that the Jacobi and Gauss–Seidel iterative processes for their solution both converge. Hence verify that the asymptotic rate of con-

vergence of the Gauss–Seidel iteration is twice that of the Jacobi iteration.

Solution

Denote the pivotal values at the points $(0, 0)$, $(\frac{1}{2}, 0)$ and $(\frac{1}{2}, \frac{1}{2})$ by a, b, c, respectively. Then the finite-difference equations at these points, in the order given, are $a - b = \text{constant}$; $-\frac{1}{4}a + b - \frac{1}{2}c = \text{constant}$, and $-\frac{1}{2}b + c = \text{constant}$. The matrix of coefficients is

$$\mathbf{A} = \begin{bmatrix} 1 & -1 & 0 \\ -\frac{1}{4} & 1 & -\frac{1}{2} \\ 0 & -\frac{1}{2} & 1 \end{bmatrix} = \mathbf{I} - (\mathbf{L} + \mathbf{U}) = \mathbf{I} - \begin{bmatrix} 0 & 1 & 0 \\ \frac{1}{4} & 0 & \frac{1}{2} \\ 0 & \frac{1}{2} & 0 \end{bmatrix}.$$

Hence the eigenvalues of the Jacobi iteration matrix $(\mathbf{L} + \mathbf{U})$ are given by the roots of the determinantal equation

$$\begin{vmatrix} -\lambda & 1 & 0 \\ \frac{1}{4} & -\lambda & \frac{1}{2} \\ 0 & \frac{1}{2} & -\lambda \end{vmatrix} = 0 = \lambda(\lambda^2 - \tfrac{1}{2}).$$

Therefore the spectral radius $\rho(J)$ is $1/\sqrt{2} < 1$, so the iteration converges.

The Gauss–Seidel iteration matrix is

$(\mathbf{I} - \mathbf{L})^{-1}\mathbf{U} =$

$$\begin{bmatrix} 1 & 0 & 0 \\ -\frac{1}{4} & 1 & 0 \\ 0 & -\frac{1}{2} & 1 \end{bmatrix}^{-1} \begin{bmatrix} 0 & 1 & 0 \\ 0 & 0 & \frac{1}{2} \\ 0 & 0 & 0 \end{bmatrix} = \begin{bmatrix} 1 & 0 & 0 \\ \frac{1}{4} & 1 & 0 \\ \frac{1}{8} & \frac{1}{2} & 1 \end{bmatrix} \begin{bmatrix} 0 & 1 & 0 \\ 0 & 0 & \frac{1}{2} \\ 0 & 0 & 0 \end{bmatrix}$$

$$= \begin{bmatrix} 0 & 1 & 0 \\ 0 & \frac{1}{4} & \frac{1}{2} \\ 0 & \frac{1}{8} & \frac{1}{4} \end{bmatrix}$$

Hence the eigenvalues of the Gauss–Seidel iteration matrix are given by the equation

$$\begin{vmatrix} -\lambda & 1 & 0 \\ 0 & (\frac{1}{4} - \lambda) & \frac{1}{2} \\ 0 & \frac{1}{8} & (\frac{1}{4} - \lambda) \end{vmatrix} = 0 = \lambda^2(\lambda - \tfrac{1}{2}),$$

showing that the spectral radius $\rho(G)$ is $\frac{1}{2}$.

As the asymptotic rate of convergence is defined by the modulus of the natural logarithm of the spectral radius, the result follows.

17. A function satisfies Poisson's equation at the points of the rectangle $0 < x < ph$, $0 < y < qh$ and has known values on its boundary. Show that the matrix of the five-point difference equations approximating this problem at the mesh points defined by $x_i = ih$, $i = 1(1)(p-1)$, and $y_j = jh$, $j = 1(1)(q-1)$, can be written in block partitioned form as

$$\mathbf{A} = \begin{bmatrix} \mathbf{B} & \mathbf{I} & & & \\ \mathbf{I} & \mathbf{B} & \mathbf{I} & & \\ & \mathbf{I} & \mathbf{B} & \mathbf{I} & \\ & & \cdot & \cdot & \cdot \\ & & & \mathbf{I} & \mathbf{B} \end{bmatrix} \text{ of order } (p-1)(q-1) \text{ where}$$

$$\mathbf{B} = \begin{bmatrix} -4 & 1 & & & \\ 1 & -4 & 1 & & \\ & 1 & -4 & 1 & \\ & & \cdot & \cdot & \cdot \\ & & & 1 & -4 \end{bmatrix} \text{ is of order } (p-1) \text{ and } \mathbf{I}$$

is the unit matrix of order $(p-1)$. (This assumes the equations are ordered row by row from left to right or right to left.)

Use the theorem on page 107, or otherwise, to show that the eigenvalues $\lambda_{i,j}$ of $\mathbf{A}$ are given by

$$\lambda_{i,j} = -4 + 2\left(\cos\frac{i\pi}{p} + \cos\frac{j\pi}{q}\right), \quad i = 1(1)(p-1), \quad j = 1(1)(q-1).$$

Deduce that the spectral radius of the corresponding Jacobi iteration matrix is

$$\tfrac{1}{2}\left(\cos\frac{\pi}{p} + \cos\frac{\pi}{q}\right).$$

Solution

As $\mathbf{B}$ is real and symmetric it has $(p-1)$ linearly independent eigenvectors $\mathbf{v}_i$ which may be taken as the eigenvectors of $\mathbf{I}$ because $\mathbf{I}\mathbf{v}_i = 1\mathbf{v}_i$. Hence the theorem may be used. The eigenvalues μ_i of B are $\mu_i = -4 + 2\cos(i\pi)/(p)$, $i = 1(1)(p-1)$. By the theorem, the eigenvalues $\lambda_{i,j}$ of **A** are

$$\lambda_{i,j} = \mu_i + 2\cos\frac{j\pi}{q}, \quad j = 1(1)(q-1).$$

Therefore

$$\lambda_{i,j} = -4 + 2\left(\cos\frac{i\pi}{p} + \cos\frac{j\pi}{q}\right).$$

The Jacobi iteration matrix is $\mathbf{D}^{-1}(\mathbf{L}+\mathbf{U})$ where $\mathbf{D} = \text{diag}(-4, -4, \ldots, -4)$ and $\mathbf{A} = \mathbf{D} - \mathbf{L} - \mathbf{U}$, i.e., $\mathbf{L}+\mathbf{U} = \mathbf{D} - \mathbf{A}$. The eigenvalues of $\mathbf{L}+\mathbf{U} = -4 - \lambda_{i,j}$. The eigenvalues of

$$\mathbf{D}^{-1}(\mathbf{L}+\mathbf{U}) = (-\tfrac{1}{4})(-4-\lambda_{i,j}) = \tfrac{1}{2}\left(\cos\frac{i\pi}{p} + \cos\frac{j\pi}{q}\right).$$

The largest value of this is

$$\tfrac{1}{2}\left(\cos\frac{\pi}{p} + \cos\frac{\pi}{q}\right).$$

18. Let μ_i represent an eigenvalue of the Jacobi iteration matrix **B** associated with the matrix **A** and let $\mathbf{v}_i$ represent the corresponding eigenvector of **B**. Prove that the Jacobi iteration matrices associated with all possible re-orderings of the equations $\mathbf{Ax} = \mathbf{b}$ have the same set of eigenvalues μ_i and corresponding eigenvectors $\mathbf{v}_i'$ where the components of $\mathbf{v}_i'$ are those of $\mathbf{v}_i$ re-ordered the same way as the equations.

Solution

Make the coefficient of every diagonal term unity. Then the equations can be written as $(\mathbf{I} - \mathbf{L} - \mathbf{U})\mathbf{x} = \mathbf{c}$ and the associated Jacobi iteration matrix is $(\mathbf{L}+\mathbf{U})$. A re-ordering of the equations and unknowns can be written as $\{\mathbf{P}(\mathbf{I}-\mathbf{L}-\mathbf{U})\mathbf{P}^T\}\mathbf{Px} = \mathbf{Pc}$, giving that $\mathbf{Px} = \{\mathbf{P}(\mathbf{L}+\mathbf{U})\mathbf{P}^T\}\mathbf{Px} + \mathbf{Pc}$ since $\mathbf{PP}^T = \mathbf{I}$. The associated Jacobi iteration matrix is $\mathbf{P}(\mathbf{L}+\mathbf{U})\mathbf{P}^T$, which is a similarity transformation of $(\mathbf{L}+\mathbf{U})$. Hence the result concerning eigenvalues. Let $\mathbf{v}'$ represent the eigenvector of $\mathbf{P}(\mathbf{L}+\mathbf{U})\mathbf{P}^T$ corresponding to the eigenvalue μ. Then $\mathbf{P}(\mathbf{L}+\mathbf{U})\mathbf{P}^T\mathbf{v}' = \mu\mathbf{v}'$. Pre-multiply by $\mathbf{P}^T$ and use $\mathbf{P}^T\mathbf{P} = \mathbf{I}$ to obtain $(\mathbf{L}+\mathbf{U})(\mathbf{P}^T\mathbf{v}') = \mu(\mathbf{P}^T\mathbf{v}')$. This equation states that the eigenvector of $(\mathbf{L}+\mathbf{U})$ corresponding to μ is $\mathbf{P}^T\mathbf{v}'$, i.e., that $\mathbf{P}^T\mathbf{v}' = \mathbf{v}$.

Premultiply by $\mathbf{P}$ to give that $\mathbf{v'} = \mathbf{Pv}$. Therefore the components of $\mathbf{v'}$ are those of $\mathbf{v}$ re-ordered the same way as the equations.

19. Given that

$$
\begin{bmatrix}
a_{11} & a_{12} & \cdot & \cdot & \cdot & a_{1n} \\
a_{21} & a_{22} & \cdot & \cdot & \cdot & a_{2n} \\
\cdot & & \cdot & & & \\
\cdot & & & \cdot & & \\
\cdot & & & & \cdot & \\
a_{n1} & a_{n2} & & & & a_{nn}
\end{bmatrix}
=
\begin{bmatrix}
l_{11} & & & & & \\
l_{21} & l_{22} & & & & \\
\cdot & & \cdot & & & \\
\cdot & & & \cdot & & \\
\cdot & & & & \cdot & \\
l_{n1} & l_{n2} & & & & l_{nn}
\end{bmatrix}
$$

$$
\times
\begin{bmatrix}
1 & u_{12} & \cdot & \cdot & \cdot & u_{1n} \\
 & 1 & u_{23} & \cdot & \cdot & u_{2n} \\
 & & \cdot & & & \\
 & & & \cdot & & \\
 & & & & \cdot & \\
 & & & & & 1
\end{bmatrix}
$$

where $l_{ij} = 0$, $i < j$ and $u_{ij} = 0$, $i > j$, show that

$$
l_{ij} = a_{ij} - \sum_{k=1}^{j-1} l_{ik} u_{kj}
$$

and

$$
u_{ij} = \frac{a_{ij} - \sum_{k=1}^{i-1} l_{ik} u_{kj}}{l_{ii}}.
$$

Develop formulae to calculate the components of the vectors $\mathbf{y}$ and $\mathbf{x}$ when the solution of the equations $\mathbf{LUx} = \mathbf{b}$ is obtained by solving $\mathbf{Ly} = \mathbf{b}$ for $\mathbf{y}$ by forward substitutions then $\mathbf{Ux} = \mathbf{y}$ for $\mathbf{x}$ by backward substitutions.

Solution

$$
a_{ij} = \sum_{k=1}^{n} l_{ik} u_{kj}
$$

$$
= \sum_{k=1}^{j-1} l_{ik} u_{kj} + l_{ij} u_{jj} + \sum_{k=j+1}^{n} l_{ik} u_{kj}.
$$

As $u_{jj} = 1$ and $u_{kj} = 0$ for $k > j$ it follows that

$$l_{ij} = a_{ij} - \sum_{k=1}^{j-1} l_{ik} u_{kj}.$$

Again,

$$a_{ij} = \sum_{k=1}^{n} l_{ik} u_{kj}$$

$$= \sum_{k=1}^{i-1} l_{ik} u_{kj} + l_{ii} u_{ij} + \sum_{k=i+1}^{n} l_{ik} u_{kj}.$$

As $l_{ik} = 0$ for $i < k$ it follows that

$$u_{ij} = \frac{a_{ij} - \sum_{k=1}^{i-1} l_{ik} u_{kj}}{l_{ii}}.$$

The ith component of $\mathbf{Ly} = \mathbf{b}$ gives that

$$b_i = \sum_{k=1}^{n} l_{ik} y_k$$

$$= \sum_{k=1}^{i-1} l_{ik} y_k + l_{ii} y_i + \sum_{k=i+1}^{n} l_{ik} u_k.$$

As $l_{ik} = 0$ for $i < k$ it is seen that

$$y_i = \frac{b_1 - \sum_{k=1}^{i-1} l_{ik} y_k}{l_{ii}}, \quad i = 1(1)n.$$

Therefore

$$y_1 = \frac{b_1}{l_{11}}, \quad y_2 = \frac{b_2 - l_{21} y_1}{l_{22}}, \quad \text{etc.}$$

The ith component of $\mathbf{Ux} = \mathbf{y}$ gives that

$$y_i = \sum_{k=1}^{i-1} u_{ik} x_k + u_{ii} x_i + \sum_{k=i+1}^{n} u_{ik} x_k.$$

As $u_{ii} = 1$ and $u_{ik} = 0$ for $i > k$, we obtain

$$x_i = y_i - \sum_{k=i+1}^{n} u_{ik} x_k, \quad i = (n-1)(1)1.$$

Therefore

$$x_n = y_n, \quad x_{n-1} = y_n - u_{n-1,n}x_n, \quad \text{etc.}$$

20.

$$\mathbf{A} = \begin{bmatrix} a_{11} & a_{12} & 0 & a_{14} & 0 \\ a_{21} & a_{22} & a_{23} & 0 & a_{25} \\ 0 & a_{32} & a_{33} & 0 & 0 \\ a_{41} & 0 & 0 & a_{44} & a_{45} \\ 0 & a_{52} & 0 & a_{54} & a_{55} \end{bmatrix}$$

(i) Show that matrix **A** is 2-cyclic.

(ii) Write down two ordering vectors for **A** in terms of the numbers 0 and 1. Is **A** consistently ordered with respect to either of them? If not, use one of them to re-order **A** into a consistently ordered matrix **B**.

(iii) Verify that **A** is consistently ordered with respect to the ordering vector $\boldsymbol{\gamma}^{(3)} = (0, 1, 2, 1, 2)$. Use $\boldsymbol{\gamma}^{(3)}$ to re-order **A** into a consistently ordered block tridiagonal matrix **C**. Write down, in matrix notation, the equation giving the re-ordering of $\mathbf{A}\mathbf{x} = \mathbf{b}$ corresponding to **C**.

Solution

(i) As in worked example 5.4, $S = \{1, 3, 5\}$, $T = \{2, 4\}$.

(ii) $\boldsymbol{\gamma}^{(1)} = (0, 1, 0, 1, 0)$, $\boldsymbol{\gamma}^{(2)} = (1, 0, 1, 0, 1)$. Consider $\boldsymbol{\gamma}^{(1)}$ and the element a_{32}. As $\gamma_3^{(1)} - \gamma_2^{(1)} = -1$, the matrix **A** is not consistently ordered with respect to $\boldsymbol{\gamma}^{(1)}$. Similarly for $\boldsymbol{\gamma}^{(2)}$ and a_{12}. Consider $\boldsymbol{\gamma}^{(2)}$. The interchange of the first and fourth components gives the vector $(0, 0, 1, 1, 1)$ which is an ordering vector for the consistently ordered matrix

$$\mathbf{B} = \mathbf{P}_{14}\mathbf{A}\mathbf{P}_{14} = \left[\begin{array}{cc:ccc} a_{44} & 0 & 0 & a_{41} & a_{45} \\ 0 & a_{22} & a_{23} & a_{21} & a_{25} \\ \hdashline 0 & a_{32} & a_{33} & 0 & 0 \\ a_{14} & a_{12} & 0 & a_{11} & 0 \\ a_{54} & a_{52} & 0 & 0 & a_{55} \end{array} \right].$$

(iii) Consider a_{12} and a_{32}. $\gamma_2^{(3)} - \gamma_1^{(3)} = 1 = \gamma_3^{(3)} - \gamma_2^{(3)}$. Similarly for the remaining non-zero elements. The interchange of the third and fourth components of $\gamma^{(3)}$ gives $(0, 1, 1, 2, 2)$ which is an ordering vector for

$$\mathbf{C} = \mathbf{P}_{34}\mathbf{A}\mathbf{P}_{34} = \begin{bmatrix} a_{11} & a_{12} & a_{14} & 0 & 0 \\ a_{21} & a_{22} & 0 & a_{23} & a_{25} \\ a_{41} & 0 & a_{44} & 0 & a_{45} \\ 0 & a_{32} & 0 & a_{33} & 0 \\ 0 & a_{52} & a_{54} & 0 & a_{55} \end{bmatrix}$$

The re-ordered equations are $(\mathbf{P}_{34}\mathbf{A}\mathbf{P}_{34})\mathbf{P}_{34}\mathbf{x} = \mathbf{P}_{34}\mathbf{b}$.

21. With the usual notation show that the eigenvalues of the Jacobi and SOR iteration matrices are respectively the roots of the equations

(i) $\det(\mu\mathbf{D} - \mathbf{L} - \mathbf{U}) = 0$, and

(ii) $\det(k\mathbf{D} - \lambda\omega\mathbf{L} - \omega\mathbf{U}) = 0$, where $k = \lambda + \omega - 1$.

Show that the matrix

$$\mathbf{A} = \begin{bmatrix} 4 & 0 & 0 & -1 \\ -1 & 4 & -1 & 0 \\ 0 & -1 & 4 & 0 \\ -1 & 0 & 0 & 4 \end{bmatrix}$$

is consistently ordered with respect to the ordering vector $(0, 1, 2, 1)$. Prove, for this particular matrix, that the eigenvalues λ of the associated SOR iteration matrix are related to the eigenvalues μ of the corresponding Jacobi iteration matrix by $(\lambda + \omega - 1)^2 = \lambda\omega^2\mu^2$. Comment on this result.

Solution

(i) The eigenvalues μ of $\mathbf{D}^{-1}(\mathbf{L} + \mathbf{U})$ are the roots of

$$\det\{\mu\mathbf{I} - \mathbf{D}^{-1}(\mathbf{L} + \mathbf{U})\} = \det \mathbf{D}^{-1}\{\mu\mathbf{D} - \mathbf{L} - \mathbf{U}\}$$
$$= \det \mathbf{D}^{-1} \det\{\mu\mathbf{D} - \mathbf{L} - \mathbf{U}\} = 0.$$

By hypothesis the elements of $\mathbf{D}$ are non-zero. Hence the result. Similarly for (ii).

Replace the non-zero elements of **A** by a_{ij} and apply the definition of consistent ordering as in exercise 20.

The eigenvalues μ are given by

$$\det \begin{bmatrix} 4\mu & 0 & 0 & -1 \\ -1 & 4\mu & -1 & 0 \\ 0 & -1 & 4\mu & 0 \\ -1 & 0 & 0 & 4\mu \end{bmatrix} = 0 = (16\mu^2 - 1)^2,$$

i.e., $\mu = \frac{1}{4}, \frac{1}{4}, -\frac{1}{4}, -\frac{1}{4}$.

The eigenvalues λ are given by

$$\det \begin{bmatrix} 4k & 0 & 0 & -\omega \\ -\lambda\omega & 4k & -\omega & 0 \\ 0 & -\lambda\omega & 4k & 0 \\ -\lambda\omega & 0 & 0 & 4k \end{bmatrix} = 0 = (16k^2 - \lambda\omega^2)^2.$$

Therefore $k^2 = \frac{1}{16}\lambda\omega^2$, where $k = \lambda + \omega - 1$. Hence the result. The matrix **A** is not block tridiagonal. Because, however, it is consistently ordered the relationship $(\lambda + \omega - 1)^2 = \lambda\omega^2\mu^2$ still holds.

References for supplementary reading

1. Albasiny, E. L. (1960) On the numerical solution of a cylindrical heat-conduction problem. *Quart. J. Mech. and Applied Math.* **13**, 374–384.
2. Bickley, W. G. (1948) Finite-difference formulae for the square lattice. *Quart. J. Mech. and Applied Math.* **1**, 35–42.
3. Bodewig, E. (1959) *Matrix Calculus*, 2nd ed., Amsterdam, North-Holland Publishing Co.
4. Buzbee, B. L. et al. (1970) On direct methods for solving Poisson's equations. *SIAM J. Numer. Anal.* **7**, 627–656.
5. Buzbee, B. L., and Dorr, F. W. (1974) The direct solution of the biharmonic equation on rectangular regions and the Poisson equation on irregular regions. *SIAM J. Numer. Anal.* **11**, 753–763.
6. Carslaw, H. S., and Jaeger, J. C. (1959) *Conduction of Heat in Solids*, 2nd ed., Clarendon Press, Oxford.
7. Courant, R., Friedrichs, K., and Lewy, H. (1928) Uber die partiellen differenzengleichungen de mathematischen Physik. *Mathematische Annalen*, **100**, 32–74.
8. Crank, J., and Nicolson, P. (1947) A practical method for numerical evaluation of solutions of partial differential equations of the heat-conduction type. *Proc. Camb Phil. Soc.* **43**, 50–67.
9. Crank, J. (1975) *Mathematics of Diffusion*, 2nd. ed., Clarendon Press, Oxford.
10. Dupont, T., Kendall, R. P., and Rashford, H. H. (1968) An approximate factorization procedure for solving self-adjoint elliptic difference equations. *SIAM J. Num. Anal.* **5**, 559–573.
11. Forsythe, G. E., and Wasow, W. R. (1960) *Finite-Difference Methods for Partial Differential Equations*, John Wiley, New York.
12. Fox, L. (1961) (Ed). *Numerical Solution of Ordinary and Partial Differential Equations*. Pergamon Press, Oxford.
13. Fox, L. (1964) *An Introduction to Numerical Linear Algebra*. Clarendon Press, Oxford.
14. George, J. A. (1972) An efficient band-oriented scheme for solving n×n grid problems. *Fall Joint Computer Conference*, AFIPS Press, New Jersey.
15. George, J. A. (1973) Nested dissection of a rectangular finite-element mesh *SIAM J. Numer. Anal.* **10**, 345–363.
16. Hageman, L. A. (1972) The estimation of acceleration parameters for the Chebyshev polynomial and the successive overrelaxation iteration methods. *AEC Res. and Dev. report WAPD—TM—1038.*
17. Hockney, R. W. (1965) A fast direct solution of Poisson's equations using Fourier analysis. *J. Assoc. Comp. Mach.* **12**, 95–113.
18. Lax, P. D. (1954) Weak solutions of non-linear hyperbolic equations and their numerical computations. *Comm. Pure Appl. Math.* **7**, 157–193.

19. Lees, M. (1966) A linear three level difference scheme for quasi-linear parabolic equations. *Maths Comp.* **20**, 516–522.

20. Lowan, A. N. (1957) The operator approach to problems of stability and convergence of solutions of difference equations and the convergence of various iteration procedures. *Scripta Mathematica, Washington, Office of Technical Services, New York.*

21. Lyusternik, L. A. (1947) A note for the numerical solution of boundary value problems for the Laplace equation and for the calculation of eigenvalues by the method of nets. *Trudy Inst. Math. Academy of Sciences of the USSR*, **20**, 49–64. (*Russian*).

22. Milne–Thomson, L. M. (1949) *Theoretical Hydrodynamics*, 2nd ed., Macmillan, London.

23. Mitchell, A. R. (1969) *Computational Methods in Partial Differential Equations.* John Wiley and Sons.

24. Noble, B. (1969) *Applied Linear Algebra.* Prentice-Hall Inc., New Jersey.

25. N.P.L. (1961) *Notes on Applied Science,* **16**, *Modern Computing Methods.* London, H.M.S.O.

26. von Neumann, J., and Richtmyer, R. D. (1950) A method for the numerical calculation of hydrodynamic shocks, *J. Appl. Phys.* **21**, 232–237.

27. O'Brien, C. G., Hyman, M. A., and Kaplan, S. (1951) A study of the numerical solution of partial differential equations. *J. Math. Phys.* **29**, 223–251.

28. Papamichael, N. (1970) Property (A) and consistent orderings in the iterative solution of linear equations. *M. Tech dissertation, Brunel University, England.*

29. Parker, I. B. and Crank, J. (1964) Persistent discretization errors in partial differential equations of parabolic type. *Computer J.* **7**, 163–167.

30. Peacemann, D. W., and Rachford, H. H. (1955) The numerical solution of parabolic and elliptic differential equations. *J. Soc. Indust. Applied Maths.* **3**, 28–41.

31. Richtmyer, R. D., and Morton, K. W. (1967) *Difference Methods for Initial-Value Problems.* Interscience Publishers.

32. Stone, H. L. (1968) Iterative solution of implicit approximations of multi-dimensional partial differential equations. *SIAM J. Numer. Anal.* **5**, 530–558.

33. The Open University. (1974) *Mathematics Course M321, Units 11–14.* The Open University Press, Milton Keynes, England.

34. Thom, A., and Apelt, C. J. (1961) *Field Computations in Engineering and Physics.* D. Van Nostrand, London.

35. Varga, R. S. (1962) *Matrix Iterative Analysis*, Prentice-Hall International, London. Prentice-Hall, New Jersey.

36. Wilkinson, J. H. (1965) *The Algebraic Eigenvalue Problem.* Clarendon Press, Oxford.

37. Wilkinson, J. H. (1963) *Rounding Errors in Algebraic Processes. N.P.L. Notes on Applied Science,* **32.** London, H.M.S.O.

8. Wilkinson, J. H. and Reinsch, C. (1971) (Ed). *Handbook for Automatic Computation, Vol 2*, Springer-Verlag, Berlin.

9. Young, D. M. (1971) *Iterative Solution of Large Linear Systems.* Academic Press, London and New York.

0. Young, D. M. (1954) Iterative methods for solving partial differential equations of elliptic type. *Trans. Amer. Math. Soc.* **76,** 92–111, 218, 242–272, 355, 394.

1. Papamichael, N. and Smith, G. D. The determination of consistent orderings for the SOR iterative method. *J. Inst. Maths. Applics.* **15,** 239–248.

Index